Advanced Computational Methods for Agri-Business Sustainability

Suchismita Satapathy
KIIT University (Deemed), India

Kamalakanta Muduli
Papua New Guinea University of Technology, Papua New Guinea

A volume in the Advances in Business Information Systems and Analytics (ABISA) Book Series

Published in the United States of America by
IGI Global
Business Science Reference (an imprint of IGI Global)
701 E. Chocolate Avenue
Hershey PA, USA 17033
Tel: 717-533-8845
Fax: 717-533-8661
E-mail: cust@igi-global.com
Web site: http://www.igi-global.com

Library of Congress Cataloging-in-Publication Data

CIP DATA PROCESSING

2024 Business Science Reference
ISBN(hc) 9798369335833 | ISBN(sc) 9798369350874 | eISBN 9798369335840

British Cataloguing in Publication Data
A Cataloguing in Publication record for this book is available from the British Library.

Advances in Business Information Systems and Analytics (ABISA) Book Series

Madjid Tavana
La Salle University, USA

ISSN:2327-3275
EISSN:2327-3283

Mission

The successful development and management of information systems and business analytics is crucial to the success of an organization. New technological developments and methods for data analysis have allowed organizations to not only improve their processes and allow for greater productivity, but have also provided businesses with a venue through which to cut costs, plan for the future, and maintain competitive advantage in the information age.

The **Advances in Business Information Systems and Analytics (ABISA) Book Series** aims to present diverse and timely research in the development, deployment, and management of business information systems and business analytics for continued organizational development and improved business value.

Coverage

- Algorithms
- Business Systems Engineering
- Forecasting
- Data Management
- Business Intelligence
- Performance Metrics
- Decision Support Systems
- Business Models
- Data Strategy
- Business Process Management

IGI Global is currently accepting manuscripts for publication within this series. To submit a proposal for a volume in this series, please contact our Acquisition Editors at Acquisitions@igi-global.com or visit: http://www.igi-global.com/publish/.

The Advances in Business Information Systems and Analytics (ABISA) Book Series (ISSN 2327-3275) is published by IGI Global, 701 E. Chocolate Avenue, Hershey, PA 17033-1240, USA, www.igi-global.com. This series is composed of titles available for purchase individually; each title is edited to be contextually exclusive from any other title within the series. For pricing and ordering information please visit http://www.igi-global.com/book-series/advances-business-information-systems-analytics/37155. Postmaster: Send all address changes to above address.

Titles in this Series

For a list of additional titles in this series, please visit: www.igi-global.com/book-series

Powering Industry 5.0 and Sustainable Development Through Innovation
Rohit Bansal (Vaish College of Engineering, India) Fazla Rabby (Stanford Institute of Management and Technology, Australia) Meenakshi Gandhi (Vivekananda Institute of Professional Studies, India) Nishita Pruthi (Maharshi Dayanand University, India) and Shweta Saini (Maharshi Dayanand University, India)
Business Science Reference • copyright 2024 • 393pp • H/C (ISBN: 9798369335505) • US $315.00 (our price)

Cases on AI Ethics in Business
Kyla Latrice Tennin (College of Doctoral Studies, University of Phoenix, USA) Samrat Ray (International Institute of Management Studies, India) and Jens M. Sorg (CGI Deutschland B.V. & Co. KG, Germany)
Business Science Reference • copyright 2024 • 342pp • H/C (ISBN: 9798369326435) • US $315.00 (our price)

Advanced Businesses in Industry 6.0
Mohammad Mehdi Oskounejad (Azad University of the Emirates, UAE) and Hamed Nozari (Azad University of the Emirates, UAE)
Business Science Reference • copyright 2024 • 278pp • H/C (ISBN: 9798369331088) • US $325.00 (our price)

Intelligent Optimization Techniques for Business Analytics
Sanjeev Bansal (Amity Business School, Amity University, Noida, India) Nitendra Kumar (Amity Business School, Amity University, Noida, India) and Priyanka Agarwal (Amity Business School, Amity University, Noida, India)
Business Science Reference • copyright 2024 • 357pp • H/C (ISBN: 9798369315989) • US $270.00 (our price)

Data-Driven Business Intelligence Systems for Socio-Technical Organizations
Pantea Keikhosrokiani (University of Oulu, Finland)
Business Science Reference • copyright 2024 • 490pp • H/C (ISBN: 9798369312100) • US $265.00 (our price)

Utilizing AI and Smart Technology to Improve Sustainability in Entrepreneurship
Syed Far Abid Hossain (BRAC University, Bangladesh)
Business Science Reference • copyright 2024 • 370pp • H/C (ISBN: 9798369318423) • US $290.00 (our price)

701 East Chocolate Avenue, Hershey, PA 17033, USA
Tel: 717-533-8845 x100 • Fax: 717-533-8661
E-Mail: cust@igi-global.com • www.igi-global.com

Table of Contents

Detailed Table of Contents

limited budgets, adverse weather, and farmer reluctance. It underscores cybersecurity, and environmental concerns, and advocates for government support, education, and awareness initiatives.

This chapter emphasizes the Find S algorithm to explore the use of cutting-edge computational techniques to improve agri-business sustainability. Precision farming, data analytics, and machine learning are all combined in "agriculture 4.0" to maximize productivity and promote sustainability. Case examples from real-world situations are provided to illustrate the usefulness of using cutting-edge computational techniques in agriculture. The chapter also covers the value of cooperation and government assistance, tackles issues related to technological adoption, and provides solutions for broader acceptance in the farming community. This chapter intends to contribute to the continuing conversation on data-driven decision-making and productivity improvement in the agricultural sector by offering insightful information to researchers, practitioners, policymakers, and stakeholders who are interested in using computational methods to improve sustainability in the industry.

In this study, a wide range of geoFigureical locations are investigated to investigate the complex relationships that exist between agricultural productivity and important environmental parameters. These elements include fluctuations in temperature, patterns of rainfall, and the application of pesticides. Through the utilization of a vast dataset that encompasses yield measures, meteorological conditions, and agricultural practices over a period of several years, we employ sophisticated statistical and machine learning techniques in order to uncover the subtle linkages that regulate crop output. The findings of our study indicate that there are substantial correlations between the outcomes of yields and particular environmental parameters. These findings show the major impact that sustainable farming practices and climate adaptation methods have on the efficiency of agricultural production. The findings highlight the significance of integrated resource management and the requirement for precision agriculture

Preface

In an era where globalization is reshaping the dynamics of agri-food markets globally, the sustainable development of agribusiness is of paramount importance. This book, *Advanced Computational Methods for Agri-Business Sustainability*, edited by Suchismita Satapathy and Kamala Kanta Muduli, delves into the complexities and opportunities presented by this evolving landscape. The creation of a single global market, free from protectionist barriers, has far-reaching implications for food security, price stability, and the resilience of food supply chains. This transformation necessitates a closer examination of the entire agribusiness spectrum, from agricultural inputs and production to the processing and distribution of food and fiber products.

Agribusiness encompasses a wide array of activities that support agricultural production, provide essential services, and facilitate the marketing, transportation, and distribution of agricultural goods. This sector not only delivers critical resources such as food, clothing, and shelter but also generates employment for millions, spanning fields like science, research, engineering, education, and government. The convergence of these diverse activities underscores the importance of addressing the environmental and resource implications of increasing agro-based production and consumption.

The drive for sustainable development in agriculture raises significant issues, such as involving small farmers in sourcing networks and complying with stringent food safety and quality regulations. This book identifies the challenges within the agricultural sectors, focusing on effective agri-food management and supply chain dynamics. It explores potential opportunities and mitigation strategies for existing issues, offering a comprehensive resource for those keen to understand the intricate realities faced by farmers and the agricultural community.

Our aim is to equip readers—regardless of their familiarity with computational and optimization techniques—with a clear understanding of the complex tasks that define modern agriculture. Through illustrative examples and detailed explanations, this book serves as a valuable tool for academicians, students, policymakers, and industry professionals. It offers insights into the current agricultural landscape and proposes initiatives for future developments, ultimately contributing to the mental well-being of the farming community.

One of the primary objectives of this book is to provide an in-depth analysis of the global occurrences of occupational stressors and their interrelationships, which can lead to depression among farmers. By highlighting these issues, we hope to prompt immediate intervention initiatives by governments and competent authorities. Additionally, the book addresses emerging aspects of global agriculture, emphasizing the role of computational and optimization techniques in managing agricultural activities effectively.

Our target audience includes professionals and researchers across various fields—agriculture, engineering, public health, management, and anthropology, among others. We also aim to support executives involved in managing expertise, knowledge, and organizational development within the farming sector. By offering a thorough understanding of current challenges and future possibilities, this book aspires to foster sustainable and resilient agricultural practices worldwide.

We invite you to explore the diverse topics covered in this book, confident that it will enrich your understanding and provide practical guidance for advancing agri-business sustainability.

Organization of the Book

Chapter 1: A Machine Learning-Based Crop Diseases Detection and Management System

This chapter introduces an innovative solution, CropGuard, aimed at addressing crop diseases through machine learning. By leveraging Convolutional Neural Networks and transfer learning, the proposed system facilitates rapid disease identification via a user-friendly mobile application. Key features include image analysis, disease identification, and real-time treatment recommendations, empowering farmers with accessible and efficient crop disease management tools.

Chapter 2: A Smart Agronomy: Deep Learning Process for Recognition and Classification Plant Leaf Diseases

Focusing on timely disease detection in agriculture, this chapter presents a deep learning approach for identifying and classifying plant leaf diseases. Utilizing the AlexNet architecture and deep learning techniques, the system automates disease recognition, offering a robust solution for efficient disease management and crop protection.

Chapter 3: Advanced Computational Forecasting for Agri-Business Supply Chain Resilience

This chapter explores the integration of advanced statistical methods and computational techniques to enhance forecasting accuracy in the agri-business sector. By employing time series analysis, predictive modeling, and data-driven insights, the chapter aims to bolster supply chain resilience by predicting market trends and demand fluctuations.

Chapter 4: Agricultural Crop Recommendations Based on Productivity and Season

Addressing the complexities of crop selection, this chapter presents a machine learning-based crop recommendation system. Leveraging SVM algorithms and comprehensive datasets, the system provides tailored crop recommendations based on productivity and prevailing seasonal conditions, offering farmers valuable insights for optimizing agricultural productivity.

Chapter 5: Artificial Intelligence (AI) Driven IoT (AIIoT)-Based Agriculture Automation

Exploring the convergence of AI and IoT in agriculture, this chapter highlights the transformative potential of AIIoT-based agriculture automation. By harnessing AI algorithms to analyze IoT data, the chapter demonstrates how automation can lead to higher output, lower expenses, and improved sustainability in agriculture.

Chapter 6: Artificial Intelligence in Agri-Business Sector: Prioritizing the Barriers through Application of Analytic Hierarchy Process (AHP)

This chapter investigates the adoption barriers of AI in agriculture, particularly in countries like India, using the Analytic Hierarchy Process methodology. By systematically identifying and prioritizing these barriers, the chapter offers insights for informed decision-making and targeted interventions to foster the adoption of AI in the agri-business sector.

Chapter 7: Barriers of Agrisupply Chain Management: During Mental and Physical Stress During Farming in Tractor

Focusing on the occupational health challenges faced by farmers, this chapter examines the ergonomic risks associated with tractor operations in agriculture. Using tools like Rapid Upper Limb Assessments (RULA) and Quick Exposure Checks (QEC), the chapter aims to identify risk factors contributing to work-related musculoskeletal disorders (WMSD) in tractor operators.

Chapter 8: Behind the Barriers; Identifying Critical Credit Access Challenges in Agri-Business Sector of India: An AHP Approach

This chapter delves into the challenges of accessing credit in the Indian agricultural industry, employing the Analytic Hierarchy Process to identify key obstacles. By offering insights for overcoming these challenges, the chapter aims to facilitate proper coordination among financial markets, agri-entrepreneurs, and government entities to promote sustainable growth in the sector.

Chapter 9: Design of Wheels of Agri-Rover for Both Dry and Wet Surfaces (Run-Way)

Exploring the design considerations for agricultural rovers, this chapter focuses on developing wheels suitable for traversing uneven, dry, and wet surfaces. Through simulation and analysis using Ansys modeling, the chapter addresses challenges related to weight, durability, and temperature resilience, aiming to optimize rover performance for year-round agricultural operations.

Chapter 10: Digitalization of SCM in the Agriculture Industry

This chapter investigates the digital transformation of agri-food supply chains, leveraging Industry 4.0 technologies and IoT. By enhancing precision farming, post-harvest processes, and supply chain management, the chapter explores how digitalization can optimize productivity, reduce waste, and promote sustainability in the agriculture sector.

Chapter 11: Harnessing Agricultural Data: Advancing Sustainability through the Application of Find S Algorithm

Focusing on the utilization of computational techniques to improve agri-business sustainability, this chapter highlights the role of data-driven decision-making in agriculture. By integrating precision farming, data analytics, and machine learning, the chapter offers insights for enhancing productivity and sustainability in the agricultural industry.

Chapter 12: Harvesting Insights Unveiling the Interplay of Climate, Pesticides, and Rainfall in Agricultural Yield Optimization

This chapter explores the complex relationships between environmental parameters and agricultural productivity. Through sophisticated statistical and machine learning techniques, the chapter aims to uncover the linkages regulating crop output, highlighting the significance of sustainable farming practices and climate adaptation methods.

Chapter 13: Identification, Classification and Grading of Crops Grain using Computer Intelligence Techniques

Addressing the challenges of crop identification and classification, this chapter investigates the application of computer intelligence techniques in agriculture. By employing machine learning algorithms, the chapter aims to develop automated systems capable of distinguishing between various crops, thereby reducing labor costs and minimizing agricultural losses.

Chapter 14: IoT, A.I., and Robotics Applications in the Agriculture Sector

This chapter explores the transformative potential of IoT, AI, and robotics in modern agriculture, emphasizing their role in enhancing productivity, resource utilization, and sustainability. By addressing challenges and highlighting future trends, the chapter offers insights for advancing sustainable farming practices in the agriculture sector.

Chapter 15: LSTM-Based Deep Learning for Crop Production Prediction with Synthetic Data

Focusing on crop production prediction, this chapter introduces a deep learning model utilizing Long Short-Term Memory (LSTM) and synthetic data. By addressing challenges such as data scarcity and dynamic complexities, the chapter aims to enhance the accuracy of crop yield predictions, thereby supporting efficient resource planning and ensuring food security.

Chapter 16: Waste Management and its Impact on Food Security

This chapter explores the intricate relationship between waste management and global food security, addressing challenges such as resource scarcity and environmental harm. By advocating for integrated policies and smart waste management rules, the chapter aims to promote resilient and healthy communities while enhancing sustainable agriculture practices.

Chapter 17: Adoption Challenges of Industry 4.0 in Agrisector and Designing a Framework to Reduce it

Investigating the barriers to Industry 4.0 adoption in the agricultural sector, this chapter proposes a framework to mitigate challenges and improve agrisector efficiency. By leveraging technologies like IoT and trend analysis, the chapter aims to enhance decision-making and promote sustainable practices in agriculture.

IN CONCLUSION

As editors of *Advanced Computational Methods for Agri-Business Sustainability*, we are proud to present a comprehensive exploration of the intersection between computational techniques and agricultural sustainability. Throughout this book, esteemed authors have delved into various facets of modern agriculture, offering innovative solutions, insights, and frameworks to address key challenges facing the agri-business sector.

From machine learning-based crop disease detection to advanced forecasting for supply chain resilience, each chapter represents a significant contribution to the field. By harnessing the power of artificial intelligence, deep learning, and IoT, these chapters showcase the transformative potential of technology in revolutionizing agricultural practices.

Moreover, our authors have not only identified barriers and challenges but have also proposed practical frameworks and solutions to overcome them. Whether it's optimizing crop recommendations, enhancing waste management practices, or addressing adoption challenges of Industry 4.0, this book offers actionable insights for stakeholders across the agricultural value chain.

As we conclude this journey, we are optimistic about the future of agriculture. By embracing computational methods, leveraging data-driven decision-making, and fostering collaboration among stakeholders, we can pave the way for a more sustainable, resilient, and prosperous agricultural sector.

We extend our gratitude to all the contributors for their invaluable insights and dedication to advancing agricultural sustainability. It is our hope that this book serves as a guiding light for researchers, practitioners, policymakers, and stakeholders striving to create a more sustainable future for agriculture worldwide.

Chapter 1
A Machine Learning-Based Crop Diseases Detection and Management System

Narendra Kumar Rao Bangole
Mohan Babu University, India

ABSTRACT

This work proposes an innovative solution to address crop diseases. The objectives include developing a machine-learning model for rapid disease identification and designing a user-friendly mobile application. The machine learning model, employing convolutional neural networks and transfer learning, is integrated into the mobile app for on-the-go disease diagnosis. Key features include image analysis, disease identification, and real-time treatment recommendations. This work termed CropGuard aims to empower farmers, regardless of technical proficiency, through accessible and efficient crop disease management. This aligns with the broader goal of sustainable agriculture by enabling timely interventions, reducing crop losses, and promoting increased productivity.

INTRODUCTION

Agriculture is always a major factor in food security and enhancing the GDP of any country. In the context of India Considering the latest data for the year 2020-2021, Agriculture is contributing 20.2% of the country's GDP. India secures second position for producing large quantities of agricultural outputs in the world. 50% of the job market in India is dependent upon Agriculture. Considering the population of India, there is an imbalance between the demand and supply. Demand is always high compared to the Supply of food products. In this situation, Crop diseases (Mondal,B., et al., 2023) have a major role in declining the rate of food production.

Crop diseases are those diseases (Rohilla. N., & Rai,M., 2021)that degrade the quality, as well as quantity of our crops like Rice paddy, wheat leaves, Cucumber leaves, tomato leaves, Strawberry leaves, and banana plants, and almost every crop production is (Harika, M., et al., 2023)affected by diseases. The common diseases are leaf spots, Bacterial leaf blight, Black rust, Blight, Charcoal rot, Fusarium, Sheath blight of rice, Brown rust, Powdery mildew, anthracnose, wilt, scab, gall, canker, dieback, and

DOI: 10.4018/979-8-3693-3583-3.ch001

so on. Several researchers have studied various crops and classified them as some harmful diseases that infect plants due to biological reasons Bacterial Diseases, Viral Diseases, Fungal Diseases, and so on. Some of the common diseases we can see are Leaf Smut and Brown spots on the Paddy leaf. There can be several reasons for crop diseases such as inappropriate Moisture and temperature differences because different crops require different levels of temperature and moisture. Wind, Frequent as well delays for rain, floods, and Chemicals such as pesticides affect soil quality, Deficiency of nutrients, and hazards.

Farmers who are the backbone of our country got troubled a lot because of such diseases. The production yield are only income source of their family. To troubleshoot these crop diseases, technologies like Machine Learning(ML) and Deep Learning(DL) is introduced. K-Clustering, and Support Vector Machine (SVM) techniques are used in classifying the diseases. At first the plant or leaves images are tested with Image Processing and Computer Vision techniques. Deep learning (Sangeevan, 2021) helps to recognize the image in detail and Machine learning classifies the different diseases into different categories for easy identification of their proper solutions, treatment, and recommendations to prevent such diseases.

PROBLEM STATEMENT

CropGuard aims to address the pressing issue of crop diseases through the development of a sophisticated Machine Learning-Based Crop Diseases Detection and Management System. The problem statement revolves around the prevalent challenges faced by farmers worldwide due to the devastating impact of plant diseases (Balafas,V., et al., 2023) on agricultural yields. Despite various existing methods, the timely and accurate identification of these diseases remains a significant hurdle, leading to substantial crop losses. CropGuard endeavors to fill this crucial gap by employing advanced machine learning algorithms to swiftly and precisely detect and manage crop diseases. The system intends to provide farmers with an intuitive tool that can identify diseases (Praveena, M., et al., 2023) early, enabling prompt and targeted interventions to mitigate their spread, thereby safeguarding crop health and ensuring optimal agricultural productivity.

SIGNIFICANCE

A Machine Learning-Based Crop disease detection (Ekanayake & Nawarathna, R. D., 2021) and Management System holds several significant advantages and benefits for agriculture and farmers:

Early Detection: One of the primary advantages is the system's ability to detect diseases in crops at an early stage. Machine learning models can ana1yze large amounts of data, including images of plants, to identify subtle signs of diseases that might not be easily noticeable to the human eye. Early detection permits for prompt intervention, preventing the range of diseases propagated and decreases crop damage.

Precision and Accuracy: Machine learning algorithms, when trained on diverse datasets, can achieve high levels of accuracy in disease identification. They can differentiate between various diseases and even distinguish them from other stressors like nutrient deficiencies or pest attacks. This precision ensures targeted and effective management strategies.

Timely Intervention: With rapid disease identification, farmers are prompted to take up actions like as applying specific treatments, adjusting irrigation or nutrient levels, or employing disease-resistant crop varieties. This proactive approach helps in controlling the spread of diseases and mitigating their impact on crop yields.

Cost and Resource Efficiency: By enabling precise and targeted interventions, a Machine Learning-Based system reduces the unnecessary use of pesticides or fungicides. Farmers can optimize resource allocation, minimizing costs associated with excessive chemical treatments while reducing environmental impact.

Empowering Farmers: Implementing such a system provides farmers with access to advanced technology. It empowers them with tools and knowledge to make informed decisions about crop health, leading to better productivity and increased incomes.

Scalability and Adaptability: Machine learning systems can continuously improve their accuracy by learning from new data. They are adaptable to different regions, crops, and evolving disease patterns, making them scalable and relevant across diverse agricultural landscapes.

Global Food Security: By safeguarding crop health and optimizing yields, a Machine Learning-Based system contributes to global food security. It helps ensure a consistent and reliable food supply, which is crucial for meeting the needs of a growing population.

LITERATURE AND COMPARATIVE STUDIES

Crop disease is a significant threat to global food safety, causing yield loss and economic damage. Early and precise illness detection is critical for adopting effective management strategies and reducing these losses. Machine learning (ML) and deep learning (DL) have are powerful tools for automated disease detection in crops, offering rapid and precise analysis (Jha P., Dembla & Dubey, 2023) of visual data like images.

Literature Review

Several research studies have been performed on ML and DL to tackle the issue of Crop disease detection and Management to increase the production of food.

1. Plant Disease Identification Using Deep Learning Approaches: This study by Hassan et al. (2019) compared various deep learning architectures like CNNs for identifying and detecting diseases in the leaves of 14 plant species. Their findings demonstrated the effectiveness of DL in achieving high accuracy for disease classification.

2. A Study on Plant Disease Detection and Classification Using Deep Learning Approaches: This work by (Mohana, et al., 2022) evaluated the performance of different DL models like VGG16, DenseNet121, and ResNet50 for classifying diseases in peanut leaves. They reported that data augmentation techniques further enhanced the accuracy of these models.

3. Machine Learning for Plant Disease Detection: An Assessment between Support Vector Machine and Deep Learning: This study by Jadon et al. (2020) compared the performance of Support Vector Machines (SVM) and DL models for detecting diseases in grape leaves. Their results showed that while both approaches achieved good accuracy, deep learning models performed slightly better, especially for complex disease cases.

Management Strategies

1. Disease prediction and forecasting: ML models can analyze historical data and weather patterns to predict disease outbreaks, allowing for proactive management interventions.

2. Precision agriculture: Combining disease detection with field-specific data enables targeted application of pesticides and fertilizers, optimizing resource use and minimizing environmental impact.

3. Decision support systems: Integrating disease detection and management strategies within user-friendly interfaces permits farmers to take knowledgeable decisions based on realtime data and recommendations.

Comparative Studies

The choice between ML and DL for crop disease detection depends on various factors, including

1. Data availability: DL models typically require larger datasets for optimal training compared to traditional ML algorithms.

Computational resources: Training DL models can be computationally expensive, requiring high-performance hardware.

2. Model interpretability: ML models are generally easier to interpret, allowing for understanding the basis of their predictions. DL models, on the other hand, can be considered "black boxes" due to their complex internal workings.

3. Human-machine interaction: Studies explore how interfaces and decision support tools can effectively communicate complex information and recommendations to farmers for optimal implementation of management strategies.

4. Comparing different management strategies: Studies evaluate the effectiveness of various methods like chemical, biological, and integrated pest management approaches based on disease type, economic feasibility, and environmental impact.

What Impact Has the Practice of Localization Contribute to the Development of Real-Time Plant Disease Methods?

Plant disease localization is a critical step in the development of real-time plant disease detection methods. To enhance localization capabilities, the use of localization datasets, image segmentation methods, and innovative models is essential. In this answer, we will explore how various research efforts have contributed to the advancement of plant disease localization. For instance, researchers have employed Convolutional Neural Networks (CNNs) equipped with a Channel-wise and Bottleneck Attention Module (CBAM) as part of the primary network in Faster RCNN. By using a CNN, such as GoogleNet, to excerpt visual features and the Selective Search Detector (SSD) structure for localization, their model demonstrated a high level of accuracy and mean average precision (Mekonnen et al., 2020). on in-field datasets.

Additionally, using Retinex for fine-tuning and integrating the Single Shot MultiBox Detector (SSD) model for detecting and localizing plant diseases in maize leaves has shown promising results. The approach involved a multilevel feature fusion model that achieved a mean average precision (mAP) score of 91.83% on the test dataset. Moreover, researchers have experimented with various image classification methods, such as MobileNetv2-YOLOv3, for detecting tomato gray leaf spot. The model demonstrated an accuracy of 93.24% on the F1-score, 91.32% on the average accuracy, and 86.98% on the IoU score. To achieve better localization, the RPN segment was used in aggregation with VGG16 and ZFnet.

The work also proposed a cascaded Mask R-CNN (MRCNN) model for segmentation loss functions of 2.0437 and a segmentation map with accuracy of 91%. Image classification methods like translations, YOLO version 2, and DarkNet-19 have been utilized in these efforts. Lastly, to address the limitations of localization datasets and labelling expenses, more advanced models and techniques are required. This includes using CenterNet2 for the citrus disease localization dataset, which achieved the highest mAP score possible at 0.914. However, there is still a lack of studies in plant disease localization due to the challenges associated with obtaining localization datasets and the cost of labelling large datasets. Therefore, further advancements and innovations are needed to overcome these challenges and enhance the accuracy of plant disease detection at the field level.

How do Adequate Public Datasets for Machine Learning and Deep Learning Work for Plant Disease Diagnosis and Identification?

Researchers in the arena of plant disease detection and identification have made significant strides in recent years, using techniques like deep learning, image classification, and advanced machine vision. For instance, Retinex, a multi-level feature fusion model, has been integrated with Single Shot MultiBox Detector (SSD) for detecting and localizing plant diseases in maize leaves. This approach has demonstrated promising outcomes, with a mean Average precision (mAP) score of 91.83% on the test dataset. Furthermore, several studies have focused on utilizing MobileNetv2-YOLOv3 and other advanced image classification methods for detecting tomato gray leaf spot. The model has shown an accuracy of 93.24% on the F1-score, 91.32% on the average accuracy, and 86.98% on the IoU score. To enhance localization, the RPN segment was combined with VGG16 and ZFnet.

In the realm of segmentation, the use of cascaded Mask R-CNN (MRCNN) has yielded impressive results. This model has achieved a segmentation map accuracy of 91% and a segmentation loss function value of 2.0437. Image classification methods like translations, YOLO version 2, and DarkNet-19 have been utilized in these efforts. Despite the progress made in this field, challenges still exist. To overcome these challenges, further advancements and innovations are required. For example, using CenterNet2 for the citrus disease localization dataset has resulted in the highest mAP score possible at 0.914. However, obtaining localization datasets and labelling expenses continue to be limitations in this domain. Additionally, more datasets are needed to train and test machine learning and deep learning models for plant disease detection and identification. Researchers have started by creating multi-crop datasets, such as PlantVillage, and specialized datasets like the Rice Dataset (Aggarwal,M., et al., 2023) and the Maize Dataset. However, the demand for diverse and extensive datasets remains high.

In conclusion, while progress has been made in the field of plant disease detection and identification, challenges and opportunities remain. These challenges include obtaining extensive datasets and overcoming labelling expenses. To capitalize on these opportunities, further advancements in deep learning and machine vision are required, along with the development of novel datasets for research and application purposes.

MACHINE LEARNING (ML) IN CROP DISEASE DETECTION AND MANAGEMENT

About Machine Learnin

Machine learning (ML) is an area of AI and computer science that emphasizes on using data and algorithms that allow AI to learn in the same way humans do, thereby increasing its precision.

Classical machine learning is commonly defined on the basis of how an algorithm makes increasingly accurate predictions. There are four main types of machine learning: supervised learning, unsupervised learning, semi-supervised learning, and reinforcement learning.

In supervised learning, the data scientists facilitate algorithms with labelled training data and permit which variables the computer should look for associations. All the algorithm's input and output are defined in supervised learning.

Unsupervised machine learning techniques data are unlabelled. They search through unlabeled data for patterns that can be used to classify data points into subgroups.

Semi-supervised learning involves feeding a little quantity of labelled training data to an algorithm. The algorithm uses this information to understand the dimensions of the data set, which it may subsequently apply to new unlabeled data.

Reinforcement learning works by creating an algorithm with a specific objective and a set of rules for achieving that goal. A data scientist will also programme the algorithm to seek positive incentives for actions that help it achieve its ultimate goal while avoiding punishments for actions that push it further away from it.

Some features of Machine Learning Algorithm is as discussed below:

Feature Engineering: Traditional ML techniques rely on engineered features extracted from images or sensor data to identify patterns associated with crop diseases. These features might include colour histograms, texture analysis, or shape descriptors.

Classical Algorithms: ML algorithms such as Support Vector Machines (SVM), Random Forests, Decision Trees, and k-nearest Neighbours (k-NN) have been used in crop disease detection. These algorithms require manually crafted features and are effective when dealing with smaller datasets.

Limited Complex Pattern Recognition: ML approaches might struggle with detecting complex patterns or variations in images due to their dependence on manually crafted features. They might not capture intricate details present in images, limiting their accuracy in disease identification.

Data Pre-processing Importance: Data pre-processing and feature engineering play a crucial role in ML-based systems. Cleaning and selecting relevant features significantly impact the model's performance.

Deep Learning in Agriculture

About Deep Learning (DL)

DL is a type of AI that communicates computers to process data in a manner encouraged by the human brain. Deep learning algorithms can recognise complex patterns in photographs, text, sounds, and other data, producing accurate insights and predictions.

Deep learning is a type of AI that encourages computers to process data in a manner inspired by the human brain. Deep learning algorithms can recognise complex patterns in photographs, text, sounds, and other data, producing accurate insights and predictions.

Benefits of Deep Learning Over Machine Learning

Machine learning systems struggle to interpret unstructured data, such as text documents, because the training dataset can contain unlimited variants. In contrast, deep learning algorithms can interpret unstructured data and generate broad observations without manual feature extraction.

Certain obstacles arise from its practical execution of DL, they are based on Large quantities of high-quality data and Computing Capacity requirement.

End-to-End Learning: DL, especially Convolutional Neural Networks (CNNs), enables end-to-end learning by automatically extracting hierarchical representations from raw images. CNNs learn feature hierarchies directly from the data, eliminating the need for handcrafted feature engineering.

Complex Pattern Recognition: DL models excel in recognizing complex patterns within images. They can capture intricate details and subtle variations in plant images, allowing for more accurate disease detection and classification.

Transfer Learning: DL models benefit from transfer learning, leveraging pre-trained networks (e.g., VGG, ResNet, Inception) on large datasets like ImageNet. This approach allows fine-tuning on smaller agricultural datasets, even with limited labelled data availability.

Big Data and Scalability: DL models thrive on large datasets. They scale well with increasing data, learning from diverse samples, and improving their performance with more training examples.

Computational Intensity: DL models are computationally intensive, requiring substantial computational resources, especially during the training phase. This aspect might pose challenges in resource-constrained environments.

Interpretability Challenges: Despite their effectiveness, DL models often lack interpretability, making it challenging to understand why a particular decision was made, which might be crucial for user trust and adoption in agriculture.

Image Dataset Availability

Several datasets are gathered from different sources with their images, and crops. Table 1 i.e. image dataset shows an overview of the available dataset.

A total of 55,303 leaf pictures were collected in the PlantVillage dataset, with a classification aim. These photos are classified into 39 types based on plant species and illnesses. The dataset contains information on 18 fungal diseases, four bacterial diseases, two mould (oomycete) diseases, two viral diseases, and one mite disease. The photos are also available in three other states: healthy, ill, and mixed.

The iBean leaf image dataset is a collection of 2,296 images, which are either healthy or affected by two different diseases, namely angular leaf spot and bean rust. The images were captured using a smartphone and the data is publicly available. The PlantLeaves dataset includes 5,503 images, divided into two categories: healthy leaves and diseased leaves. The dataset is available for download and use.

Table 1. Image dataset

Dataset	Crop	Images	Classes	Type
CropDeep	Various Crops	32,147	32	C
IP102	Corn	11,000	103	C
PlantVillage	Multiple	55,303	39	C
iBean	Bean	2,296	4	C
PlantLeaves	Multiple	5,503	23	C
PlantDoc	Multiple	3,345	19	C

The PlantLeaves dataset as in Figure1 includes 4,503 images, divided into two categories: healthy leaves and diseased leaves. The dataset is available for download and use. The PlantDoc dataset comprises 3,345 images with a classification target. The dataset covers 14 plant species and 19 classes of diseases. Because the dataset is open source, we can readily obtain it for benchmarking. Apple leaf, Apple rust leaf, Apple Scab Leaf, Bell pepper leaf, Bell pepper leaf spot, Blueberry leaf, Cherry leaf, Corn leaf blight, Corn Grey leaf spot, Corn rust leaf, Grape leaf, Grape leaf black rot are all PlantDoc dataset types. Potato leaf, Potato leaf early blight, Potato leaf late blight, Peach leaf, Peach leaf, Strawberry leaf, Tomato leaf, Tomato leaf bacterial spot, Tomato leaf early blight leaf, Tomato leaf late blight leaf, Tomato leaf mosaic virus, Tomato leaf two-spotted spider mites leaf, Tomato leaf yellow virus leaf, and Tomato leaf yellow virus leaf. There are 8,851 annotations, with an average of 3.4 annotations per image.

Figure 1. Visualization of the plantdoc dataset with image annotations

CHALLENGES AND FUTURE DIRECTIONS

The challenges for making the application for crop disease detection are going to be tough. We have to train a large number of datasets. Some of the Challenges are:

Data collection and quality: Collecting high-quality labelled images of crop diseases for training ML models is a time-consuming and resource-intensive process. This may hinder the widespread adoption of ML-based systems in the agricultural sector.

Limited annotated datasets: While recent studies have demonstrated the potential of ML and DL techniques in plant disease detection, the lack of large, annotated datasets can be a significant barrier to the implementation of such systems.

Complexity of diseases: The complexity of crop diseases can make it challenging to accurately identify and classify those using ML algorithms. Additionally, distinguishing between different disease stages or variants can be a significant hurdle.

Privacy concerns: In an era of growing data privacy concerns, ensuring the security and confidentiality of farmers' sensitive data while utilizing ML techniques can be a challenging task.

Future Direction

The future direction of this research can involve exploring more complex and adaptive ML techniques. Solving such big issues for the farmers may lead us to the following directions.

i. Innovative strategies for collecting and annotating high-quality datasets, such as crowdsourcing or leveraging drone imagery, can be explored to enhance the capabilities of ML-based crop disease detection systems.

ii. Integrating systems with decision support algorithms can enhance the effectiveness of ML-based crop disease detection systems by providing actionable insights and recommendations for efficient crop disease management.

iii. Developing scalable and efficient computing infrastructures can be crucial to the successful implementation of ML-based systems in the agricultural sector, as these systems may require substantial computational resources to process large datasets and make accurate predictions.

Finally, fostering collaborations between ML researchers, precision agriculture experts, and farmers can lead to more practical and tailored solutions for real-world applications of ML-based crop disease detection and management systems.

ARCHITECTURE

Figure 2. Overall workflow diagram

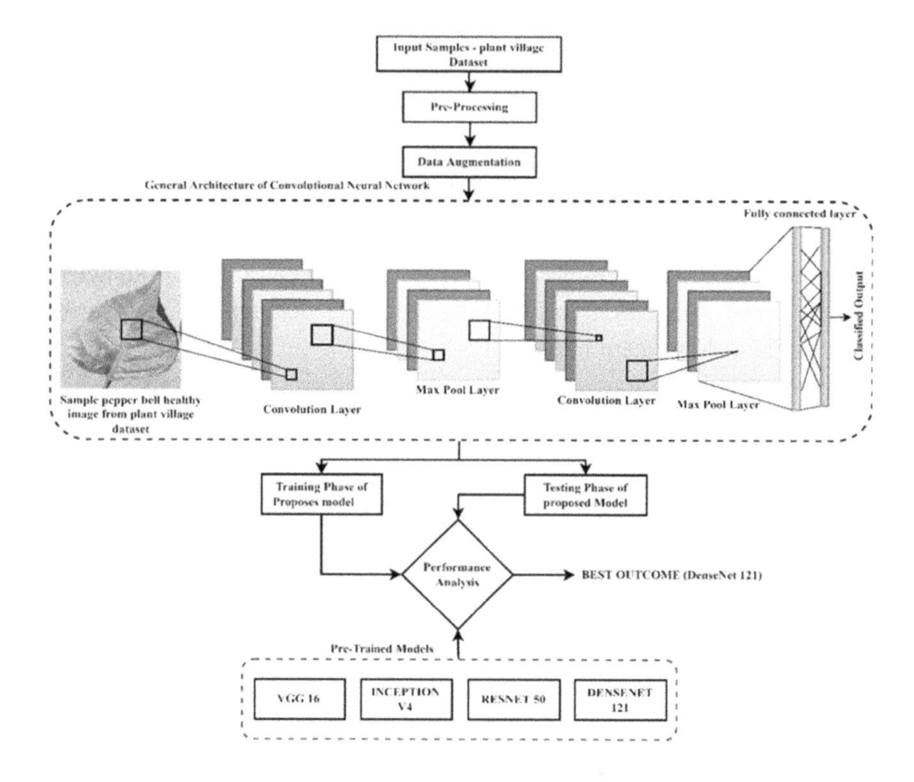

The architecture as in Figure 2 is discussed below. First, a large dataset of crop disease images will be collected and labelled. These images will serve as the foundation for the development of the machine learning models.

Second Step is Pre-Processing where the images are well-maintained for further processing. Such as maintaining the size and clarity of the image. No blurring and noisy errors should be there. Such things are maintained in these steps.

A feature extraction stage will follow. Here, advanced DL models such as Convolutional Neural Networks (CNNs) and Transfer Learning models like ResNet or VGG will be employed to automatically extract relevant features from the images.

The extracted features will then be used as input for an ML classifier, which will be trained to identify the presence and severity of various crop diseases. Techniques such as Logistic Regression, Support Vector Machines, and Neural Networks can be utilized for this purpose.

The classifier's performance will be evaluated using metrics such as accuracy, precision, recall, and F1 score. This will provide valuable insights into the effectiveness of the ML models in detecting crop diseases.

A user-friendly interface will be developed, allowing farmers and other stakeholders to easily upload images of their crops and receive disease detection results. This interface can be implemented using mobile app development frameworks like Flutter and Django.

IMPLEMENTATION

Algorithm Pseudo Format

The following are the stepwise algorithm steps to predict the disease:

1. Import necessary libraries: numpy, pandas, os, matplotlib, seaborn, cv2, tensorflow, tqdm, sk-learn, and keras.
2. Define the list of disease types and the directory where the training data is located.
3. Create a list of training data by looping through each disease type and file in the training directory.
4. Convert the training data list into pandas dataframe and randomize the order of the training set.
5. Plot a histogram of the frequency of each disease type in the training set.
6. Display images for different species using a function that takes in the species name, number of rows, and number of columns as input.
7. Define functions to read and resize images.
8. Preprocess the training data by resizing and normalizing the images and converting the disease IDs into categorical variables.
9. Split the training set into training and validation sets.
10. Define a function to build the DenseNet121 model with custom layers.
11. Compile the model with the Adam optimizer and categorical cross-entropy loss function.
12. Define callbacks for learning rate reduction and model checkpointing.
13. Train the model using the training and validation sets, with data augmentation.
14. Evaluate the model on the validation set.
15. Plot the accuracy and loss over the course of training.
16. Define a function to predict the disease for a given input image.
17. Load an image, preprocess it, and predict the disease.
18. Display the input image and the predicted disease

Sequence diagram is represented in Figure 3.

The implementation involves the following detailed steps:

1. Data Preparation:

- **Gather a dataset:** Choose a suitable dataset with labelled images of healthy and diseased crops (e.g., PlantVillage, PlantDoc).
- **Pre-process images:**
  - Resize images to consistent dimensions.
  - Normalize pixel values for better model convergence.
  - Apply data augmentation techniques (random cropping, flipping, and rotations) to increase dataset size and variety.
- **Split data:** Divide the dataset into training, validation, and testing sets for model training, evaluation, and final testing.

Figure 3. Sequence diagram

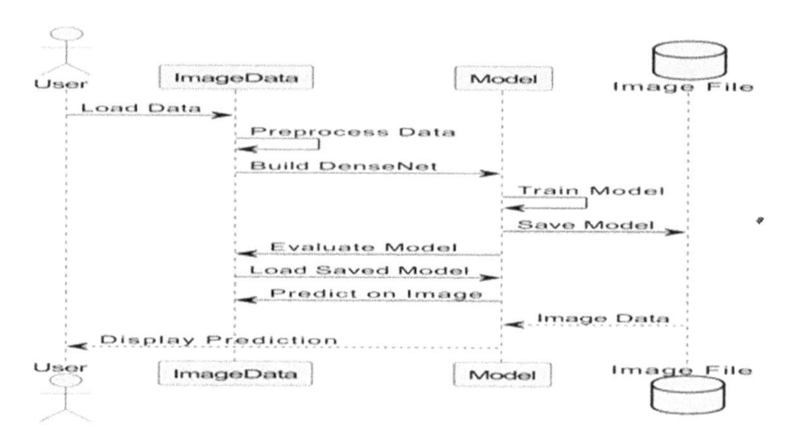

2. Model Selection and Training:

- **Choose a machine learning or deep learning model:**
    - Traditional ML: Support Vector Machines (SVMs), Random Forests, or K-Nearest Neighbours (KNN) can be effective for smaller datasets or simpler tasks.
    - Deep learning: Convolutional Neural Networks (CNNs) are highly successful for image-based tasks, especially with large datasets.
- **Train the model:**
    - Use a suitable library like Tensor Flow or PyTorch.
    - Define model architecture and hyper parameters.
    - Feed training data to the model for learning patterns.
    - Monitor training progress and adjust hyper parameters as needed.

Figure 4. Overview of the training accuracy of the different classification models

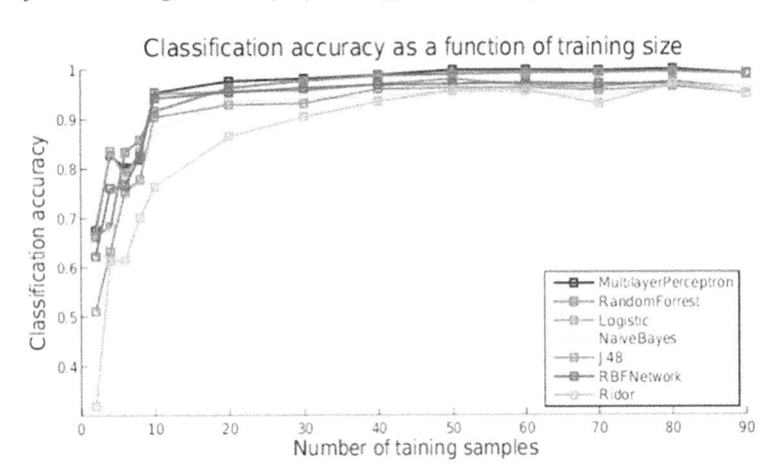

The graph provided as in Figure 4 is a plot of training accuracy against the number of epochs. An epoch is basically defined as a single forward and backward pass of all the training samples in the dataset. The training accuracy is the proportion of correct predictions made by the model during training. We have trained and evaluated eighteen CNN model architectures.

In the case of crop disease detection, the task is to classify an image into one of several possible classes, each representing a different type of crop disease.

The model being used here is a deep learning model. It's worth noting that while the specific model used isn't provided, it could be any one of the popular models like AlexNet, VGG16, VGG19, DenseNet, ResNet, ResNext, ShuffleNet, MobileNet, etc. Each of these models has a unique architecture, designed to capture specific patterns or features in the input data.

The accuracy curve represents the learning process of the model. The curve initially rises as the model starts to learn the underlying patterns in the training data. However, as the model learns more and more about the data, it begins to make better and more accurate predictions.

There are several important takeaways from this graph:

The accuracy curve suggests that the model can learn from the training data.

The training accuracy plateaus around 0.8 - 0.9. This implies that the model is unable to further improve its performance on the training data, suggesting a phenomenon known as overfitting.

A further increase in training epochs beyond the point of diminishing returns would not improve the model's performance.

In summary, this graph provides a clear visual representation of the model's learning process and indicates that the model has successfully learned the underlying patterns in the training data, reaching a high level of accuracy. However, the presence of overfitting suggests that there may be some areas for improvement.

3. Disease Detection:

- **Load the trained model:**

- Load the saved model weights and configuration.

- **Pre-process new images:** Apply the same pre-processing steps used during training.
- **Make predictions:**
    - Pass pre-processed images to the model.
    - Obtain model predictions, indicating the probabilities of different diseases.

4. Management Strategies:

- **Implement decision logic:**
    - Define actions based on predicted diseases.
- **Consider:**
    - Pesticide or fungicide application (if necessary).
    - Cultural practices (e.g., crop rotation, pruning).
    - Integrated pest management (IPM) strategies.
- **Present recommendations:**
    - Provide clear and actionable information for farmers.
    - Consider a user-friendly interface or decision support system.

5. Evaluation and Refinement:

- **Measure performance:**
    - Evaluate model accuracy, precision, recall, and F1-score on the testing set.
- **Refine model:**
    - Adjust model architecture, hyperparameters, or training data to improve performance.
- **Monitor system effectiveness:**
    - Track disease incidence and crop health over time to assess system impact.

6. Additional Considerations:

- **Hardware and software:**
    - Ensure appropriate computing resources for model training and deployment.
- **Real-time monitoring:**
    - Explore integration with mobile devices or sensors for field-level monitoring.
- **Integration with agricultural systems:**

- Consider linking with farm management software for comprehensive decision-making

7. Packages used in Implementation:
   1. **Operating Systems:** This module provides functions for interacting with the operating system, like listing files and directories.
   2. **Pandas:** It offers high-performance, easy-to-use data structures and data analysis tools. Think of it as an advanced spreadsheet.
   3. **matplotlib.pyplot**: This module allows you to create various visualizations like plots, charts, and histograms.
   4. **seaborn**: Built on top of matplotlib, seaborn offers a higher-level interface for making statistical graphics more visually appealing.
   5. **cv2:** This module (OpenCV) focuses on computer vision tasks like image processing and object detection.
   6. **Tensorflow**: This popular library is used for machine learning and deep learning applications.
   7. **sklearn.metrics**: This submodule of scikit-learn provides metrics for evaluating machine learning models, like accuracy and precision.
   8. **keras.models**: This submodule allows you to create and manage neural network models in keras.
8. Platform Information:

- Flutter is an open source framework by Google for building beautiful, natively compiled, multi-platform applications from a single codebase.
- Flask is a micro web framework written in Python. It is classified as a microframework because it does not require particular tools or libraries.
- Kaggle is a data science competition platform and online community of data scientists and machine learning practitioners
- Flutter is basically for the frontend part where it easily integrates with Flask for backend development.
- It supports popular libraries and frameworks such as TensorFlow, PyTorch, and scikit-learn, making it versatile for various machine learning tasks.
- Windows, Linux, macOS, and Android are the platforms.

CONCLUSION

This study aims to explore the application of machine learning (ML) and deep learning (DL) techniques in quality agriculture production, particularly in plant disease detection and classification. The scope of our research encompasses both classification and object detection methodologies. We categorize all relevant studies into their respective classes to provide a comprehensive overview using a novel classification scheme. This approach enables a clear understanding of the current state of knowledge in the field.

In addition to discussing the research methodologies, we have figured out different available datasets for classifying and detecting plant disease. Our analysis highlights these datasets suitability for classification and object detection tasks.

The future direction of this research can involve exploring more complex and adaptive ML techniques. Solving such big issues for the farmers may lead us to the following directions.

- Innovative strategies for collecting and annotating high-quality datasets, such as crowd-sourcing or leveraging drone imagery, can be explored to enhance the capabilities of ML-based crop disease detection systems.
- Integrating systems with decision support algorithms can enhance the effectiveness of ML-based crop disease detection systems by providing actionable insights and recommendations for efficient crop disease management.
- Developing scalable and efficient computing infrastructures can be crucial to the successful implementation of ML-based systems in the agricultural sector, as these systems may require substantial computational resources to process large datasets and make accurate predictions.

Finally, fostering collaborations between ML researchers, precision agriculture experts, and farmers can lead to more practical and tailored solutions for real-world applications of ML-based crop disease detection and management systems.

REFERENCES

Aggarwal, M., Khullar, V., & Goyal, N. (2023). Exploring classification of rice leaf diseases using machine learning and deep learning. *3rd International Conference on Innovative Practices in Technology and Management (ICIPTM)*, (pp. 1-6). IEEE. 10.1109/ICIPTM57143.2023.10117854

Balafas, V., Karantoumanis, E., Louta, M., & Ploskas, N. (2023). Machine learning and deep learning for plant disease classification and detection. *IEEE Access : Practical Innovations, Open Solutions*, 11, 114352–114377. 10.1109/ACCESS.2023.3324722

Ekanayake, & R. D. Nawarathna. (2021). Novel deep learning approaches for crop leaf disease classification: A review, *International Research Conference on Smart Computing and Systems Engineering (SCSE)*, (pp. 49-52). IEEE. .10.1109/SCSE53661.2021.9568324

Harika, S., Sandhyarani, G., Sagar, D., & Reddy, G. V. S. (2023). *Image-based black gram crop disease detection. International Conference on Inventive Computation Technologies (ICICT)*, Lalitpur, Nepal. 10.1109/ICICT57646.2023.10134027

Jha, P. (2023). Comparative analysis of crop diseases detection using machine learning algorithm. *Third International Conference on Artificial Intelligence and Smart Energy (ICAIS)*. IEEE. 10.1109/ICAIS56108.2023.10073831

Mekonnen, N., & Burton, S. (2020). Machine learning techniques in wireless sensor network based precision agriculture. *Journal of the Electrochemical Society*, 167(3), 037522. 10.1149/2.0222003JES

Mondal, B., Bhushan, M., Dawar, I., Rana, M., Negi, A., & Layek, S. (2023). Crop disease prediction using machine learning and deep learning: an exploratory study. *International Conference on Sustainable Computing and Smart Systems (ICSCSS)*, (pp. 278-283). IEEE. 10.1109/ICSCSS57650.2023.10169612

Praveena, M., Dubisetty, V. B., Varaprasad, K. V., Rama, M., Vadana, P. S., & Sai, T. S. R. (2023). An in-depth analysis of deep learning and machine learning methods for identifying rice leaf diseases. *4th International Conference on Smart Electronics and Communication (ICOSEC)*. IEEE. 10.1109/ICOSEC58147.2023.10276335

Rohilla, N., & Rai, M. (2021). Advanced machine learning techniques used for detecting and classification of disease in plants: a review. *3rd International Conference on Advances in Computing, Communication Control and Networking (ICAC3N)*. IEEE. 10.1109/ICAC3N53548.2021.9725616

Sangeevan. (2021). Deep learning-based pesticides prescription system for leaf diseases of home garden crops in Sri Lanka. *International Research Conference on Smart Computing and Systems Engineering (SCSE)*. IEEE. .10.1109/SCSE53661.2021.9568308

Chapter 2
A Smart Agronomy:
Deep Learning Process for Recognition and Classification Plant Leaf Diseases

Chandra Prabha Ramakrishnappa
http://orcid.org/0000-0003-4236-2782
BMS Institute of Technology and Management, India

Seema Singh
BMS Institute of Technology and Management, India

ABSTRACT

The main part of the agriculture process is the timely detection of leaf diseases to have a healthy growth. In routine implementation, the identification of diseases is realized either by manual or laboratory testing. Physical testing involves few expertise and results could vary from individuals which can result in false interpretation while the latter requires extra time and might not be able to deliver the production, due to which the spread of disease gradually increases. Hence an automated system is required for the identification and classification of the disease. This chapter intends leaf sickness detection and recognition by applying deep learning for two data split ratios. The classification task is performed using Alex-net, a pre-trained architecture. The data set has three categories of leaf disease, namely, bacterial leaf blight, brown spot, and leaf blast, each consisting of 40 infected images. The proposed architecture classifies the diseases into three categories. The comparison study for various performance metrics—such as recall, precision, and specificity—is measured.

INTRODUCTION

Agriculture is a major source of national wealth in many countries. Crop diseases are important causes of decreased quantity and quality of production; thus, recognising plant diseases is critical. Disease signs can appear in several areas of the plant; nevertheless, plant leaves are frequently utilised to identify diseases. Early and correct diagnosis is an important first step in reducing losses caused by plant diseases. An inaccurate diagnosis might lead to poor management decisions, such as using the wrong chemical application, potentially resulting in additional health loss and production decline. The unaided

DOI: 10.4018/979-8-3693-3583-3.ch002

eye method is a classic way of illness identification that involves a lot of labour, is prone to human error, takes a long time, and is not suitable for big fields.

An underdeveloped growth in plants due to various plant diseases causes a dreadful effect on the profit and yield in society. Due to this underdeveloped growth, financial damage valued all over the world is up to $20 billion per year. (Ahmed, K., Shahidi, T. R., Alam, S. M. I., & Momen, S. (2019)). Accurate identification is very challenging due to various geographical conditions and other reasons. Also, the conventional methods that are used in most farming areas largely depend on specialists in the domain, experience, skill set, and manual testing. Many of the methods are costly, consume more time, and are labor-intensive with the pain of identifying precisely (Ahmed, K., Shahidi, T. R., Alam, S. M. I., & Momen, S. (2019)). Hence, a quick, accurate plant disease classifier is greatly essential to support the agriculturalists which consecutively leads to the growth of occupational and ecosystem agriculture. There are many issues such as animate or abiotic causes, Nutrient deficiency, leaf disease, Microorganisms, and environmental changes that constrain plant growth.

Figure 1. Rice leaf diseases

Figure 1 illustrates a few rice leaf categories Here the plant leaf disease is one of the factors to hinder the plant growth c identification model using the pre-trained - deep learning algorithms designed to answer the listed issues Sladojevic, S., Arsenovic, M., Anderla, A., Culibrk, D., & Stefanovic, D. (2016).A few rice leaf infections such as rice blast, rice brown spot, rice sheath blight, rice bacterial leaf blight, rice bacterial sheath rot, rice bakanae disease, rice sheath rot, and rice bacterial wilt (Rahman, C. R., Arko, P. S., Ali, M. E., Khan, M. A. I., Apon, S. H., Nowrin, F., & Wasif, A. (2020)).

In the proposed design the three-leaf disease namely Leaf smut, Bacterial blight, and Leaf Blast is considered for the classification. Brown Spot and Leaf Blast are prominent leaf infections. Bacterial Leaf Blight and brown are taken as the most projecting and unsafe rice leaf infection. Leaf smut is triggered by fungal Pyricularia grisea. The leaf blades have small black linear lesions and leaf tips may turn grey and dry. Bacterial blight is caused by the bacteria Xanthomonas oryzae. It is elongated lesions near the leaf tips and margins and gradually turns white to grey due to fungal attack. Brown spot is caused by the fungus Helminthosporiose. The structure has round to oval designed lesions and a dusky auburn colored on rice leaves (Ahmed, K., Shahidi, T. R., Alam, S. M. I., & Momen, S. (2019)).

The rice quality and profit decrease due to these leaf diseases. An effective and accurate diagnosis will increase the profit and quality. Effective diagnosis includes monitoring of the diseases, their occurrences, and frequencies through manual observation.

Figure 2. Different leaf disease categories

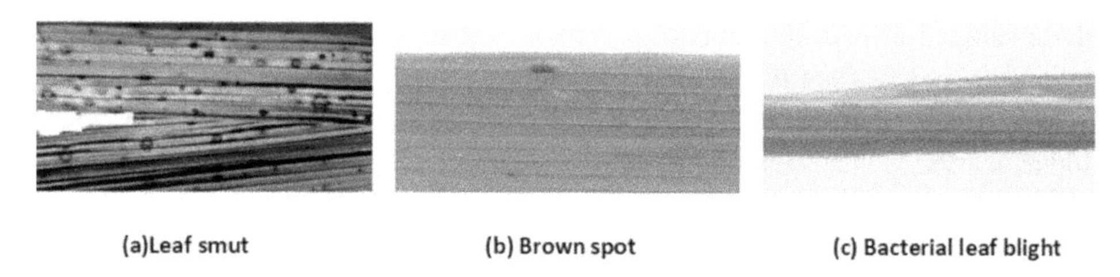

Figure 2 illustrates the different leaf categories. In manual observation, farmers need to monitor the plant at definite time intervals which consume most of the time and delay in classification. This tandem of tasks becomes very challenging when huge farming is are considered using supplementary time and additional employment charges. Primary detection of the infected plants, their on-time treatment, and future preparation strategies can be scheduled to avoid the diseases to have a good yield by reducing crop losses. Despite the time consumption, the correction leaf diseases cannot be identified due to a lack of expertise in the agriculture domain. Manual testing is a tough job to identify the disease. So automatic detection is required to support the planters in predicting the disease effectively in a short time, so the crop losses can be avoided and further their profit and yields can boost as per their expectation.

Machine learning is a field in modern computing. A lot of research has been done to make machines intelligent. Learning is a natural human behaviour that has been extended to include robots. Various approaches have been suggested to do this. Traditional machine learning techniques are used in a wide range of applications. Researchers have made numerous efforts to increase the accuracy of machine learning systems. Another dimension was considered, which led to the deep learning notion. Deep learning is a subset of machine learning. So yet, few applications of deep learning have been investigated. This will undoubtedly address challenges in a variety of new application domains and subdomains through the use of deep learning.

Machine learning can be utilized in the following fields: (a) crop management, which includes predicting yield, identifying diseases, weeds, and improving crop quality; (b) livestock management, which involves producing livestock and ensuring animal welfare; (c) water management; and (d) soil management.

The proposed works present a simple architecture deep learning-based system for disease identification and classification for the mentioned named rice leaf diseases. Various performance metrics are also measured for different data split ratios.

LITERATURE SURVEY

Javidan, S. M., Banakar, A., Vakilian, K. A., & Ampatzidis, Y. (2023) suggests to diagnose and classify grape leaf diseases a novel image processing algorithm and multi-class support vector machine (SVM) are applied feature dimension reduction and features are performed by applying principal component analysis (PCA) and relief feature selection respectively. Yu, H., Liu, J., Chen, C., Heidari, A. A., Zhang, Q., Chen, H., & Turabieh, H. (2021) proposes K-means clustering and a deep-learning model for accurate diagnosis. This paper explores the effect of several k values (2, 4, 8, 16, 32, and 64) and models

(VGG-16, ResNet18, Inception v3, VGG-19, and the enhanced deep learning model. Bari, B. S., Islam, M. N., Rashid, M., Hasan, M. J., Razman, M. A. M., Musa, R. M., & Majeed, A. P. A. (2021) proposes a faster region-based convolutional neural network (Faster R-CNN) for rice plant disease detection with an accuracy of 99.25. Sujatha, R., Chatterjee, J. M., Jhanjhi, N. Z., & Brohi, S. N. (2021) proposes performance is better for disease classification accuracy (CA) compared to that of ML Shin, J., Mahmud, M. S., Rehman, T. U., Ravichandran, P., Heung, B., & Chang, Y. K. (2022) suggests that an advanced machine vision system will improve the whole agriculture management system and give farmers with useful recommendations and insights into decision-making. Sangeetha, R., Logeshwaran, J., Rocher, J., & Lloret, J. (2023) proposes that improved machine vision systems will improve the overall agriculture management system and give farmers with valuable recommendations and insights into decision-making.

K. Ahmed et al. (2019) discusses the different machine-learning-based algorithms such as KNN, Naive Bayes, Decision Tree, and Logistic Regression that can be applied with the needed pre-processing process for classification. Out of all, the Decision tree algorithm provides an accuracy of over 97% after 10-fold cross-validation. Sladojevic et al. (2016) present their work in the classification of thirteen varieties of plant diseases by applying a deep CNN fine-tuned model achieving an accuracy of 96.3%. The recognition of the rice disease with VGG16 with an accuracy of 99.53% with the test data set is attained (Preetom S. Arko et al.(2020)). The k-means clustering technique can also be useful in identifying the infected area of the plant (S. Archana, K., & Sahayadhas, A., 2018). J. P. Shah, H. B. Prajapati, and V. K. Dabhi, (2016) present the different segmentation techniques applied to the infected leaf and different classifiers for the detection. (Islam, T., Sah, M., Baral, S., & RoyChoudhury, R. (2018)) aims Naive Bayes, a simple classifier. This work deliberates a technique of identifying and categorizing paddy leaf disease using the percentage of RGB value of the infected portion giving an accuracy greater than 89% in classification.

The authors (Pinki, F., Khatun, N., & Islam, S.M, 2017) present the image pre-processing with K-means clustering segmentation that separates the infected areas. Based on features like (color, texture, and shape) the disease's overall classification accuracy of 92.06% is achieved by the support vector machine classifier

The Histogram of an Oriented Gradient is also applicable to extract the features and classify the categories using the random forest where the plant disease can be identified (Maniyath, S.R, 2018).

Prajapati, Harshadkumar B et al. (2017) presents the work related to finding accurate attributes using K-means clustering with centroid feeding for segmentation of the infected part of a leaf image. The training and testing accuracy obtained is 93.33% and 73.33% by applying the SVM classifier for multiclass. A cross-validation of 5-fold and 10 fold is performed to achieve an accuracy of 83.80% and 88.57% respectively for the features of color, shape, and texture. The authors present the classification work to recognize the leaf infection in tomato plants with an accuracy of 97.28% with the model ResNet with stochastic gradient descent (SGD), the model used a batch size of 16 with iterations of 4992 (Keke Zhang et al. 2020)). A classification model incorporating the CV algorithm, RPN algorithm, and TL algorithm is also applied to resolve the detection of plant disease identification in the tough atmosphere (Yan Guo et al., 2020).

The authors K. Pradhan et al (2019) present the use of Alexnet to extract the features and an SVM classifier to classify the plant disease based on the different split ratios. The accuracy is limited to 91.37%.

The authors use the deep convolutional neural to classify into three stages golden apple snail infested normal, and unhealthy (Lucas G.B., Campbell C.L., 1992)). Based on the survey, machine learning, and the neural network has influenced its role in agriculture in recognizing plant disease. According to a survey, the work is limited in using the existing retrained simple Deep convolution neural network with an augmentation process. The model recommended assurances of the robust convolutional neural network and decreased the number and excellence necessities of the convolutional neural network for the selected data set and found good outcomes.

DATA RESOURCES AND METHODOLOGY

Image Acquisition

Materials were collected from the UC Irvine Machine Learning Repository. The data set consists of three categories of leaf infections. 1) Bacterial Leaf 2) Blight Brown Spot, 3) Leaf Blast with each 40 in jpeg. The dataset was developed by manually sorting out sick leaves into dissimilar disease classes. The farmers were consulted to name the infections for trial leaves. The infected leaves were named in their native language by the agriculturists. Later, by referring to professionals in the agriculture field, English names were assigned to those diseases. The images were acquired in direct sunlight with a bleached background. The images were reduced to the desired resolution for processing.

Building a Classifier Model

The use of conventional machine learning is avoided due to inadequately labelled datasets to train a model and poor performance in classification. Transfer learning is a very main step during the adjustment of the existing network (Hussain, E., Mahanta, L. B., Das, C. R., & Talukdar, R. K. 2020). Based on the limited availability of the data sets and technical computation resources, construction of the network from the base is not possible. Hence pre-trained already existing CNN is fine-tuned with the specific own dataset. The fine-tuning of the available pre-trained network on a labelled large-scale natural image dataset is referred to as transfer learning. Fine-tuning includes weight adjustments and resetting the last layers of the network with back propagation to familiarise deep attributes to the new datasets.

Accordingly, transfer learning is followed for classifying the leaf disease into three categories.

In the proposed work, the last fully connected layers for pre-trained models are updated with a modified fully connected layer having three output nodes that represent the three classes. Model training was conducted with a stochastic gradient descent optimizer for 10 epochs, and 120 iterations with a batch size of 120 images.

Image Augmentation

Data augmentation is a data-space approach to the issue of insufficient data. A group of methods known as "data augmentation" work to improve the quantity and caliber of training datasets so that deeper learning models can be constructed on them (Shorten, C., & Khoshgoftaar, T. M., 2019).

The Deep learning DCNNs model outputs a very effective and accurate classification of images. To avoid the over-fitting issue, the networks must work with huge data during the training process (Szegedy, C., Liu, W., Jia, Y.Q., Sermanet, P., Reed, S. and Anguelov, D. (2015)).

(a) (b) (c)

Figure 3. Augmented images for different data splits (a) 50%of training set and 50% of test data (b) 60%of training set and 40% of test data (c) 70% of training set and 30% of test data

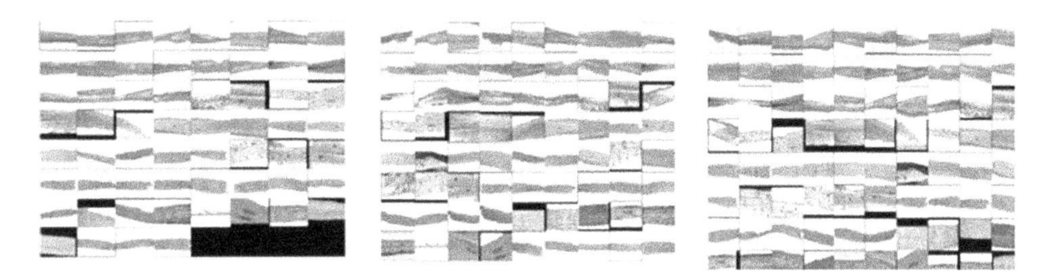

The raw data is restricted due to unavailability, and the augmentation process such as image rotation, and image translation has been performed. Figure 3(a),3(b),3(c) illustrates the augmented shot for various splits.

Deep convolution neural networks architecture

The deep learning models are biologically inspired to provide good performance (Krizhevsky, A., Sutskever, I. and Hinton, G. (2012)). A classic DCNN consists of four different types of layers namely convolutional layer, activation layers, pooling layer, and the last fully connected layers (Krizhevsky, A., Sutskever, I. and Hinton, G. 2012). The network which is simple to design and easy to train is AlexNet. The Alex net is an extremely capable DCNN that has the following stages 1) convolution layers, 2) pooling layers, 3) Rectified Linear Unit (ReLU) layers, and 4) fully connected layers connected in cascaded form.

The input to the model is of the size of 227*227*3 and the output part has possibilities of three classes that are performed by softmax layer function.

Requirements

The software platform used is MATLAB. The execution task is executed by Central Processing Unit (CPU) with the feature of /i5/7th/1TB/8GB/2GB.

RESULTS AND DISCUSSION

The classification model is trained and tested by an image dataset of 120 images. The different leaf diseases used are Bacterial Leaf, Blight Brown Spot, and Leaf Blast with each 40. Accuracy and error rate are not only the terms applied to evaluate the model. The other evaluation parameters are applied in this work to evaluate the model. The parameters used are precision, recall, specificity, and F1 score. Also, the average of the three-leaf categories is computed to know the performance of the model. The model has tested for a data split ratio of 50:50, 60:40, and 70:30. In the 50:50 data split ratio, the testing and training images are 60 each. In the data split of 60:40 and 70:30, the number of training and testing images are 72 and 48, 84 and 36 respectively. The fine-tuned model automatically extracts the features and is passed to deep CNN for further classification. The images are classified into three labels Bacterial Leaf Blight, Brown Spot, and Leaf Blast.

The evaluation parameters applied to the model are precision, recall F1, and Specificity, which is applied for individual leaf categories, and weighted average precision, weighted average recall, and weighted average F1 are measured for the overall leaf categories.

a) Results for data set with equal number of training and testing data set.

Figure 4. Training process for the data split of 50:50 with validation accuracy of 88.33%

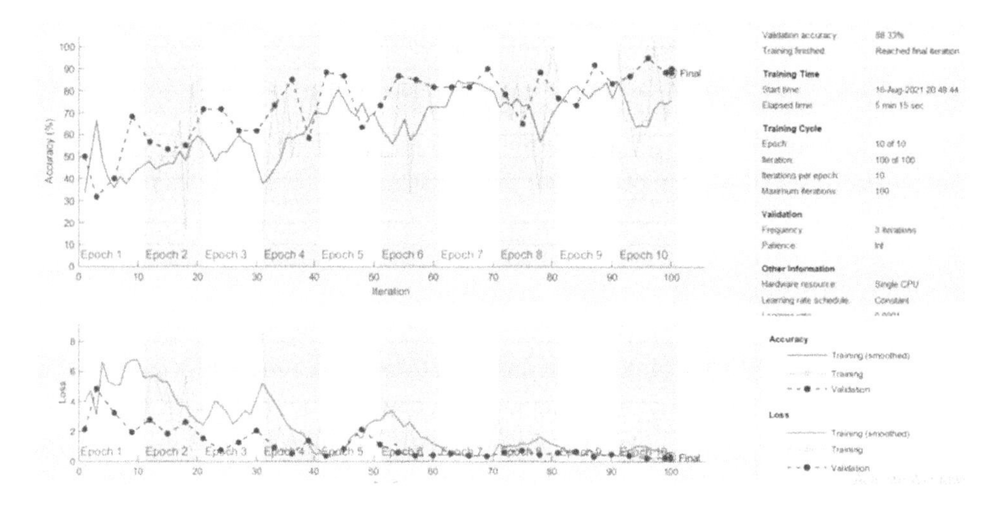

Figure 5. (a) Confusion matrix for the training data set; (b) Confusion matrix for the testing data

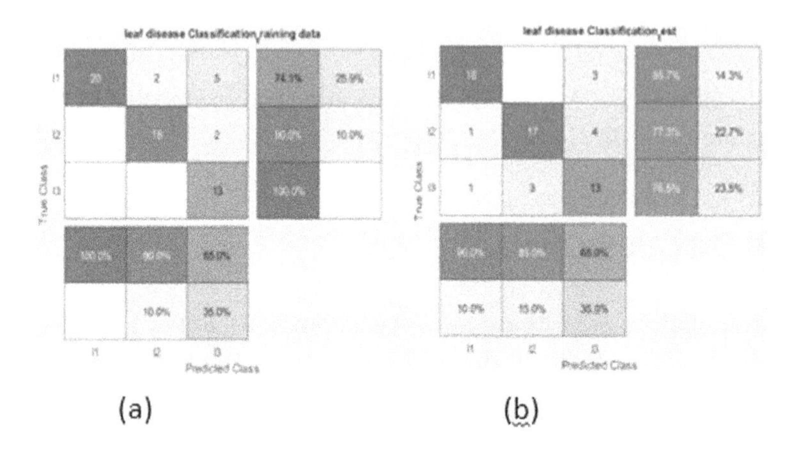

The plot training process and the confusion matrix for equal data set split ratio are represented in Figure 4 and Figure 5.

Figure 6. Individual leaf category evaluation parameter for the data split of 50:50

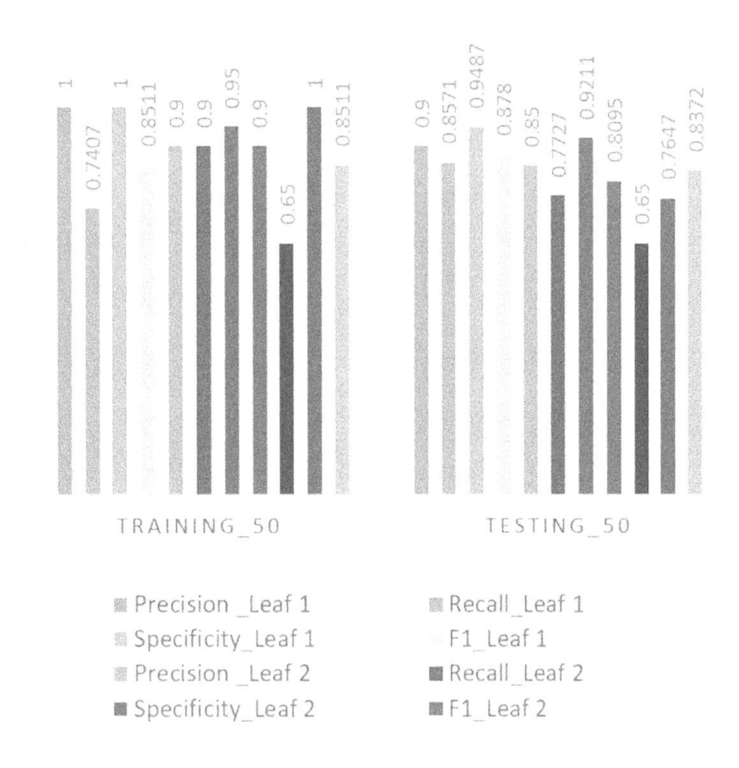

The evaluation parameter of recall, precision, and specificity for individual leaf categories is as in Figure 6.

The values of weighted average parameters for the 50% training images and 50% test images are figured in Figure 7.

For the data split of 50:50, the weighted average precision, weighted average recall, and weighted average F1 for the training data set are 89.08%,85.00%, and 85.37% whereas for the test images, it is 82.92%,80.9%, and 80.32% respectively.

Figure 7. Weighted average evaluation parameters for the data split of 50:50

b) Results for data set with 60% training images and 40% testing images

Figure 8. Training process for the data split of 60:40 with a validation accuracy of 98.6%

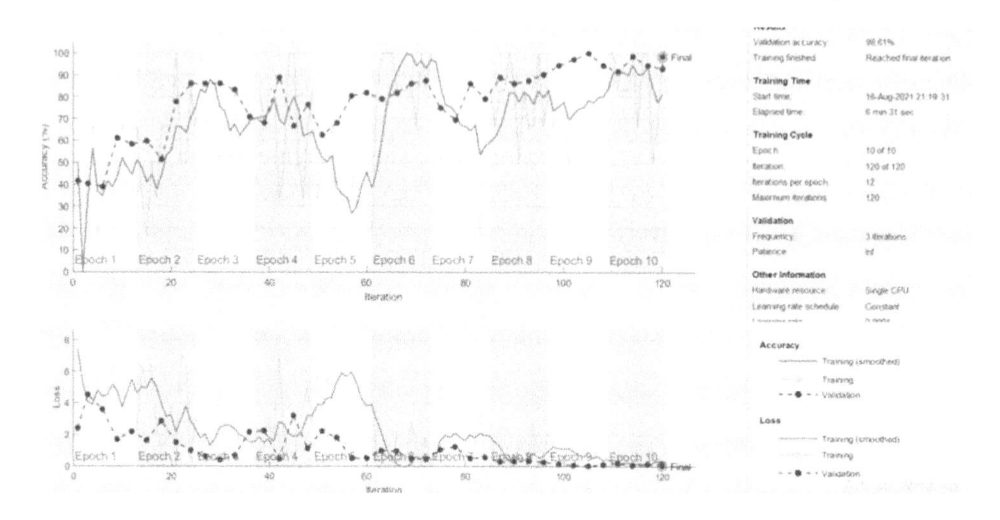

(a) (b)

Figure 9. (a) Confusion matrix for 60% of the training data set, (b) Confusion matrix for 40% of the testing data

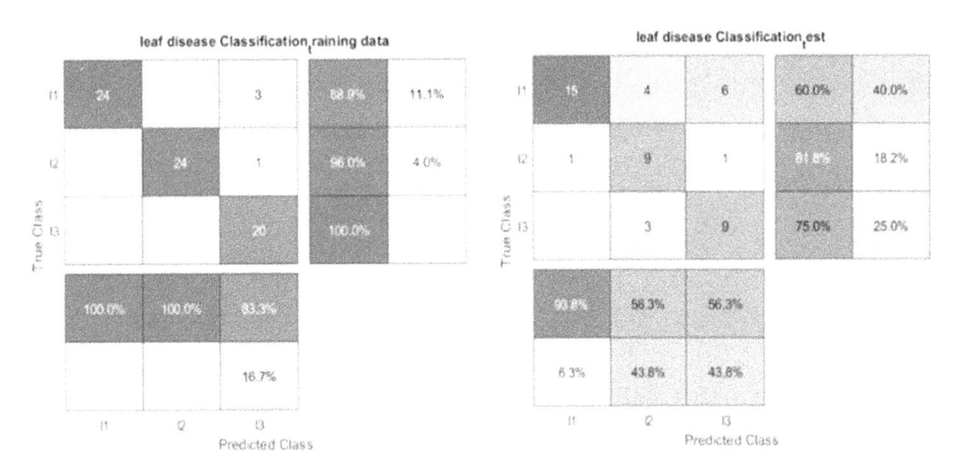

The training process and confusion matrix are illustrated in Figure 8 and Figure 9.

Figure 10. Individual leaf category evaluation parameter for 60% of training images and 40% of testing images

The individual leaf evaluation parameters are represented in Figure 10.

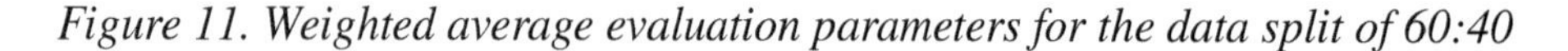

Figure 11. Weighted average evaluation parameters for the data split of 60:40

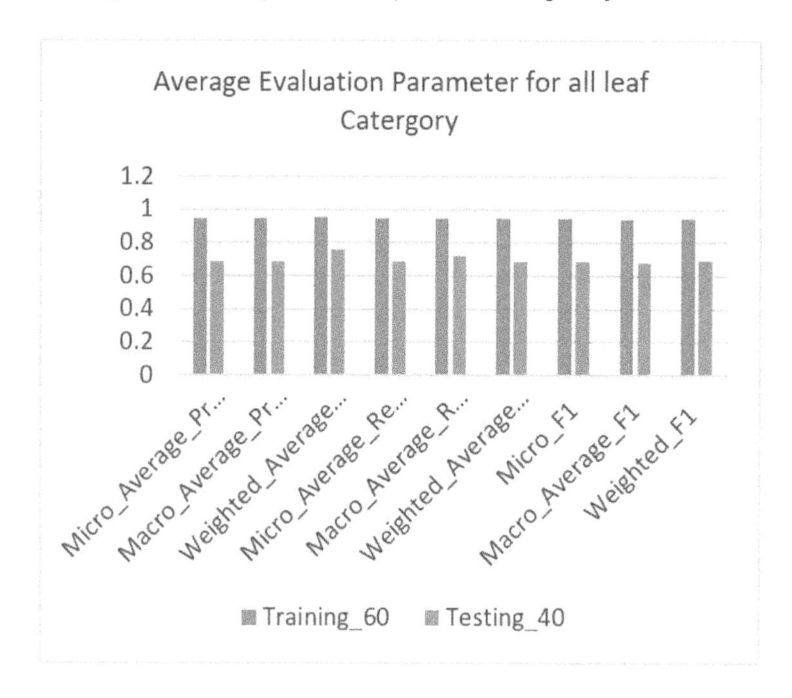

For the data split of 60:40, the weighted average precision, weighted average recall, and weighted average F1 for the training data set are 95.37%,94.44%, and 94.56% whereas for the test images, it is 75.78%,68.75%, and 69.46% respectively and same is depicted in Figure 11.

c) Results for data set with 70% training images and 30% testing images

Figure 12. Training process for the data split of 70:30 with validation accuracy of 97.62%

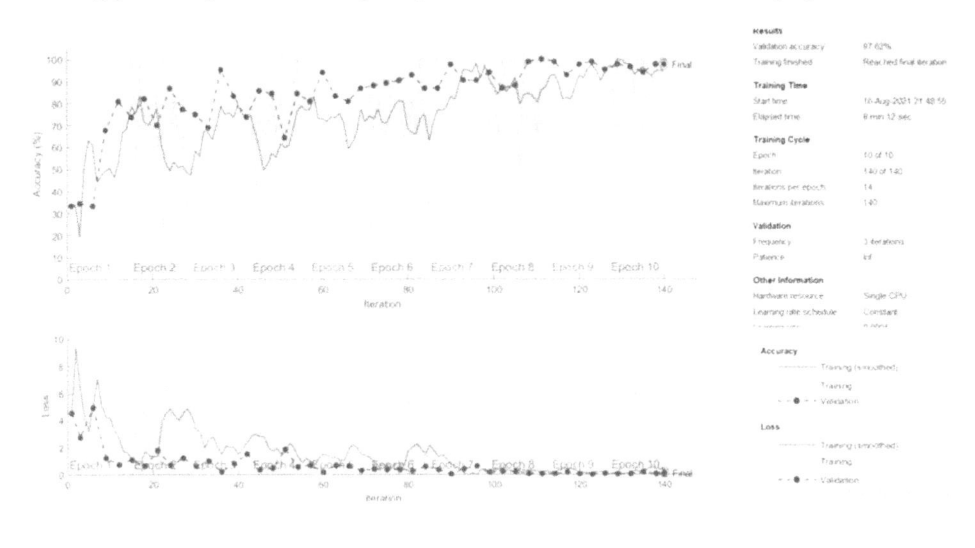

(a) (b)

Figure 13. (a) Confusion matrix for 70% of the training data set; (b) Confusion matrix for 30% of the testing data

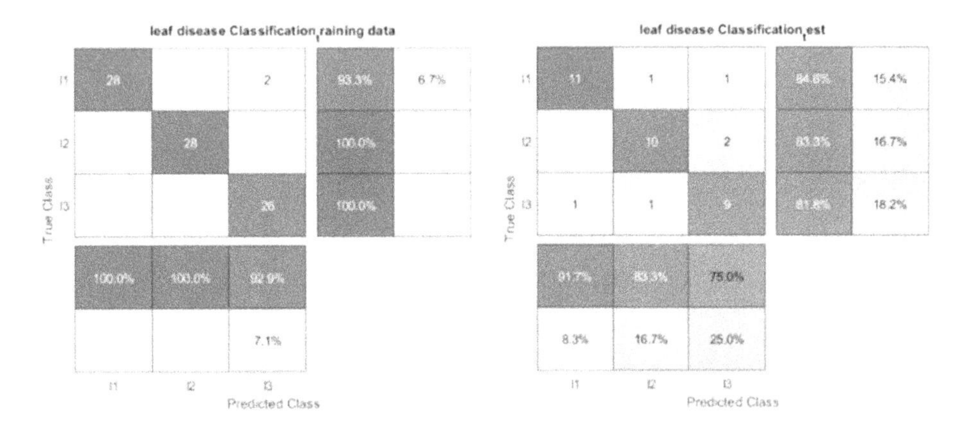

Figure 14. Individual leaf category evaluation parameter for 70% of training images and 30% of testing images

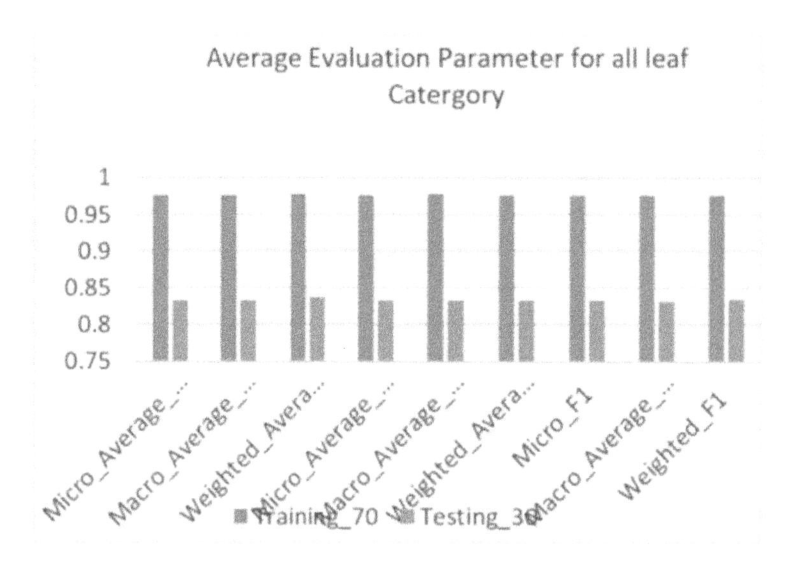

Figure 15. Weighted average evaluation parameters for the data split of 70:30

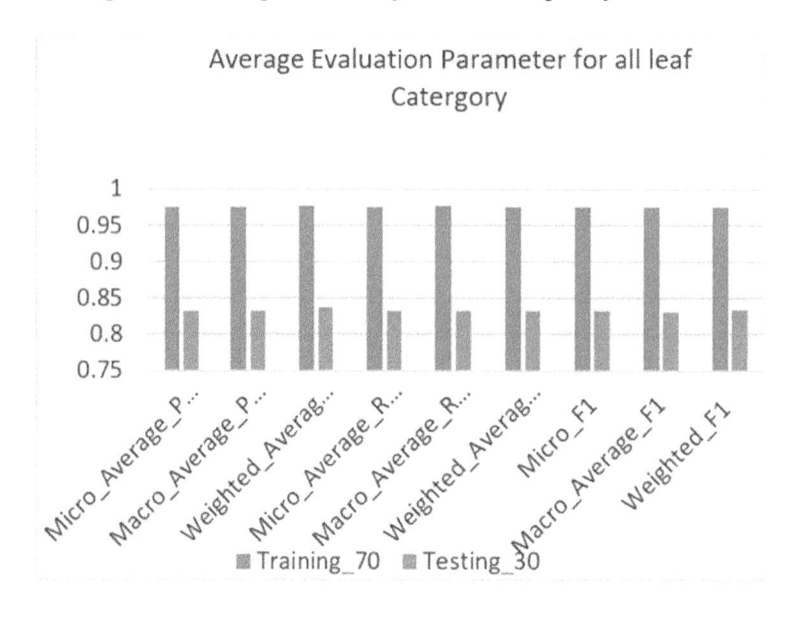

For the data split of 70:30, the training process confusion matrix is represented in Figure 12 and Figure 13, the weighted average precision, weighted average recall, and weighted average F1 for the training data set are 97.79%, 97.62%, and 97.62% whereas for the test images, it is 83.88%, 83.33%, and 83.47% respectively and the same is highlighted in Figure 14 and Figure 15

Figure 16. Comparison of performance metric for the different data split ratio

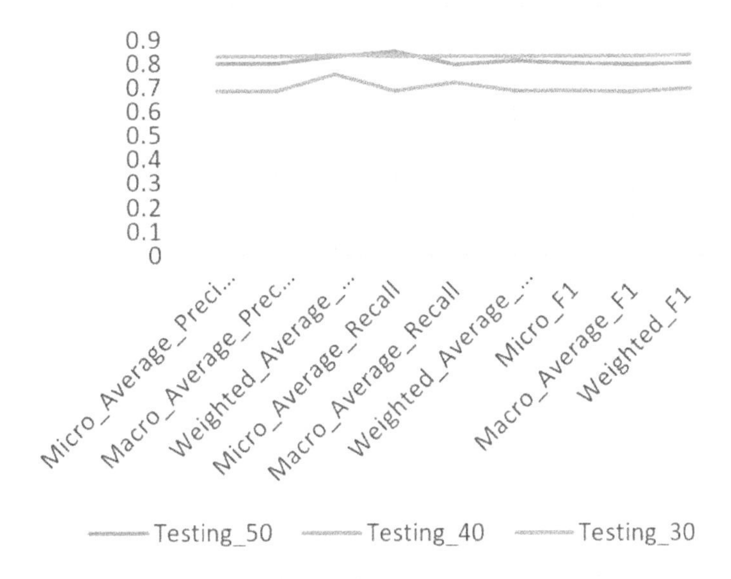

The performance metric comparison as mentioned in Figure 16 depicts that when the network is trained with a large number of data sets (70%), it can provide a good classification rate.

CONCLUSION

Agriculture is one of the economic sources for society. To provide a good yield to the farmer a timely identification of disease is very much needed. The model proposed gives us the classification accuracy results with trained augmented images.

The weighted average precision, weighted average recall, and weighted average F1 for the training data set are 89.08%, 85.00%, and 85.37% for the data split of 50:50, while these values are 82.92%, 80.9%, and 80.32% for the test data

The weighted average precision, weighted average recall, and weighted average F1 for the training data set are 95.37%, 94.44%, and 94.56% for the data split of 60:40, while they are 75.78%, 68.75%, and 69.46% for the test data

The weighted average precision, weighted average recall, and weighted average F1 for the training data set are 97.79%, 97.62%, and 97.62% for the data split of 70:30, respectively, while for the test data, they are 83.88%, 83.33%, and 83.47%.

The pre-trained fine-tuned model for the rice plant disease classification gives results for the data split ratio of 50:50, 60:40, and 70:30 data split.

The 70:30 data split is performing well. This study is limited to use of single pre-trained deep learning network; no deployment to hardware unit is employed. The approach can still be improved to yield a significant classifier that divides the classification task into discrete classes with good precision by modifying the hyper-parameters of the architectures.

REFERENCES

Ahmed, K., Shahidi, T. R., Alam, S. M. I., & Momen, S. (2019, December). Rice leaf disease detection using machine learning techniques. In *2019 International Conference on Sustainable Technologies for Industry 4.0 (STI)* (pp. 1-5). IEEE. 10.1109/STI47673.2019.9068096

Anthimopoulos, M., Christodoulidis, S., Ebner, L., Christe, A., & Mougiakakou, S. (2016). Lung pattern classification for interstitial lung diseases using a deep convolutional neural network. *IEEE Transactions on Medical Imaging*, 35(5), 1207–1216. 10.1109/TMI.2016.253586526955021

Archana, K. S., & Sahayadhas, A. (2018). Automatic rice leaf disease segmentation using image processing techniques. *Int. J. Eng. Technol, 7*(3.27), 182-185.

Atila, Ü., Uçar, M., Akyol, K., & Uçar, E. (2021). Plant leaf disease classification using EfficientNet deep learning model. *Ecological Informatics*, 61, 101182. 10.1016/j.ecoinf.2020.101182

Dhaka, V. S., Meena, S. V., Rani, G., Sinwar, D., Ijaz, M. F., & Woźniak, M. (2021). A survey of deep convolutional neural networks applied for prediction of plant leaf diseases. *Sensors (Basel)*, 21(14), 4749. 10.3390/s2114474934300489

Ferentinos, K. P. (2018). Deep learning models for plant disease detection and diagnosis. *Computers and Electronics in Agriculture*, 145, 311–318. 10.1016/j.compag.2018.01.009

Gill, A., Kaur, T., & Devi, Y. K. (2023). 38 A review for identification and detection of plant disease using machine learning. *Recent Advances in Computing Sciences:Proceedings of RACS 2022*, 216. Research Gate.

Guo, Y., Zhang, J., Yin, C., Hu, X., Zou, Y., Xue, Z., & Wang, W. (2020). Plant disease identification based on deep learning algorithm in smart farming. *Discrete Dynamics in Nature and Society*, 2020, 1–11. 10.1155/2020/2479172

Hussain, E., Mahanta, L. B., Das, C. R., & Talukdar, R. K. (2020). A comprehensive study on the multi-class cervical cancer diagnostic prediction on Pap smear images using a fusion-based decision from ensemble deep convolutional neural network. *Tissue and Cell, 65*, 101347.10.1016/j.tice.2020.101347

Islam, T., Sah, M., Baral, S., & Choudhury, R. R. (2018, April). A faster technique on rice disease detectionusing image processing of affected area in agro-field. In *2018 Second International Conference on Inventive Communication and Computational Technologies (ICICCT)* (pp. 62-66). IEEE. 10.1109/ICICCT.2018.8473322

Jogekar, R. N., & Tiwari, N. (2021). A review of deep learning techniques for identification and diagnosis of plant leaf disease. *Smart Trends in Computing and Communications:Proceedings of SmartCom 2020*, (pp. 435-441). Springer. 10.1007/978-981-15-5224-3_43

Kamilaris, A., & Prenafeta-Boldú, F. X. (2018). Deep learning in agriculture: A survey. *Computers and Electronics in Agriculture*, 147, 70–90. 10.1016/j.compag.2018.02.016

Krizhevsky, A., Sutskever, I., & Hinton, G. E. (2012). Imagenet classification with deep convolutional neural networks. *Advances in Neural Information Processing Systems*, 25.

Krizhevsky, A., Sutskever, I., & Hinton, G. E. (2017). ImageNet classification with deep convolutional neural networks. *Communications of the ACM*, 60(6), 84–90. 10.1145/3065386

Lu, J., Tan, L., & Jiang, H. (2021). Review on convolutional neural network (CNN) applied to plant leaf disease classification. *Agriculture*, 11(8), 707. 10.3390/agriculture11080707

Patil, R. R., & Kumar, S. (2020). A Bibliometric Survey on the Diagnosis of Plant Leaf Diseases using Artificial Intelligence. *Library Philosophy and Practice*, 1-26.

Pinki, F. T., Khatun, N., & Islam, S. M. (2017, December). Content based paddy leaf disease recognition and remedy prediction using support vector machine. In *2017 20th international conference of computer and information technology (ICCIT)* (pp. 1-5). IEEE. 10.1109/ICCITECHN.2017.8281764

Prajapati, H. B., Shah, J. P., & Dabhi, V. K. (2017). Detection and classification of rice plant diseases. *Intelligent Decision Technologies*, 11(3), 357–373. 10.3233/IDT-170301

Rahman, C. R., Arko, P. S., Ali, M. E., Khan, M. A. I., Apon, S. H., Nowrin, F., & Wasif, A. (2020). Identification and recognition of rice diseases and pests using convolutional neural networks. *Biosystems Engineering*, 194, 112–120. 10.1016/j.biosystemseng.2020.03.020

Ramesh, S., Hebbar, R., Niveditha, M., Pooja, R., Shashank, N., & Vinod, P. V. (2018, April). Plant disease detection using machine learning. In *2018 International conference on design innovations for 3Cs compute communicate control (ICDI3C)* (pp. 41-45). IEEE. 10.1109/ICDI3C.2018.00017

Sangeetha, R., Logeshwaran, J., Rocher, J., & Lloret, J. (2023). An improved agro deep learning model for detection of Panama wilts disease in banana leaves. *AgriEngineering*, 5(2), 660–679. 10.3390/agriengineering5020042

Shah, J. P., Prajapati, H. B., & Dabhi, V. K. (2016, March). A survey on detection and classification of rice plant diseases. In *2016 IEEE International Conference on Current Trends in Advanced Computing (ICCTAC)* (pp. 1-8). IEEE. 10.1109/ICCTAC.2016.7567333

Shin, J., Mahmud, M. S., Rehman, T. U., Ravichandran, P., Heung, B., & Chang, Y. K. (2022). Trends and prospect of machine vision technology for stresses and diseases detection in precision agriculture. *AgriEngineering*, 5(1), 20–39. 10.3390/agriengineering5010003

Shorten, C., & Khoshgoftaar, T. M. (2019). A survey on image data augmentation for deep learning. *Journal of Big Data*, 6(1), 1–48. 10.1186/s40537-019-0197-0

Shrivastava, V. K., Pradhan, M. K., Minz, S., & Thakur, M. P. (2019). Rice plant disease classification using transfer learning of deep convolution neural network. *The International Archives of the Photogrammetry, Remote Sensing and Spatial Information Sciences*, 42(W6), 631–635. 10.5194/isprs-archives-XLII-3-W6-631-2019

Sladojevic, S., Arsenovic, M., Anderla, A., Culibrk, D., & Stefanovic, D. (2016). Deep neural networks based recognition of plant diseases by leaf image classification. *Computational Intelligence and Neuroscience*, 2016, 2016. 10.1155/2016/328980127418923

Sujatha, R., Chatterjee, J. M., Jhanjhi, N. Z., & Brohi, S. N. (2021). Performance of deep learning vs machine learning in plant leaf disease detection. *Microprocessors and Microsystems*, 80, 103615. 10.1016/j.micpro.2020.103615

Swain, S., Nayak, S. K., & Barik, S. S. (2020). A review on plant leaf diseases detection and classification based on machine learning models. *Mukt shabd, 9*(6), 5195-5205.

Szegedy, C., Liu, W., Jia, Y., Sermanet, P., Reed, S., Anguelov, D., & Rabinovich, A. (2015). Going deeper with convolutions. In *Proceedings of the IEEE conference on computer vision and pattern recognition* (pp. 1-9). IEEE.

Tugrul, B., Elfatimi, E., & Eryigit, R. (2022). Convolutional neural networks in detection of plant leaf diseases. *Revista de Agricultura (Piracicaba)*, 12(8), 1192.

Zhang, K., Wu, Q., Liu, A., & Meng, X. (2018). Can deep learning identify tomato leaf disease? *Advances in Multimedia*, 2018, 2018. 10.1155/2018/6710865

Chapter 3
Advanced Computational Forecasting for Agri-Business Supply Chain Resilience

Kali Charan Rath
Department of Mechanical Engineering, GIET University, India

Lakshmi Prasad Panda
http://orcid.org/0000-0002-9185-2967
Government College of Engineering, Kalahandi, India

N. V. Jagannadha Rao
School of Management Studies, GIET University, Gunupur, India

Gopal Krushna Mohanta
Department of Mechanical Engineering; GIET University, Gunupur, India

Anmol Panda
http://orcid.org/0009-0001-8826-7302
GIET University, India

ABSTRACT

This chapter focuses on using advanced statistical methods to improve predictions in the agri-business sector. It integrates cutting-edge computational techniques and statistical models to address supply chain disruptions in agriculture. The main goal is to create a robust forecasting framework that predicts market trends, demand fluctuations, and enhances supply chain resilience. The novelty lies in combining advanced statistical methodologies like time series analysis, predictive modeling, and data-driven insights for a comprehensive approach. This aims to improve supply chain management in agri-business by fostering adaptability and resilience in changing market conditions.

DOI: 10.4018/979-8-3693-3583-3.ch003

INTRODUCTION

In India, the agri-business sector faces daunting challenges amidst global dynamics. Climate change brings erratic weather, shifting rainfall, and rising temperatures, directly affecting agricultural productivity. Resource scarcity, especially water, land, and energy, further strains farming practices, particularly in water-stressed and degraded regions, increasing farmers' vulnerability [Carayannis et al. (2018); Ulvenblad et al. (2020)] . Market volatility, influenced by domestic and international factors, poses additional uncertainty, impacting the profitability and sustainability of agri-businesses throughout the value chain.

India urgently needs to address food security amid a burgeoning population of over 1.3 billion people, with surging demand for food staples, straining the agricultural sector and necessitating adaptable supply chains to balance production and consumption amidst fluctuating patterns and evolving preferences [Kumar et al. (2023); Pandey & Pandey (2023) ; Shetty (2018)].Supply chain disruptions, environmental uncertainties, and risks from extreme weather events, pest outbreaks, and disease epidemics compound challenges in the agri-business sector, hindering efficient movement of goods and exacerbating vulnerability[Belhadi et al. (2024) ; Xu et al. (2021)].

Advanced computational forecasting strengthens India's agri-business supply chains by leveraging data analytics, AI, and predictive modeling for proactive decision-making, optimizing resource allocation, logistics, and demand fluctuations, enhancing operational efficiency and profitability [Kagalkar et al. (2023) ; Roy et al. (2023) ; Sarkar et al. (2023)].

Utilizing advanced computational forecasting, this chapter delves into enhancing coordination within India's supply chain, benefiting smallholder farmers with timely guidance and market insights. Novelty of the work reflects by integrating cutting-edge methodologies with agricultural expertise, it proposes tailored solutions employing AI and machine learning to tackle climate change, resource scarcity, and market fluctuations, thus bolstering the resilience of India's agri-business.

This chapter highlights how advanced computational forecasting empowers smallholder farmers in India by providing timely information and tools. It emphasizes inclusive strategies and collaborative partnerships among academia, industry, and government to drive innovation and address challenges in the agricultural sector. Its contextualized approach offers actionable insights for building resilience and fostering sustainable development.

METHODOLOGY

This study's methodology rigorously examines how advanced computational forecasting strengthens agri-business supply chain resilience in India. It combines theoretical frameworks, empirical analysis, and practical applications for a comprehensive understanding of the research problem.

A comprehensive literature review focuses on theoretical foundations, conceptual frameworks, and empirical evidence of advanced computational forecasting in agri-business supply chains, emphasizing computational modeling, predictive analytics, machine learning, and optimization. Assessing their relevance in the Indian agricultural context illuminates their potential to address supply chain challenges. Case studies offer real-world insights into computational forecasting's impact, spanning diverse commodities, supply chains, and regions, while empirical analysis evaluates performance in predicting demand, optimizing inventory, and managing disruptions, using statistical techniques and computational tools to derive actionable insights for improved supply chain efficiency and resilience.

Collaborative workshops facilitate knowledge exchange and solution co-creation among stakeholders in India's agri-business sector. They bring together diverse groups to discuss trends, share best practices, and explore opportunities for innovation. Through interactive sessions, participants tackle challenges in adopting advanced computational forecasting techniques to enhance supply chain resilience [Osumba et al. (2021)].

A comprehensive framework, informed by literature, case studies, and stakeholder insights, is developed to address key drivers and barriers to implementing advanced forecasting in Indian agri-business. Validated through expert review and workshops, this framework provides actionable recommendations for enhancing computational capabilities and supply chain resilience [Siddh et al. (2021); Joshi et al. (2023)].

Overally, the methodology employs a multi-method approach that combines qualitative and quantitative research methods and participatory techniques stakeholder engagement strategies to generate actionable insights and recommendations for advancing computational forecasting capabilities and enhancing supply chain resilience in the Indian agri-business sector.

AGROBUSINESS SCHEMES IN INDIA

India has implemented a range of agribusiness schemes aimed at fostering growth, sustainability, and resilience in the agricultural sector. These schemes encompass initiatives such as the Pradhan Mantri Krishi Sinchayee Yojana (PMKSY) for enhancing water use efficiency, the National Agricultural Market (e-NAM) to facilitate transparent trading, and the Pradhan Mantri Fasal Bima Yojana (PMFBY) providing crop insurance coverage. Additionally, schemes like Paramparagat Krishi Vikas Yojana (PKVY) promote organic farming, while Rashtriya Krishi Vikas Yojana (RKVY) supports agricultural development projects at the state level. Other key initiatives include the National Horticulture Mission (NHM) for holistic horticultural growth, the Pradhan Mantri Kisan Samman Nidhi (PM-KISAN) offering direct income support to farmers, and efforts under the Atma Nirbhar Bharat Abhiyan to bolster self-reliance in agriculture [Bhargav (2017) ; Prabha et al. (2016) ; Das et al. (2020) ; Negia & Kumarb (2020)].

Other few schemes are: Atma Nirbhar Bharat Abhiyan ; Mission for Integrated Development of Horticulture (MIDH); Soil Health Card Scheme; National Food Security Mission (NFSM); Interest Subvention Scheme for Agriculture Loans ; Pradhan Mantri Matsya Sampada Yojana (PMMSY); Dairy Entrepreneurship Development Scheme (DEDS); National Livestock Mission (NLM) ; Sub-Mission on Agricultural Mechanization ; Pradhan Mantri Kisan Sampada Yojana (PMKSY); Agri Export Policy; Mega Food Parks Scheme; Startup India Scheme

These schemes collectively aim to improve productivity, enhance farmer livelihoods, and ensure food security while fostering sustainable agricultural practices across the country.

The Benefits of These Schemes

These schemes in India aim to boost agricultural productivity, ensure financial security, promote sustainability, enhance market access, provide income support, encourage innovation, strengthen infrastructure, foster exports, and ensure food security while improving rural livelihoods.

a) Pradhan Mantri Krishi Sinchayee Yojana (PMKSY): PMKSY enhances water efficiency in agriculture through irrigation projects and micro-irrigation, increasing productivity, stabilizing farmer incomes, and promoting sustainable water use for agricultural development.

Farmers from states like Maharashtra, which face frequent droughts, can benefit from watershed management projects that conserve rainwater for agricultural use.

b) National Agricultural Market (e-NAM): e-NAM is an online platform connecting agricultural buyers and sellers, reducing costs, eliminating middlemen, and improving market efficiency for fairer prices and less wastage.

Farmers from states like Punjab, known for their surplus production of grains, can use e-NAM to access markets beyond their immediate vicinity, thereby fetching better prices for their crops.

c) Pradhan Mantri Fasal Bima Yojana (PMFBY): PMFBY offers crop insurance against natural calamities, pests, and diseases, ensuring financial stability for farmers and encouraging modern agricultural practices for improved yields and confidence in farming.

Farmers from states like Uttar Pradesh, which often face floods and erratic weather patterns, can benefit from PMFBY by mitigating their financial losses during crop failures.

d) Paramparagat Krishi Vikas Yojana (PKVY): PKVY incentivizes organic farming through financial support for organic inputs and practices, fostering sustainable agriculture, preserving biodiversity, and enhancing soil fertility for healthier ecosystems and safer food production with increased market opportunities for organic produce.

Farmers from states like Sikkim, which has been declared as the first organic state in India, can avail benefits from PKVY by transitioning to organic farming methods.

e) Rashtriya Krishi Vikas Yojana (RKVY): RKVY aids states in agricultural projects to boost productivity, sustainability, and competitiveness. With flexible funding, it fosters innovation, technology adoption, and infrastructure development, leading to increased productivity, income, and rural development.

Farmers from states like Madhya Pradesh, known for their innovative agriculture practices, can benefit from RKVY by accessing funds for projects that improve agricultural productivity and sustainability.

f) National Horticulture Mission (NHM): NHM supports holistic horticulture growth by aiding development, infrastructure, and market connections. It creates jobs, boosts rural income, and improves nutrition while conserving biodiversity and natural resources.

Farmers from states like Himachal Pradesh, famous for its apple orchards, can benefit from NHM by receiving support for infrastructure development and market linkages to sell their produce.

g) Pradhan Mantri Kisan Samman Nidhi (PM-KISAN): PM-KISAN provides small and marginal farmers with 6,000 annually, enhancing their income and livelihood resilience. This support promotes investment in agricultural inputs, reduces rural poverty, and stimulates economic growth through increased rural consumption.

Farmers from states like Bihar, where a significant portion of the population depends on agriculture for livelihood, can benefit from PM-KISAN by receiving direct financial assistance.

h) Atma Nirbhar Bharat Abhiyan: Atma Nirbhar Bharat Abhiyan promotes agricultural self-reliance through marketing, infrastructure, and credit reforms. It fosters a favorable policy environment, encourages private sector participation, and strengthens agricultural value chains, leading to a more resilient and sustainable sector capable of meeting India's diverse economic needs.

Farmers from states like Rajasthan, where lack of infrastructure hampers agricultural growth, can benefit from Atma Nirbhar Bharat Abhiyan by gaining access to better market facilities and credit support.

i) Mission for Integrated Development of Horticulture (MIDH): MIDH supports horticulture development through interventions like area expansion and post-harvest management. By integrating value chains, it enhances productivity and market access, promoting diversification and rural prosperity through increased employment and income.

Farmers from states like Karnataka, known for its diverse horticulture produce, can benefit from MIDH by receiving support for expanding their horticulture activities and accessing better markets.

j) Soil Health Card Scheme: The Soil Health Card Scheme provides farmers with soil health cards, aiding informed decisions on fertilizer use. Through soil testing, it optimizes inputs, boosts fertility, and increases yields sustainably. This saves costs, reduces pollution, and conserves soil health for long-term agricultural sustainability.

Farmers from states like Punjab, where intensive agriculture has led to soil degradation, can benefit from Soil Health Card Scheme by adopting soil conservation practices based on the card recommendations.

k) National Food Security Mission (NFSM): NFSM enhances food security by increasing production of rice, wheat, and pulses, targeting small farmers for inclusive growth.

Farmers from states like Uttar Pradesh, where rice and wheat are major crops, can benefit from NFSM by receiving support for adopting modern technologies and practices to enhance crop yields.

l) Interest Subvention Scheme for Agriculture Loans: The Interest Subvention Scheme for Agriculture Loans offers interest subsidy to farmers, reducing credit costs, fostering investment in agricultural inputs and technologies, and boosting productivity, income, and rural development.

Farmers from states like Gujarat, where access to credit is crucial for agricultural growth, can benefit from interest subvention scheme by availing loans at lower interest rates.

m) Pradhan Mantri Matsya Sampada Yojana (PMMSY): PMMSY aims to boost fish production and aquaculture via infrastructure development, technology adoption, and value addition, promoting growth, employment, income generation, food security, nutrition, entrepreneurship, and fisheries development.

Farmers from coastal states like Andhra Pradesh, known for their fisheries sector, can benefit from PMMSY by receiving support for modernizing their fish farming practices and accessing better markets.

n) Dairy Entrepreneurship Development Scheme (DEDS): DEDS supports modern dairy farms and processing plants, fostering entrepreneurship, enhancing productivity, creating jobs, and stimulating growth in the dairy sector.

Farmers from states like Gujarat, famous for its dairy cooperatives, can benefit from DEDS by setting up modern dairy farms and processing units, thereby increasing their income from dairy activities.

o) National Livestock Mission (NLM): NLM aims to enhance livestock productivity, improve livelihoods, ensure food security, and promote climate resilience in the livestock sector.

Farmers from states like Rajasthan, where livestock rearing is a significant source of income, can benefit from NLM by receiving support for improving the productivity and health of their livestock.

p) Sub-Mission on Agricultural Mechanization: The Sub-Mission on Agricultural Mechanization facilitates farm machinery adoption for small and marginal farmers, enhancing productivity and rural livelihoods.

Farmers from states like Punjab, known as the 'granary of India,' can benefit from agricultural mechanization by reducing labor costs and increasing farm productivity.

q) Pradhan Mantri Kisan Sampada Yojana (PMKSY): PMKSY aids food processing infrastructure, cold chain facilities, and value addition, benefiting farmers in states like Maharashtra by improving processing and adding value to their produce.

r) Agri Export Policy: This policy aims to double agricultural exports and integrate Indian farmers with global value chains through policy support and infrastructure development.

Farmers from states like Punjab, which produce surplus grains, can benefit from Agri Export Policy by accessing international markets and fetching better prices for their produce.

s) Mega Food Parks Scheme: This scheme facilitates the establishment of mega food parks with state-of-the-art infrastructure for food processing, storage, and value addition.

Farmers from states like Uttar Pradesh, with a large agricultural base, can benefit from Mega Food Parks Scheme by accessing modern processing facilities and reducing post-harvest losses.

t) Startup India Scheme: This scheme encourages entrepreneurship in the agri-business sector by providing financial support, mentorship, and incubation facilities to startups and innovators.

Farmers from states like Telangana, known for their innovative agricultural practices, can benefit from Startup India Scheme by receiving support for developing and scaling up agri-business ventures.

These above schemes illustrate about farmers from different states can avail benefits from various government schemes and initiatives aimed at promoting agricultural growth, sustainability, and livelihood enhancement.

Agri Business Model and Rural Development

An agro-business model is a framework or plan that outlines the various aspects of operating a business in the agricultural sector. It encompasses activities related to the production, processing, marketing, and distribution of agricultural products and services. Agro-business models are designed to generate profits while also contributing to the development of rural areas.

Here's a breakdown of the key components of an agro-business model:

a) **Production:** Agricultural production encompasses cultivating crops, raising livestock, and managing resources like land and water, involving decisions on what, where, and how to produce commodities.
b) **Processing and value addition:** After harvesting, agricultural products undergo processing such as milling, drying, packaging, and preserving to enhance value or prepare for consumption or further use.
c) **Marketing and distribution:** Agro-business models encompass marketing and distribution strategies, involving target market identification, branding, distribution channel establishment, and sales management.
d) **Financial management:** Effective financial management is essential for agro-business success, involving budgeting, cost analysis, pricing strategies, cash flow management, financing, and performance monitoring.
e) **Human resources and skills development:** Agro-business success hinges on skilled workforce management, addressing hiring, training, retention, and fostering innovation and continuous learning.
f) **Environmental sustainability:** Sustainable agriculture practices are vital for agro-businesses, minimizing environmental impacts and ensuring long-term viability through methods like organic farming, water and soil conservation, and biodiversity preservation.
g) **Community engagement and social responsibility:** Agro-business models must account for social and community impacts by engaging with local communities, respecting land rights, promoting fair labor practices, and contributing to local development initiatives.

Importance of agro-business models for rural development:

a) **Income generation:** Agro-businesses offer rural residents opportunities for income through employment, entrepreneurship, or product sales, reducing poverty and enhancing livelihoods in rural areas.
b) **Employment opportunities:** Agro-businesses, relying on substantial labor for farming, processing, transportation, and marketing, combat unemployment and underemployment in rural areas by creating job opportunities.

c) **Infrastructure development:** Agro-businesses drive investment in rural infrastructure like roads, storage facilities, processing plants, and marketplaces, benefiting both the business and the broader community through improved access to services and markets.
d) **Value addition and market access:** Agro-businesses add value to agricultural products through processing and marketing, enabling farmers to fetch higher prices and improving market access while reducing post-harvest losses.
e) **Technology transfer and innovation:** Agro-businesses drive agricultural innovation and knowledge transfer, improving productivity and efficiency in rural areas, thus fostering overall economic development.
f) **Sustainable development:** Agro-business models that prioritize environmental sustainability contribute to the conservation of natural resources and the preservation of ecosystems. This ensures the long-term viability of agriculture and supports the well-being of rural communities.

So, agro-business models play a vital role in driving rural development by creating economic opportunities, improving livelihoods, fostering innovation, and promoting sustainable agriculture practices. They serve as engines of growth that benefit both the agricultural sector and the broader rural economy.

Case Study: Amul Dairy Cooperative

Amul is a renowned dairy cooperative in India, owned and managed by rural milk producers. It operates on a cooperative model where farmers collectively own the dairy and participate in its management. Founded in 1946 in the state of Gujarat, Amul has grown into one of the largest dairy organizations in the world.

a) **Production:** Amul's agro-business model begins with dairy production by small-scale rural farmers who own cows and buffaloes. These farmers supply milk to Amul collection centers located in villages across Gujarat and other states.
b) **Processing and Value Addition:** Once collected, the milk is transported to Amul's dairy processing plants, where it undergoes pasteurization, homogenization, and other processing stages to produce various dairy products such as milk, butter, ghee, cheese, yogurt, and ice cream. Value addition occurs through branding, packaging, and product differentiation.
c) **Marketing and Distribution:** Amul has a robust marketing and distribution network that reaches both urban and rural markets. It employs a multi-tiered distribution system involving wholesalers, retailers, and cooperatives, ensuring wide accessibility of its products across India.
d) **Financial Management:** Amul operates on a cooperative business model where profits are shared among its member farmers after deducting operational expenses. Financial management is overseen by elected representatives from the farming community who serve on the board of directors.
e) **Human Resources and Skills Development:** Amul provides training and capacity-building programs to its member farmers on best dairy farming practices, animal husbandry, and quality control measures. This helps to improve the productivity and livelihoods of rural dairy farmers.
f) **Environmental Sustainability:** Amul promotes sustainable dairy farming practices among its member farmers, including fodder cultivation, waste management, and water conservation initiatives. It also adheres to environmental regulations in its processing plants to minimize pollution.

g) **Community Engagement and Social Responsibility:** Amul actively engages with rural communities by providing healthcare services, education programs, and infrastructure development initiatives. It empowers women through its women's dairy cooperative societies and supports local community development projects.

Figure 1. Milk production (liters) vs. revenue (in INR)

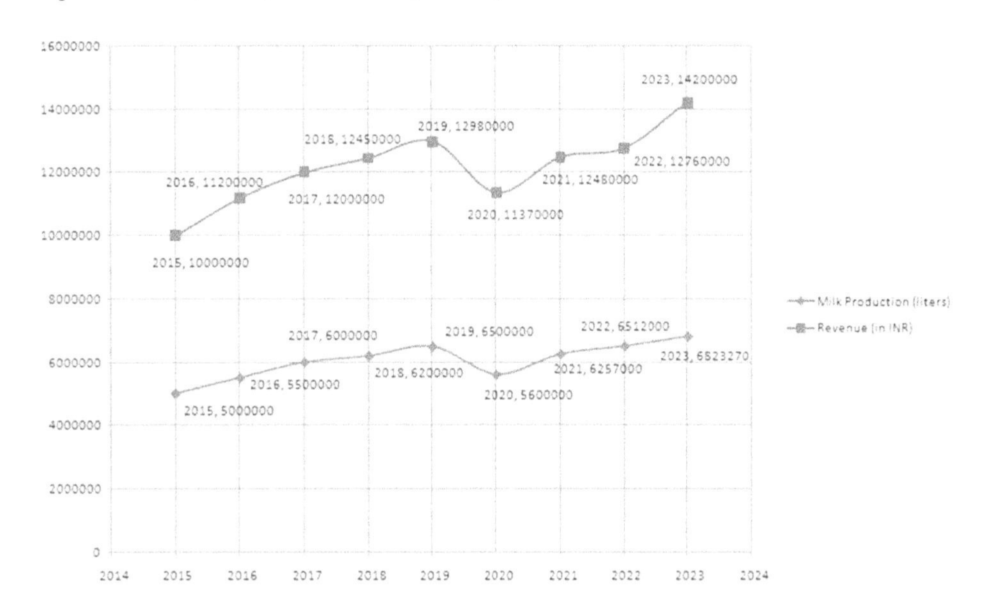

Impact on Rural Development

a) Economic Empowerment: Amul has empowered millions of rural farmers by providing them with a stable source of income through dairy farming.
b) Employment Generation: The dairy cooperative model has created numerous employment opportunities in rural areas, both directly and indirectly, through milk collection, processing, distribution, and retailing.
c) Rural Development: Amul's presence has led to the development of infrastructure, social institutions, and economic activities in rural regions, contributing to overall rural development.
d) Poverty Alleviation: By enabling small-scale farmers to earn a fair income from milk production, Amul has played a significant role in alleviating poverty in rural communities.

Amul Dairy Cooperative serves as a remarkable example of an agro-business model that not only generates profits but also promotes rural development in India. Through its cooperative structure, value chain integration, and commitment to sustainability, Amul has transformed the lives of millions of rural farmers and contributed to the socio-economic development of rural areas across the country.

Case Study on Mushroom Cultivation for Rural Development

Agribusiness plays a crucial role in India's rural development by providing livelihood opportunities, enhancing agricultural productivity, and ensuring food security. Mushroom cultivation stands out as a promising sector within agribusiness due to its relatively low investment requirements, high yield potential, and compatibility with small-scale farming. This case study explores the impact of mushroom cultivation on rural development in India, focusing on a specific agribusiness venture.

The case study centers on a mushroom cultivation initiative launched in a rural village in Maharashtra, India. The initiative aimed to empower local farmers, particularly women, by introducing them to mushroom cultivation as an alternative source of income.

Agribusiness Model

a) **Training and capacity building:** The initiative began with training sessions conducted by agricultural experts to educate farmers about mushroom cultivation techniques, including substrate preparation, spawning, casing, and harvesting. The training also covered aspects of marketing, quality control, and financial management.
b) **Infrastructure development:** The project provided essential infrastructure such as mushroom spawn, growing containers, composting units, and drying facilities. Farmers were assisted in setting up low-cost mushroom houses using locally available materials.
c) **Market linkages:** To ensure market access and fair prices for the harvested mushrooms, the project facilitated linkages with local markets, supermarkets, restaurants, and food processing companies. Additionally, efforts were made to create awareness among consumers about the nutritional benefits of mushrooms.
d) **Financial support:** Financial assistance in the form of subsidies, grants, or low-interest loans was provided to farmers for initial investment in infrastructure, spawn procurement, and other operational expenses.

Table 1. Mushroom cultivation status

Aspect	Description	Data/Details
Production Rank	National Ranking	2nd (as of 2023)
Production Share	Percentage of National Production	9.89%
Major Cultivated Varieties	Popular Species	Button Mushroom (Agaricus bisporus), Oyster Mushroom (Pleurotus spp.), Milky Mushroom (Calocybe indica)
Investment Range (per unit)	Small-scale	10,000 - 50,000
	Medium-scale	50,000 - 2,00,000
	Large-scale	2,00,000 - 10,00,000+
Profit Margin	Average	30% - 50%

continued on following page

Table 1. Continued

Aspect	Description	Data/Details
Policy Impact	Government Schemes	• Maharashtra State Horticulture Mission (MSHM) provides subsidies for mushroom farm setup and training. • National Horticulture Board (NHB) offers financial assistance and technical support. • NABARD and other banks provide loans for mushroom farming ventures.
	Impact on Profit	• Subsidies reduce initial investment costs. • Training programs improve cultivation practices and yield. • Loans provide access to capital for expansion.
Challenges	Obstacles faced by farmers	• Access to quality spawn and compost. • Fluctuations in market prices. • Lack of proper storage and transportation facilities.
Potential Solutions	Initiatives to address challenges	• Government support for spawn and compost production units. • Establishment of cold storage facilities and market linkages. • Training programs on post-harvest management and marketing.

Impact on Rural Development

a) **Economic empowerment:** Mushroom cultivation provided a viable source of additional income for rural households, particularly women, who often face limited employment opportunities. The steady income generated from mushroom sales helped improve their standard of living and contributed to poverty alleviation.
b) **Skill development:** Through training and hands-on experience in mushroom cultivation, farmers acquired new skills and knowledge, enhancing their agricultural expertise and employability. This led to a more diversified and resilient rural economy.
c) **Social cohesion:** The project fostered community cohesion by promoting collective decision-making, knowledge sharing, and mutual support among farmers. Collaborative efforts in mushroom cultivation strengthened social ties and solidarity within the village.
d) **Environmental sustainability:** Mushroom cultivation is relatively eco-friendly, as it utilizes agricultural by-products such as crop residues and organic waste as substrates. By promoting sustainable farming practices, the initiative contributed to environmental conservation and resource optimization.

Various challenges and overcomes of this model:

a) Limited access to quality spawn:

Challenge: Farmers may face difficulties in accessing high-quality mushroom spawn, which is essential for successful cultivation.

Overcome: Collaborate with government agencies, research institutions, or private suppliers to ensure consistent availability of quality spawn. Establish local spawn production units or training programs to empower farmers to produce their own spawn.

b) Fluctuations in market demand:

Challenge: Market demand for mushrooms can be volatile, leading to income instability for farmers.

Overcome: Diversify product offerings by exploring value-added products like dried mushrooms, mushroom powder, or pickled mushrooms. Build long-term partnerships with buyers, such as restaurants, supermarkets, and food processing companies, to secure stable market outlets.

c) Pest and disease management:

Challenge: Mushroom cultivation is susceptible to pests, diseases, and contamination, leading to crop losses.
Overcome: Implement strict hygiene protocols and proper sanitation measures in mushroom houses. Train farmers in integrated pest management techniques, including biological control methods and use of natural pesticides. Regular monitoring and prompt intervention are crucial to prevent and manage pest outbreaks effectively.

d) Limited technical knowledge and skills:

Challenge: Farmers may lack adequate technical knowledge and skills required for successful mushroom cultivation.
Overcome: Provide comprehensive training and capacity-building programs covering all aspects of mushroom cultivation, including substrate preparation, spawn inoculation, cropping techniques, and post-harvest handling. Offer hands-on training sessions, demonstration plots, and mentorship programs to facilitate experiential learning and skill development.

e) Infrastructure constraints:

Challenge: Limited access to infrastructure such as suitable growing houses, composting units, and drying facilities can hinder the scalability of mushroom cultivation.
Overcome: Implement low-cost, scalable infrastructure solutions using locally available materials. Encourage community participation in infrastructure development by organizing collective construction efforts or providing financial support for building essential facilities. Explore innovative technologies such as low-cost polyhouses or controlled environment systems to optimize production efficiency and minimize resource requirements.

f) Access to finance:

Challenge: Lack of access to affordable credit and financial resources may impede farmers' ability to invest in mushroom cultivation.
Overcome: Facilitate access to financial services such as microcredit, subsidies, grants, or revolving funds tailored to the needs of smallholder farmers. Collaborate with financial institutions, NGOs, or government agencies to design inclusive financing schemes with flexible repayment terms and low-interest rates. Empower farmers with financial literacy training and assistance in preparing business plans to secure funding for their ventures.

g) Market linkages and value chain integration:

Challenge: Limited market linkages and inefficient value chains can hinder farmers' ability to capture value from their mushroom produce.

Overcome: Strengthen market linkages by establishing direct procurement arrangements with buyers, cooperatives, or marketing federations. Facilitate value chain integration through vertical integration or partnership with downstream players such as processors, distributors, or exporters. Promote branding, packaging, and quality certification initiatives to enhance product competitiveness and market access.

The following graph shows one example of the production of mushroom and % shares of various states of India during the recent past year.

Figure 2. Production of mushroom by various states in Indian

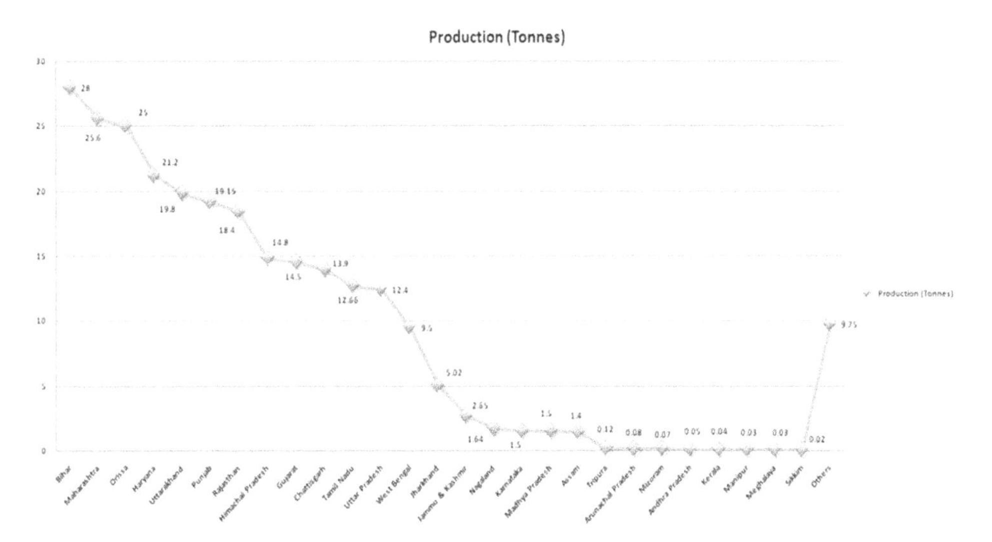

Figure 3. Percent share of various states for mushroom production in Indian

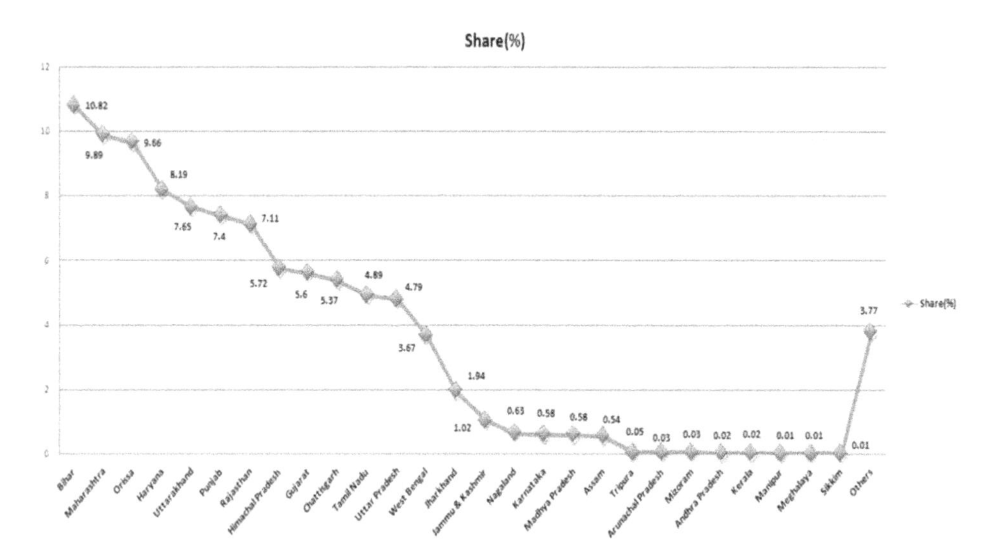

From the output graph, it is observed that:

a) Production analysis:

Bihar has the highest production with approximately 28 tons.
Other states with relatively high production include Maharashtra, Orissa, Haryana, and Uttarakhand.
States like Sikkim, Manipur, Meghalaya, Kerala, Andhra Pradesh, Mizoram, Arunachal Pradesh, and Tripura have very low production, each contributing less than 0.1 tons.

b) Share analysis:

Bihar also has the highest share percentage among all states, with approximately 10.82%.
The top five states with the highest share percentages are Bihar, Maharashtra, Orissa, Haryana, and Uttarakhand.
Sikkim, Manipur, Meghalaya, Kerala, Andhra Pradesh, Mizoram, Arunachal Pradesh, and Tripura have negligible shares, each contributing less than 0.05%.

c) Comparison:

Bihar, Maharashtra, Orissa, Haryana, and Uttarakhand are not only the top producers but also have significant shares, indicating their dominance in production and distribution.
States like Jharkhand, Jammu & Kashmir, Nagaland, Karnataka, Madhya Pradesh, and Assam have moderate production but relatively lower shares, suggesting potential areas for improvement in distribution or utilization.

The plot provides a clear visual comparison between production and share, allowing for easy identification of states with high production but lower shares or vice versa.

d) Overall trend:

There is a variation in both production and share among different states, reflecting the diverse agricultural landscape of India.

The graph highlights the disparity in agricultural productivity and distribution across states, which could be influenced by factors such as climate, soil fertility, infrastructure, and government policies.

In summary, the analysis of the output graph provides insights into the production and share distribution of various states in India during the specified period, helping stakeholders in agriculture and policymaking to identify areas for intervention and improvement.

Suggestions for better mushroom production

Followings are few suggestions for better mushroom production:

a) **Quality Spawn:** Start with high-quality spawn (seed material) sourced from reputable suppliers. Ensure that the spawn is healthy, free from contaminants, and of the desired mushroom species.
b) **Substrate Selection:** Choose a suitable substrate for mushroom cultivation based on the mushroom species you intend to grow. Common substrates include composted agricultural waste (such as straw, sawdust, or agricultural residues) or synthetic substrates. Ensure the substrate is well-composted, free from pests and diseases, and properly pasteurized or sterilized to eliminate competitors.
c) **Optimal Environmental Conditions:** Maintain optimal environmental conditions for mushroom growth, including temperature, humidity, light, and air circulation. Different mushroom species have specific requirements, so it's essential to research and monitor these parameters closely.
d) **Hygiene and Sanitation:** Maintain strict hygiene and sanitation practices throughout the mushroom cultivation process. Clean and disinfect all equipment, growing containers, and growing areas to prevent contamination by pathogens, pests, and competing organisms.
e) **Proper Spawning and Casing:** Ensure proper spawning and casing procedures. Spawning involves mixing the spawn with the substrate, while casing involves covering the spawn-substrate mixture with a layer of casing material (such as peat moss or vermiculite) to promote fruiting.
f) **Humidity Control:** Maintain high humidity levels within the growing environment to promote mushroom formation and growth. This can be achieved through regular misting or the use of automated humidification systems.
g) **Ventilation:** Ensure adequate air exchange and ventilation within the growing area to prevent the buildup of carbon dioxide and maintain optimal oxygen levels for mushroom growth.
h) **Harvesting Techniques:** Harvest mushrooms at the right stage of maturity to maximize yield and quality. Use sharp, clean knives or scissors to cut mushrooms from the substrate, taking care not to damage surrounding mushroom pins or mycelium.
i) **Post-Harvest Handling:** Handle harvested mushrooms carefully to minimize bruising and damage. Store mushrooms in a cool, humid environment to extend shelf life and maintain quality.
j) **Continuous Learning and Improvement:** Stay informed about the latest techniques, research, and developments in mushroom cultivation. Experiment with new methods, varieties, and substrates to optimize production and quality.

By following these suggestions and paying close attention to detail throughout the cultivation process, you can improve mushroom production and achieve higher yields of high-quality mushrooms. Additionally, seeking guidance from experienced growers, attending workshops, or training programs, and joining mushroom cultivation associations or forums can provide valuable insights and support for successful mushroom cultivation.

The increased investment in mushroom cultivation likely contributed to the improved production and profitability observed in later years. However, careful monitoring and management are necessary to address the fluctuations and ensure sustainable growth in the future. Additionally, analyzing the reasons behind the fluctuations, such as market dynamics, weather conditions, or operational issues, can provide insights for strategic decision-making in mushroom cultivation ventures.

By addressing various challenges through proactive measures and strategic interventions, agribusiness ventures focused on mushroom cultivation can enhance their resilience, sustainability, and impact on rural development in India. Collaboration among stakeholders, including government agencies, research institutions, NGOs, private sector entities, and local communities, is essential for implementing effective solutions and unlocking the full potential of mushroom cultivation as a catalyst for inclusive growth and poverty alleviation.

AGRI-BUSINESS SUPPLY CHAIN SYSTEM IN INDIA

Here's a short, detailed breakdown of the agri-business supply chain system in India with examples of key players and processes:

a) **Input supply:** Input supply refers to the provision of essential resources and materials necessary for agricultural production, including seeds, fertilizers, pesticides, machinery, and equipment.

- Seed Companies: Companies like Mahyco, Nuziveedu Seeds, and Advanta Seeds provide high-quality seeds adapted to various agro-climatic conditions.
- Fertilizer Manufacturers: Companies such as IFFCO, Coromandel International, and Rashtriya Chemicals & Fertilizers produce and distribute fertilizers like urea, DAP, and complex fertilizers.
- Pesticide Suppliers: Bayer Crop Science, Syngenta, and UPL Limited supply pesticides, herbicides, and insecticides to protect crops from pests and diseases.
- Agricultural Machinery Companies: Examples include Mahindra & Mahindra, TAFE (Tractors and Farm Equipment Limited), and Escorts Limited, which manufacture and distribute tractors, harvesters, and other farm machinery.

b) **Farmers:** Smallholder farmers across India cultivate crops like rice, wheat, pulses, cotton, and vegetables using inputs acquired from seed companies, fertilizer manufacturers, and pesticide suppliers.

c) **Produce collection and aggregation:** Produce collection and aggregation refer to the process of gathering agricultural products from farms, local markets, or collection centers and consolidating them for further processing, distribution, or sale in the market.

- Mandis (Wholesale Markets): Farmers bring their produce to local mandis, such as the Azadpur Mandi in Delhi or the Vashi Market in Mumbai, where traders and wholesalers purchase agricultural commodities in bulk.
- Collection Centers: Companies like ITC e-Choupal and BigBasket operate collection centers where farmers can sell their produce directly or through farmer cooperatives.

d) **Processing and value addition:** Processing and value addition involve converting raw agricultural produce into processed goods through various techniques such as cleaning, sorting, grading, packaging, and sometimes transforming them into higher-value products. These activities enhance the quality, shelf-life, and marketability of agricultural commodities.

- Milling Companies: Companies like ITC, Cargill, and Adani Wilmar operate rice and wheat mills, processing grains into rice, flour, and other value-added products.
- Food Processing Units: Parle Agro, PepsiCo India, and Nestle India process fruits, vegetables, and other agricultural products into juices, snacks, and packaged foods.

e) **Distribution and logistics:** Distribution and logistics encompass the planning, organization, and execution of activities involved in transporting, storing, and delivering agricultural products from producers to consumers or intermediary points within the supply chain. This includes managing transportation routes, storage facilities, inventory, and ensuring timely and efficient delivery of goods to meet market demands.

- Cold Storage and Warehousing: Companies such as Snowman Logistics and Dev Bhumi Cold Chain operate cold storage facilities and warehouses to store perishable agricultural products.
- Transportation Providers: Transport companies like Transport Corporation of India (TCI) and Blue Dart provide logistics services for transporting agricultural goods from production areas to markets.

f) **Marketing and retail:** Marketing and retail involve strategies and activities aimed at promoting and selling agricultural products to consumers. This includes branding, advertising, pricing, packaging, and distribution through various retail channels such as supermarkets, grocery stores, online platforms, farmers' markets, and direct sales to consumers.

- Supermarkets and Hypermarkets: Retail chains like Reliance Fresh, Big Bazaar, and Spencer's Retail sell a wide range of fresh produce and packaged food products to consumers.
- Online Platforms: Companies like Grofers, BigBasket, and Amazon Pantry offer online platforms for purchasing groceries and agricultural products.

g) **Export and international trade:** Export and international trade refer to the commercial activities involved in selling agricultural products to foreign markets. This includes the negotiation, sale, shipment, and delivery of goods across international borders. Exporters manage various aspects such as compliance with trade regulations, documentation, logistics, and payment terms to facilitate the smooth flow of agricultural products to overseas customers.

Exporters: Companies such as Adani Agri Logistics, Olam International, and Louis Dreyfus Company export agricultural commodities like rice, spices, fruits, and vegetables to international markets.

h) **Regulatory framework:** The regulatory framework comprises laws and oversight ensuring compliance with quality, safety, and trade standards in the agricultural sector.

Regulatory bodies like the Food Safety and Standards Authority of India (FSSAI) and the Ministry of Agriculture and Farmers Welfare set standards and regulations related to food safety, quality, and trade.

i) **Technology integration:** Technology integration involves incorporating digital tools, innovations, and solutions into various stages of the agricultural supply chain to improve efficiency, productivity, traceability, and decision-making processes.

- Farm Management Apps: Apps like AgriApp, CropIn, and FarmERP provide farmers with tools for crop management, weather forecasting, and market information.
- IoT Devices: Companies like Intello Labs and Ninjacart use IoT devices for monitoring crop conditions, inventory management, and supply chain traceability.

The agri-business supply chain system in India offers numerous benefits across the agricultural sector [Kagalkar et al. (2023); Roy et al. (2023)]. Firstly, it enhances efficiency by streamlining processes from input supply to distribution, reducing wastage, and ensuring timely delivery of agricultural products to markets. Secondly, it promotes economic growth by providing employment opportunities, particularly in rural areas, and facilitating trade both domestically and internationally, thus contributing to the country's GDP. Additionally, the supply chain system enables farmers to access quality inputs, technology, and market information, empowering them to improve productivity and incomes. Furthermore, by integrating technology and modern practices, it enhances transparency, traceability, and food safety standards, thereby ensuring consumer trust and confidence in agricultural products. Overall, a well-functioning agri-business supply chain system plays a pivotal role in the sustainable development of India's agriculture sector, fostering resilience, competitiveness, and food security.

CONCLUSION

In conclusion, the agricultural sector and its associated agribusiness supply chain in India are fundamental to sustainable development, playing crucial roles in ensuring food security, economic stability, and environmental conservation. Through the implementation of sustainable practices and technological advancements, agriculture maximizes resource efficiency while minimizing negative environmental impacts, thereby enhancing resilience to climate change and fostering efficient supply chains.

Agribusiness, as an integral part of the agricultural sector, drives innovation, supports smallholder farmers, and promotes inclusive growth, thereby contributing to social development and economic prosperity. The integration of various stakeholders, processes, and technologies within the agribusiness supply chain facilitates the smooth flow of agricultural products from farm to market, benefiting farmers, consumers, and the overall economy.

However, persistent challenges such as infrastructural limitations, market volatility, and regulatory constraints pose obstacles to the efficiency and sustainability of the supply chain. Addressing these challenges necessitates collaborative efforts among government, industry, and civil society to invest in infrastructure, improve market access, promote innovation, and strengthen regulatory frameworks.

By addressing these challenges and capitalizing on emerging opportunities, the agribusiness supply chain in India can further enhance its contribution to the growth and development of the agricultural sector. This will ensure continued food security, prosperity, and resilience for future generations, while also preserving natural resources and fostering sustainable development.

Future Scope of Work

Looking ahead, there are significant opportunities for further enhancing the agri-business supply chain system in India. Continuing efforts should focus on leveraging advancements in technology such as artificial intelligence, blockchain, and IoT to optimize supply chain processes, enhance traceability, and improve decision-making. Moreover, there is a need to prioritize investments in infrastructure, particularly cold storage facilities, transportation networks, and digital connectivity, to reduce post-harvest losses, improve market access, and enhance value chain efficiency. Additionally, initiatives aimed at promoting sustainable agricultural practices, enhancing farmer income, and addressing climate change impacts should be integrated into supply chain strategies to ensure long-term viability and resilience. Collaborative partnerships between public and private sectors, as well as engagement with smallholder farmers and rural communities, will be essential to drive inclusive and sustainable growth within the agri-business supply chain. By embracing innovation, sustainability, and inclusivity, the future of the agri-business supply chain in India holds tremendous potential to transform the agriculture sector and contribute to the nation's socio-economic development goals.

REFERENCES

Belhadi, A., Kamble, S., Subramanian, N., Singh, R. K., & Venkatesh, M. (2024). Digital capabilities to manage agri-food supply chain uncertainties and build supply chain resilience during compounding geopolitical disruptions. *International Journal of Operations & Production Management*. Advance online publication. 10.1108/IJOPM-11-2022-0737

Bhargav, S. (2017). Agricultural Marketing in Growth of Rural India. International Journal of Management. *IT and Engineering*, 7(7), 306–319.

Carayannis, E. G., Rozakis, S., & Grigoroudis, E. (2018). Agri-science to agri-business: The technology transfer dimension. *The Journal of Technology Transfer*, 43(4), 837–843. 10.1007/s10961-016-9527-y

Das, A. (2020). An appraisal of agribusiness industries in india and their market growth scenario. *Indian Journal of Economics and Development*, 16(2s), 496–499.

Joshi, S., Singh, R. K., & Sharma, M. (2023). Sustainable agri-food supply chain practices: Few empirical evidences from a developing economy. *Global Business Review*, 24(3), 451–474. 10.1177/0972150920907014

Kagalkar, S., Agashe, A., Paralkar, T. A., & Deogaonkar, A. (2023). Narrative Synthesis of the Economic Impact of Agricultural Supply Chain and Distribution Networks on Output: Economic Impact of Agricultural Supply Chains. International Journal of Professional Business Review: Int. *J. Prof. Bus. Rev.*, 8(6), 22.

Kumar, M., Raut, R. D., Jagtap, S., & Choubey, V. K. (2023). Circular economy adoption challenges in the food supply chain for sustainable development. *Business Strategy and the Environment*, 32(4), 1334–1356. 10.1002/bse.3191

Negia, C. S., & Kumarb, S. (2020). *Promoting agro-based industry in India (issues and challenges)*. INTERNATIONAL JOURNALOF TRADE & COMMERCE-IIARTC. 10.46333/ijtc/9/1/27

Osumba, J. J., Recha, J. W., & Oroma, G. W. (2021). Transforming agricultural extension service delivery through innovative bottom–up climate-resilient agribusiness farmer field schools. *Sustainability (Basel)*, 13(7), 3938. 10.3390/su13073938

Pandey, P. C., & Pandey, M. (2023). Highlighting the role of agriculture and geospatial technology in food security and sustainable development goals. *Sustainable Development (Bradford)*, 31(5), 3175–3195. 10.1002/sd.2600

Prabha, R. K., Rai, B. N. J. P., & Singh, S. R. (2016). Role of government schemes in Indian agriculture and rural development. Indian Agriculture and Farmers, 92-102.

Roy, S., Ghosh, S., Beck, C. D., & Sinha, A. P. (2023). Supply chain traceability: A case of blockchain modelling application to agro-business product in India. *International Journal of Sustainable Agricultural Management and Informatics*, 9(4), 295–319. 10.1504/IJSAMI.2023.134067

Sarkar, N. C., Mondal, K., Das, A., Mukherjee, A., Mandal, S., Ghosh, S., & Huda, S. (2023). Enhancing livelihoods in farming communities through super-resolution agromet advisories using advanced digital agriculture technologies. *Journal of Agrometeorology*, 25(1), 68–78.

Shetty, H. (2018). Food security through agricultural sustainability: In Indian context. *Int. J. Res. Eng. Sci. Manage*, 1, 685–691.

Siddh, M. M., Soni, G., Jain, R., Sharma, M. K., & Yadav, V. (2021). A framework for managing the agri-fresh food supply chain quality in Indian industry. *Management of Environmental Quality*, 32(2), 436–451. 10.1108/MEQ-05-2020-0085

Ulvenblad, P., Barth, H., Ulvenblad, P. O., Ståhl, J., & Björklund, J. C. (2020). Overcoming barriers in agri-business development: Two education programs for entrepreneurs in the Swedish agricultural sector. *Journal of Agricultural Education and Extension*, 26(5), 443–464. 10.1080/1389224X.2020.1748669

Xu, Z., Elomri, A., El Omri, A., Kerbache, L., & Liu, H. (2021). The compounded effects of COVID-19 pandemic and desert locust outbreak on food security and food supply chain. *Sustainability (Basel)*, 13(3), 1063. 10.3390/su13031063

Chapter 4
Agricultural Crop Recommendations Based on Productivity and Season

A. V. Senthil Kumar
http://orcid.org/0000-0002-8587-7017
Hindusthan College of Arts & Science, India

Aparna M.
Hindusthan College of Arts & Science, India

Amit Dutta
All India Council for Technical Education, India

Samrat Ray
IIMS, India

Hakikur Rahman
http://orcid.org/0000-0002-2132-1298
Presidency University, Bangladesh

Shadi R. Masadeh
Isra University, Jordan

Ismail Bin Musirin
Universiti Teknologi MARA, Malaysia

Manjunatha Rao L.
National Assessment and Accreditation Council, India

Suganya R. V.
VISTAS, India

Ravisankar Malladi
http://orcid.org/0000-0002-8250-6595
Koneru Lakshmaiah Education Foundation, India

Uma N. Dulhare
http://orcid.org/0000-0002-4736-4472
Muffakham Jah College of Engineering and Technology, India

ABSTRACT

This chapter aims to develop an agricultural crop recommendation system leveraging the power of machine learning algorithms. The proposed system takes into account crop productivity and prevailing season as crucial factors in making appropriate crop suggestions. The authors proposed the SVM algorithm, which was trained and evaluated on a comprehensive dataset comprising historical agricultural data with diverse features such as climate variables, soil properties, and geographical factors. The data was further segmented based on seasonal patterns to provide crop recommendations tailored to specific timeframes. The models' performance was evaluated using standard metrics, and an ensemble approach was considered to enhance the system's robustness. Ultimately, the developed system offers farmers and agricultural experts a valuable tool for making informed decisions, optimizing crop selection, and

DOI: 10.4018/979-8-3693-3583-3.ch004

increasing overall agricultural productivity

INTRODUCTION

Machine Learning

Machine learning is a subfield of artificial intelligence (AI) that focuses on developing algorithms and models that enable computers to learn and improve their performance on a specific task without being explicitly programmed. The fundamental idea behind machine learning is to allow computers to recognize patterns, make decisions, and solve problems based on data rather than relying on explicit instructions from programmers. In order to find patterns and correlations in the data, machine learning algorithms learn from past data and experiences. These patterns and relationships can then be used to predict the future, categorize new data, or improve decision-making procedures. Machine learning is widely used in various fields, including natural language processing, computer vision, recommendation systems, autonomous vehicles, finance, healthcare, and more. With the availability of large datasets, powerful computing resources, and advances in algorithms, machine learning continues to make significant contributions to solving complex problems and driving innovations across industries.

Recommender System

A recommender system is a type of information-collecting system designed to suggest relevant items to users based on their preferences, interests, and past behavior. These systems are commonly used in various online platforms to enhance user experience by providing personalized recommendations, thereby increasing user engagement and satisfaction. Recommender systems include a wide range of applications, including e-commerce, online streaming services, social media, and content platforms. Other advanced techniques and hybrid approaches may also be used, combining elements of collaborative filtering, content-based filtering, and additional factors like context, demographics, and popularity. Recommender systems have become an essential part of many online platforms, helping users discover new content, products, and services they are likely to enjoy while also benefiting businesses by increasing user engagement and driving sales. However, designing an effective recommender system involves addressing challenges such as data sparsity, cold start problems (when new users/items have limited data), and ensuring fairness and diversity in recommendations.

Knowledge Discovery in Databases

Knowledge Discovery in Databases (KDD) is the process of extracting useful and actionable knowledge from large volumes of data. It is an interdisciplinary field that combines techniques from databases, machine learning, statistics, and data mining to discover patterns, trends, relationships, and insights that are hidden in the dated is a critical process in data-driven decision-making and plays a different role in various applications, including customer relationship management, fraud detection, market analysis, healthcare informatics, and scientific research. It is important to note that the KDD process is not a one-time activity; it often involves an iterative approach as new data is collected, and new knowledge is discovered, leading to continuous improvement and refinement of insights.

BACKGROUND

Deep Learning for Crop Yield Prediction: A Systematic Literature Review

AlexandrosOikonomidis,CagtayCatal and Ayalewkassahun (2022)Deep Learning has been applied to the crop yield prediction problem as suggested in this work; however, a thorough overview of the studies is lacking. Thus, the purpose of this work is to present a review of the most recent state-of-the-art use of deep learning for agricultural production prediction. To find and evaluate the most pertinent studies, we conducted a Systematic Literature Review (SLR). 456 appropriate studies were found, and after applying selection and quality assessment criteria to the relevant research, we chose 44 primary studies for further review. The main goals, the target crops, the algorithms used, the features employed, and the data sources utilized were all thoroughly analyzed and synthesized in the primary searches. We found that the most widely used method, Convolutional Neural Network (CNN), performs best if it comes to Root Mean Square Error (RMSE). The absence of a large training dataset, which increases the potential of over fitting and, ultimately, lowers model performance in real-world scenarios, is one of the biggest obstacles. Given that this field's academics frequently concentrate on the significance of unexplored study subjects, it is beneficial to highlight both the present obstacles and the potential for further research. Our findings on this study are helpful to practitioners who wish to create innovative crop yield prediction models for their use as well as to researchers in this field. The difficulties are significant for researchers in this subject because they will be aware of these problems before creating their models. This SLR study addresses various difficult tasks that practitioners face when developing new agricultural yield prediction models. Thus, choosing the methods and model parameters calls for careful analysis based on the literature. To conduct a systematic review, we obtained 456 appropriate studies for this reason. To the best of our knowledge, there hasn't been a published systematic literature review on the use of deep learning to predict crop production. Certain SLR publications and some standard research papers on crop yield prediction, for example, do not concentrate on the use of deep learning in agricultural yield prediction; instead, they discuss the use of classic machine learning. There is currently no SLR paper that focuses on the application of deep learning in agricultural yield prediction, thus we must distinguish between shallow and deep learning in this case. In this regard, the current study represents a ground-breaking endeavor that paves the way for a methodical examination of the state-of-the-art understanding of the development of Deep Learning-based techniques for crop production prediction. Different studies employed different deep-Learning techniques and algorithms. Our findings indicate that most articles used supervised learning. This outcome resulted from the widespread use of CNN to estimate crop yield.

Data Mining and Wireless Sensor Network for Agriculture Pest/Disease Predictions

A.K. Tripathy et.al. (2021) has proposed in this paper Information exactness horticulture angles, especially the pest/disease administration, require energetic crop-weather information. An exploration was conducted in a semiarid locale to get the crop-weather-pest/disease relations utilizing remote tactile and field-level observation information on closely related and forbid bug (Thrips) – illness (Bud Rot) flow of groundnut trim. Information mining strategies had been used to flip the records into useful information/ knowledge/relations/trends and relationships of the crop-weather-pest/disease continuum. These elements gotten from the information mining strategies and prepared through scientific models were approved with comparing observation information. Comes about gotten from 2009 & 2010 kharif seasons (rainstorm) and 2009-10 & 2010-11 rabi seasons (post rainstorm) information may be utilized to create a genuine to close real-time choice bolster framework for pest/disease expectations. In a long time, there has been a huge disease zone of edit illnesses and creepy crawly bother as well as the degree of its reality, which caused colossal financial misfortunes to the laborers. Climate plays a vital part in rural generation. Oilseed crops are more prevalent in weather-based fragile agriculture systems (semi-arid regions). Among the oilseed plants groundnut (peanut) vegetation are susceptible to assault by means of pests/diseases to a plenty large extent than many different crops. The disease occurs with the incidence ranging from 0- 98%. A try has been made to recognize the hidden relationships between the most prevailed disorder (BNV) / pest (Thrips) and climate parameters of the Groundnut crop. The crop weather-pest/disease dynamics and hidden members of the family had been got and quantified by the use of DM techniques. The statistical method collectively with regression mining based totally correlations helped in growing multivariate regression mannequin that has been used to advance an empirical prediction mannequin (noncumulative) to difficulty the forecast for populace buildup, initiation & severity of pest /disease. Apart from this, a cumulative prediction mannequin has been developed (which was determined to be greater correct than the non-cumulative one) and examined the usage of two season's data. This will assist in taking strategic selections to retail the crop from pest/disease impacts and enhance the crop yields.

An Analysis of Agricultural Soils by Using Data Mining Techniques

Palepu et.al. (2021) has proposed in this paper the software of Data mining methods in agriculture specifically on soils can revise the scenario of pledge making and enhance cultivation yields in a higher way. The evaluation of soils plays an essential function for decision making on countless problems associated to agriculture field. This paper offers about the position of information mining in point of view of soil evaluation in the area of agriculture and additionally confers about various statistics mining methods and their associated work through numerous authors in context to soil analysis domain. The information mining methods are very up-to-the-minute in the vicinity of soil analysis. In the modern-day days of society, information mining is used in a huge areas and many off-the-shelf information mining tools, strategies and strategies are on hand and sphere of have an effect on facts mining utility software's are reachable, however information mining in agricultural soil datasets is a comparatively a childish lookup field. These days mining thought and methods used to resolve agriculture problems. In this paper it has been mentioned about how information mining methods are utilized in agriculture field. Globally, day to day the requirement of meals is escalating; consequently, the agricultural scientists, farmers, government, and researchers are tiresome to put more strive and use strategies in agriculture for enhancement in production. As an effect, the facts generated in the area of agricultural information improved day via day. As the diploma of information enlarges, it requires instinctive way for this information to be mined and analyzed when needed. Even at present, a very solely some farmers are truly making the use of the new methods, equipment, and methods in agriculture for higher production. Data mining can be classified into two sorts such that one is descriptive any other one is predictive. Descriptive records mining considers the present data, that is uncooked records and then summarizes it summarized. Descriptive mining represents the traits of previous occasions and approves us to analyze how they have an effect on the future. The basis of predictive mining relies upon probabilities, it is used to predict future based totally on the values regarded from regarded results. Forecasting includes the usage of the variables or discipline in the database to estimate nameless results. Agriculture is the most vast software vicinity especially in the growing international locations like India. Use of statistics technological know-how in agriculture can exchange the state of affairs of choice making and farmers can yield in higher way. Data mining performs an important function for choice making on a number of troubles associated to agriculture field.

Analyzing Soil Data Using Data Mining Classification Techniques

Rajeswari and Arunesh(2020) proposed this system. The goal of the work is to predict soil kind the usage of records mining classification techniques. Methods/Analysis: Soil kind is estimated by the use of facts mining classification strategies such as JRip, J48 and NaiveBayes. These classifier algorithms are utilized to extract the know-how from soil facts and two sorts of soil are viewed such as Red and Black. The JRip mannequin can produce greater reliable consequences of this information and the Kappa Statistics in the forecast had been increased. Application/Improvement: For fixing the troubles in Big Data, environment friendly techniques can be created that make use of Data Mining to decorate the exactness of classification of massive soil record sets. Data Mining (DM) turns into famous in the subject of agriculture for soil classification, barren region administration and crop and pest management. Assessed the range of affiliation strategies in DM and utilized them in the database of soil science to predict the significant relationships and furnished affiliation policies for one-of-a-kind soil kinds in agriculture. Similarly, agriculture prediction, disorder detection and optimizing the pesticides are analyzed with the use of several information mining strategies earlier2. In3 analyzed the J48 classification algorithm with excessive accuracy to predict the soil fertility rate. In4 investigated the use of more than a few DM methods for understanding discovery in agriculture zones and added extraordinary well-known shows for information discovery in the structure of Association Rules, Clustering, Classification and Correlation. In5 envisioned the soil fertility training the use of classification strategies have been Naïve Bayes, J48 and K-Nearest Neighbor algorithms. In6 used Adopted facts mining strategies to estimate crop yield analysis. Multiple Linear Regression (MLR) techniques are used to discover the linear relationship between based and unbiased variables. K-Means clustering strategy used to be additionally used to structure 4 clusters thinking about Rainfall as a key parameter. Decision tree, Bayesian Network records mining strategies and the non-linear processes have been implemented. Optimization based Bayesian Network strategy was once regarded as higher than non-linear. In8 analyzed the digital magnitude of soil fertility and the crop administration elements to predict the maize yields and decide the yield variability and the hole between farmers. Classification and regression tree evaluation had been used to predict the result. In9 investigated two complete strategies to calculate the manufacturing associated yield hole and soil fertility associated nutrient balance.

Data Analytics for Crop Management: A Big Data View

Nabila Chergui and MohandTaharKechadi (2022) has proposed in this system the latest advances in Information and Communication Technologies have a large influence on all sectors of the economic system worldwide. Digital Agriculture seemed as the end result of the democratization of digital units and advances in artificial talent and information science. Digital agriculture created new strategies for making farming greater productive and environment friendly whilst respecting the environment. Recent and state-of-the-art digital units and facts science allowed the series and evaluation of good-sized quantities of agricultural datasets to assist farmers, agronomists, and experts recognize higher farming duties and make higher decisions. In this paper, we are existing a systematic assessment of the utility of records mining methods to digital agriculture. We introduce the crop yield administration method and its aspects whilst limiting this learning about crop yield and monitoring. After figuring out the predominant classes of statistics mining methods for crop yield monitoring, we talk about panoply of current works on the use of facts analytics. This is accompanied by means of a typical evaluation and dialogue on the influence of large records on agriculture. DA, (also referred to as digital farming or clever farming)1, is a current strategy that makes use of digital and clever units [sensors, cameras, satellite, drones, the Global Positioning System (GPS)] in conjunction with Data Mining (or information analytics) to enhance productiveness and to optimize the use of resources. Digital Agriculture (DA) comes as a response to the growing demand for enhancing productiveness whilst lowering farming operational costs. DA can be utilized in nearly all agricultural fields. For instance, in crop production: DA permits the correct administration of crops, which consists of fields, wasteland, crop, pest, and irrigation management, soil classification, etc. In Animal production: DA permitted monitoring the animal over its complete existence cycle, its meal quantity, fitness manipulate and safety from diseases, and so on. Fishery, Animal Husbandry, farm animals and dairy farming are some examples. Digital agriculture (DA) is a data-driven method that exploits the hidden records inside the amassed information to achieve new insights; reworking the farming practices from intuitive-based decision-making to informed-based decision-making. DA depends on environment friendly statistics series practices, environment friendly information instruction and storage techniques, environment friendly statistics analytics, and environment friendly deployment and exploitation of the won insights to make ultimate farming selections. There are quite a few different challenges and barriers that want to be addressed, amongst them are lack of data, lack of skills, and lack of maturity and requirements so that it can be adopted and deployed rapidly and easily. In this study, we discover techniques that deal with the whole system of information mining; from information series to understanding deployment. We discover this procedure from a massive statistical view, with an extra focal point on crop monitoring and administration in an try to recognize the challenges that DA is presently facing.

Spiking Neural Networks for Crop Yield Estimation Based on Spatiotemporal Analysis of Image Time Series

Pritam Bose and Nikola Kasabov (2020) has proposed in this paper affords spiking neural networks (SNNs) for a long way off sensing spatiotemporal contrast of picture time series, which make use of the pretty parallel and low-power-consuming neuromorphic hardware buildings possible. It offers the improvement and checking out of a methodological framework that makes use of the spatial accumulation of time sequence of Moderate Resolution Imaging Spectro radiometer 250-m selection data and historical crop yield records to educate an SNN to make nicely timed prediction of crop yield. The lookup work additionally consists of an evaluation of the most beneficial wide variety of elements wanted to optimize the consequences from our experimental statistics set. CROP manufacturing performs an essential function in meal protection and monetary improvement of a country. In previous years, the fluctuation of crop yield in China attracted an awesome situation in the financial system and even led to the meal disaster of the entire country. In this paper, we have introduced an SNN mannequin to make well timed predictions of crop yield. The mannequin is the use of spatial accumulation of blocks of MODIS-NDVI 250-m decision information and historic crop yield data. The proposed mannequin introduces for the first time SNNs as important methods in far flung sensing, in our case here—for spatiotemporal records modeling, analysis, and land use/crop prediction. Compared to the other conventional methods, SNN performed substantially better. The SNN yield estimation model's stability and accuracy are the product of numerous key features, including: 1) SNN's ability to apply time-dependent machine learning rules to capture spatiotemporal patterns from spatiotemporal data; 2) SNN model interpretability for comprehending the data and the processes that generated them; 3) the integrated use of remotely sensed data along with historical statistical information, with parameters retrieved from satellite images being combined by the main crop growing season; and 4) the precise division of the study area based on agricultural knowledge and meticulous sample data selection. The first ten days of June were the wheat's ripe time, thus we only used the NDVI to predict winter wheat yields from the first week of March to the middle ten days of April. It was important to the winter wheat production in this area that we were able to anticipate the winter wheat yield accurately approximately 40 days before harvest time, or during the booting–heading stage of the crop. Every year, around the same period from the beginning of March to the middle of April, the NDVI photos were gathered. The pictures are a five-day interval composite of ten days' worth of data.

Smart Farming System Using Data Mining

Chandak et.al. (2021) has proposed in this paper Smart farming system is an autonomous & sophisticated mechanism, which will aid in the growth of agriculture yield by applying hi-tech agriculture techniques without human intervention. The paper represents an overview of recent smart farming software solutions. The suggested system uses data mining techniques and information from satellite imagery, the Internet, and soil testing reports that are inserted into already-existing databases. To make decisions based on weather awareness, it skillfully uses clustering algorithms to track crop growth stages, ensure appropriate water use, decide which fertilizer to apply based on crop stage, and choose which pesticide to apply to shield crops from disease and insect attack. By strategically controlling farm operations, this method can raise field output. The largest use of water worldwide is agriculture, with irrigation making up around 70% of total usage. The home and industrial sectors account for 10% and 20%, respectively, even though these percentages fluctuate appreciably throughout countries. As the populace is growing day by way of day, the demand for meals is on the increase too. There are positive extra elements that affect greater crop yield, such as environmental elements like erratic climate prerequisites main to crop loss, farmers lack knowledge in embracing more modern applied sciences that can be used for enhancement of gross earnings from agriculture. Despite all such problems, agriculture is a cardinal supply of employment and performs a key function in the socio-economic improvement of India. So, to enhance the condition, we can make use of technological know-how in a smarter way. To make this viable we want extra productivity from farming. The home and industrial sectors account for 10% and 20%. Without increased efficiencies, agricultural water consumption is anticipated to expand globally by around 20% in a few years. In the proposed system, Data mining is used for all facts mapping & processing. Data Mining is about discovering policies in data. The technological know-how of information mining is narrowly linked to statistics storage and is intertwined with database administration systems. The clever farming machine will additionally alert the farmers about crucial weather prerequisites which will once more make each feasible anomaly to be sustained. Water is a limited resource and its conservation is the biggest crisis nowadays, but using this system will aid in the proper utilization of water & no wastage or under-oversupply. Summarizing all the smart farming systems is an ideal solution for future farming.

Krishimantra: Agricultural Recommendation System

Vikas Kumar and Vishal Dave (2021) has proposed in this system with the evolution of Web 2.0, ICT has emerged as an important want of human beings. There is a hole between the farmers and the understanding of agricultural experts. ICT can fill the hole between farmers and experts. In this paper, we have proposed a semantic net based total structure to generate agricultural recommendations, the usage of spatial records and agricultural understanding bases. Our know-how base acts as an area professional and will ship hints to the farmers based totally on local weather stipulations and geographic data. We have proven experimental outcomes as a section of the implementation of our proposed architecture. A farmer sends a question to the question engine, to get records for a particular crop. The query may also be associated with GIS data, crop know-how base or both. The final result of the question is displayed on a cell device. India has the fourth biggest agricultural area in the world and offers the predominant ability of livelihood for over 58.4% of India's population. Indian agriculture is going through one of the essential challenges of deceleration in agriculture growth. The main motive for the deceleration in agricultural increase is declining funding in agricultural lookup and improvement mixed with the inefficiency of establishments offering inputs and offerings along with rural credit score and extension services. Climate change is one of the vital economic and environmental challenges of our time. Climate has varying effects on agriculture and precision farming. In developing economies, significant percentages of the population are still untouched by the revolution of new technologies and are unaware of such advances. In most of the villages, farmers nevertheless focus on historical farming practices. Moreover, due to changes in climate parameters like temperature, rainfall, humidity, sunny days, and soil moisture etc., agricultural yield is affected severely. Hence, there is a need for enabling technologies to work together and generate recommendations for farmers based on climatic parameters in the form of spatial data and knowledge repositories in the form of ontologies. This work is a step toward filling the gap between farmers and agriculture experts by implementing an information system which will make use of geographical data and agricultural domain knowledge bases. We have implemented the information system partially and shown the initial results. Most of the research and implementation work needs to be done, to realize the architecture. For the complete realization of our proposed architecture, we need to develop ontologies for cotton and spatial data, user Interfaces, integration modules and building SPARQL queries.

Sustainable Development: The Role of GIS and Visualization

Latuet.al. (2021) Has proposed in this system the bodily surroundings in which we stay and on which our preserve with existence relies have confined capacity, consequently, human beings need to no longer think about it as a useful resource to be exploited for temporary maximized profits. If we are to experience a secure and profitable existence on the planet Earth then we should use the constrained sources at our disposal accurately by being precise stewards of our very own environment. Our surroundings have been dominated by using our moves and the want to higher manipulate it has been debated broadly and vigorously in current years. In many creating countries, which include the insular island countries of the Pacific, financial improvement and environmental conservation are frequently in conflict. This paper argues that choice makers, specifically in typical societies, bypass professional recommendations furnished by means of Geographic or Spatial Information Systems (GIS, or SIS). The functionality of GIS has been prolonged to consist of modeling and visualization of terrain points so that selection makers will be capable no longer solely to see the cutting-edge country of the assets that they managed however additionally see the influences of their selections and the future nation of these resources. The paper examines the influence of monetary improvement things to do on the coastal ecosystems in exemplar growing countries, in the Pacific, and proposes GIS-Visualization techniques for transferring past subsistence and financial improvement aspirations to socially, economically, and environmentally sustainable improvement activities. Resources of monetary cost are scarce and are critically threatened with irreversible depletion in the Small Island Developing States (SIDS) of the Pacific. Resources which took nature many years to nurture are now obliterated in days and these which had been considered with the aid of the islanders as being successful of infinite exploitation are now being annihilated at an alarming rate. With constrained areas of land, populace densities vary from 8.2 people per rectangular kilometer in the island of Niue to 587 of us per rectangular kilometer in Nauru, and a comparable projected rate of populace growth of 3% every year in the first decade of the twenty-first century, SIDS in the Pacific face such acute troubles of land shortages and populace stress as in distinctive small island global places spherical the world.

A Brief Survey of Data Mining Techniques Applied to Agricultural Data

Hetal Patel and Dharmendra Patel (2014) has proposed this paper as with many other sectors the amount of agriculture data is increasing on a daily basis. However, the software of statistics mining techniques and strategies to find out new insights or information is a fantastically novel lookup area. In this paper, we furnish a quick evaluation of a range of Data Mining methods that have been utilized to mannequin facts from or about the agricultural domain. The Data Mining strategies utilized on Agricultural records consist of k-means, bi clustering, ok nearest neighbor, Neural Networks (NN) Support Vector Machine (SVM), Naive Bayes Classifier and Fuzzy means. As can be viewed the appropriateness of information mining methods is to a sure extent decided by way of the exceptional kinds of agricultural statistics or the issues being addressed. Agriculture is the backbone of the Indian economy. Nearly two-thirds of its population directly depend on agriculture for its livelihood. Even though large areas in India have been brought under irrigation, only one-third of the cropped part is irrigated. The productivity of agriculture is very low. So as the demand for meals is increasing, researchers, farmers, agricultural scientists, and authorities are making an attempt to put greater effort into and strategies for extra production. And as a result, the agricultural information will increase day by way of day. As the quantity of records increases, it requires an involuntary way for these statistics to be extracted when needed. Still today, very few farmers are using the new methods, equipment, and approaches of farming for higher production. Data mining can be used for predicting the future developments of agricultural processes. Descriptive records mining duties signify the regular houses of the statistics in the database whilst predictive records mining is used to predict express values primarily based on patterns decided from recognized results. Prediction entails the use of some variables or fields in the database to predict unknown or future values of different variables of interest. As long way as the facts mining method is concerned, in most instances predictive statistics mining strategy is used. The Predictive statistics mining method is used to predict future crops, climate forecasting, pesticides, and fertilizers to be used, income to be generated and so on.

A Survey on Data Mining Techniques for Crop Yield Prediction

Medar.et.al.(2020) has proposed in this paper presents a variety of crop yield prediction techniques and the usage of information mining techniques. The Agricultural device is very complicated because it offers a giant information state of affairs which comes from a range of factors. Crop yield prediction has been a subject of activity for producers, consultants, and agricultural-associated organizations. Data mining science has acquired high-quality growth with the speedy improvement of laptop science, and synthetic intelligence. Data Mining is a rising lookup area in agriculture crop yield analysis. Data Mining is the procedure of figuring out the hidden patterns from a giant quantity of data. Yield prediction is a very essential agricultural trouble that needs to be solved primarily based on the accessible data. The hassle of yield prediction can be solved using statistical mining techniques. Data Mining is the method of extracting beneficial and necessary data from massive units of data. Data mining in agriculture subject is a noticeably novel lookup field. Yield prediction is a very necessary agricultural problem. Data mining techniques The training data must be taken from past data and utilized for training, which is necessary to understand how to categorize yield forecasts for the future. Crop models and decision tools are increasingly used in the agricultural field to improve production efficiency. Crop forecasting is the practice of projecting crop yields and production several months ahead of time, usually before the harvest itself occurs. Based on meteorological and agronomic data, several indices are derived. With the enhancement of statistics mining technologies, in particular, these barring any premises or human beings subjective, facts mining can be utilized in many areas.

Crop and Yield Prediction Model

ShreyaBhanose and KalyaniBogawar (2020) has proposed in this paper A certain, methodical strategy is required in the agricultural sector to forecast crop yields and assist farmers in making the right choices that will improve farming quality. Due to the lack of a crop knowledge base, predicting the best crops is very hard. Crop prediction is a productive strategy for raising income and improving farming quality. In the realm of data mining, using the data clustering method is an effective way to extract important information and provide predictions. Numerous strategies have been used thus far, both for crop prediction and other purposes. A crop forecast model helps farmers make the right choices. A crop prediction is a big hassle that occurs. A farmer had an interest in grasp how an awful lot produce he was going to expect. Traditionally farmers determine this primarily based on an everlasting journey for unique yield, plant life and climate conditions. Character at once thinks about produce prediction as an alternative than regarding crop prediction. If the right crop is predicted then yield will be better. Problem of crop and yield prediction the usage of modified k-means clustering algorithm thereby developing higher revenue for berry farmers. Clustering is the system of grouping the information into training or groupings, so that objects inside a cluster have excessive similarity in settlement to every different however are surprisingly multiple to objects in choice clusters. A bunch of information objects can be dealt with at the same time in the course of the time that you crew and so may also appear a basic of facts compression. Unlike category, clustering is an effective capacity for partitioning the series of records into groups based totally on facts likeness and then ascribe labeling to the distinctly small variety of groups. Clustering is an unsupervised study as it no longer remembers predefined lessons and types labeled education examples. The fantastic clusters will rely on how dense it is. So, a cluster having a greater variety of factors is a cluster of proper quality.

The Impact of Data Analytics in Crop Management Based on Weather Conditions

Rani. et.al (2020)has proposed in this paper agriculture is the most essential software area, in particular in creating international locations like India. When choosing the number of troubles in the agricultural sector, statistics mining performs a critical role. The manner of information mining objectives is to extract records from the present facts set and convert them into a singular, human-readable layout for future use. Because local weather can have a substantial influence on crop productivity, crop administration in a precise agriculture area is influenced using its climatic conditions. Good crop administration can be made viable with the assistance of real-time climate data. Automation of considerable statistical extraction in the pursuit of expertise and developments is made viable through the use of statistics and communications technology. This makes it less difficult to extract information immediately from digital sources, switch it to an invulnerable digital machine of documentation, limit production costs, expand yield, and elevate market prices. It used to be additionally found how records mining aids in the evaluation and prediction of beneficial patterns from enormous, dynamically altering climatic data. Fuzzy logic, synthetic neural networks, genetic algorithms, choice trees, and guide vector machines have all been using researchers and engineers in agricultural and organic engineering to learn about the soil, climate, and water regimes that affect crop boom and pest administration in agriculture.

Machine Learning in Agriculture: A Review

Liakos et.al.(2018) has proposed in this paper computing device mastering has emerged with large information applied sciences and high-performance computing to create new possibilities for records intensive science in the multi-disciplinary Agri-technologies domain. In this paper, we current a complete assessment of lookup committed to functions of desktop gaining knowledge of in agricultural manufacturing systems. The works analyzed have been classified in (a) crop management, inclusive of functions on yield prediction, sickness detection, weed detection crop quality, and species recognition; (b) farm animals management, together with functions on animal welfare and cattle production; (c) water management; and (d) soil management. The filtering and classification of the introduced articles reveal how agriculture will gain from computing device studying technologies. By making use of computer studying to sensor data, farm administration structures are evolving into actual time synthetic Genius enabled packages that supply prosperous suggestions and insights for farmer selection assist and action. Agriculture performs a essential function in the international economy. Pressure on the agricultural gadget will increase with the persevering with growth of the human population. Agri-technology and precision farming, now additionally termed digital agriculture, have arisen as new scientific fields that use facts severe processes to power agricultural productiveness whilst minimizing its environmental impact. The facts generated in contemporary agricultural operations is furnished by way of a range of distinctive sensors that allow a higher perception of the operational surroundings (an interplay of dynamic crop, soil, and climate conditions) and the operation itself (machinery data), main to greater correct and quicker choice making.

Web Based Recommendation System for Farmers

Shinde et.al. (2020) has proposed in this paper an agricultural is nonetheless the use of typical approaches of guidelines for agriculture. Currently guidelines for farmers are based totally on mere one to one interplay between farmers and professionals and distinctive professionals have exceptional recommendations. Recommendation can be furnished to farmers the use of previous agricultural things to do with assist of information mining principles and the market style can be merged with it to grant optimized effects from recommender. The paper proposes the use of information mining to grant tips to farmers for crops, crop rotation and identification of splendid fertilizer. The System can be used by way of farmers on net as properly on android primarily based cellular gadgets. Agriculture is a top occupation in India from a long time and as a result performs a necessary function in an Indian economy. India is an agricultural us of a with 2d very best land place of greater than 1.4 million square-kilometers beneath cultivation. India possesses a fantastic possible to be a superpower in the area of agriculture. Agriculture promotes poverty upliftment and rural development. Agriculture is India's largest monetary area and employed 52.1% of complete work pressure in 2009-10. As of 2011, India had a massive and various agricultural sector, accounting, on average, for about 16% of GDP and 10% of export earnings. agricultural is carried out from a while and consequently we have a prosperous series of agricultural previous records which can used for recommendation. Data mining strategies and algorithms can be used for recommending single crops and pattern of flowers for crop rotation. However, to reap optimized and valid outcomes machine wants to be in non-stop mastering which can be achieved via which include trendy datasets in the system. The paper proposes the use of records mining strategies to furnish tips to farmers for crops, crop rotation and identification of excellent fertilizer. The consequences from the suggestion gadget are optimized with appreciation to parameter consideration.

EXISTING SYSTEM

A higher level of efficiency should be possible with more people and land, but it is not possible. Farmers used to rely on word-of-mouth, but current climate conditions make this impossible. Data used to get insights into Agro-facts are created by agricultural elements and parameters. Some advances in agriculture sciences are driven by the growth of the IT industry, providing farmers with high-quality agricultural information. In the contemporary context, it is desirable to have the intelligence to apply modern technical methods in the sector of agriculture. Using the data, machine learning techniques create a well-defined model that aids in prediction-making. It is possible to find solutions for agricultural problems such as crop prediction, rotation, water and fertilizer requirements, and protection. Due to the atmosphere-altering climate, it is necessary to have a high approach to help crop harvesting be less difficult and help farmers with their yield manufacturing and management. This might make agriculture better for aspiring farmers. A farmer can receive a set of recommendations to help with crop output with the help of data mining. Crops are suggested for implementation of this technique according to their climatic characteristics and quantity. The development of valuable extraction from agricultural databases is made possible by data analytics. After the Crop Dataset was evaluated, crops were recommended depending on season and productivity.

Advantages of Crop Recommendation

- Intelligent crop recommendation systems can help farmers to select the best crops to grow based on a variety of factors, including crop productivity, the prevailing season, climate variables, soil properties, and geographical factors. This can lead to significant increases in crop yields and profits.
- It can help farmers to reduce their risk by providing them with information about the risks associated with growing different crops in different conditions. This can help farmers to avoid making costly mistakes.
- The crop recommendation systems can help farmers to save time and effort by automating the crop selection process. This frees up farmers to focus on other tasks, such as crop management and marketing.

RESEARCH METHODOLOGY

In this work, we suggest an approach that makes use of SVM algorithms to become aware of climate best and predict the most appropriate crop for cultivation. We reflect on the consideration of crop and climate information as inputs to our algorithm, and our technique additionally suggests a fantastic fertilizer for the anticipated crop. The check effects exhibit that our approach precisely predicts crop decision and yield, which can radically gain farmers. We consider the overall performance of every algorithm and evaluate them to make sure that we are the usage of the most positive approach. We additionally take steps to make sure that our records are dependable and accurate. Our find out about indicate that machine learning can be a beneficial device for predicting crop yield and supporting farmers make knowledgeable choices The machine starts evolving by accumulating and integrating quite a number of facts sources, such as historic crop overall performance data, soil fitness indicators, climate patterns, and unique seasonal trends. Through cautious statistics preprocessing and characteristic engineering, the device ensures that the entered records are strong and informative for the studying algorithms. The personal interface of the proposed machine is designed to be straightforward and reachable to farmers of all backgrounds. Farmers can without problems enter their location, soil characteristics, and different applicable data, receiving crop hints tailor-made to their unique conditions. The gadget will additionally constantly research and adapt to new data, making sure that its pointers continue to be updated with altering conditions, technology, and agricultural practices.

Input Data

The input data for the agricultural crop recommendation system consists of various agricultural and environmental factors relevant to crop productivity. This data is collected from different sources, including historical records, weather stations, soil testing, and satellite imagery. The input data typically includes information such as geographic location (latitude and longitude or name), seasonal information, climate variables (temperature, precipitation, and humidity), soil properties (pH level, nutrient content), altitude, and other relevant factors that influence crop growth and yield.

Data Preprocessing

Dataset preprocessing is a fundamental step in preparing the entered information for teaching the computer to gain knowledge of models. It includes cleansing the facts to deal with lacking values, outliers, and inconsistencies. This may also consist of strategies like imputation, outlier detection, and records normalization or standardization. Preprocessing ensures that the records are in an appropriate layout and devoid of mistakes or anomalies, enabling the fashions to analyze successfully and make correct predictions.

Feature Selection

The machine learning algorithm cannot work on the same datasets. We need to transform the data set into a corresponding algorithm to prepare the proper input for the specific algorithm and maximize the classifier accuracy. We use different feature extraction techniques. The value of class attributes is changed from numeric values to alphabetic values. We make clusters of the performance class to the relevant range.

Crop Recommendation Using the SVM

Once the SVM mannequin has been trained, it can be used to generate crop guidelines for new locations. To do this, the mannequin is honestly fed the aspects of the new place and it predicts crop productivity. The plants with the best possible envisioned productiveness are then encouraged by the farmer.

System Testing

System testing is a crucial phase in the development process to ensure the agricultural crop recommendation system functions correctly and meets its intended objectives. Several types of testing are essential for a robust system:

1. Unit Testing: This involves testing individual components or functions of the system in isolation to verify their correctness and proper functionality.
2. Integration Testing: Integration testing checks the interactions between different modules or components of the system to ensure they work seamlessly together.
3. Functional Testing: This type of testing validates whether the system functions as expected and produces accurate crop recommendations based on the provided input.
4. Performance Testing: Performance testing assesses the system's responsiveness and scalability, ensuring it can handle multiple users simultaneously and provide prompt responses.

System Implementation

The device implementation segment includes deploying the agricultural crop suggestion device for realistic use via farmers and agricultural experts. Here are the key steps in gadget implementation:

1. Deployment Environment: Set up the required hardware and software program infrastructure to host the system. This should contain cloud servers or devoted on-premise servers.

2. User Interface: Develop and finalize the effortless interface the place where customers can enter applicable facts and get hold of crop recommendations.
3. Model Integration: Integrate the educated desktop getting-to-know fashions (SVM) into the device to allow crop prediction based totally on entered data.
4. Database Integration: Implement a database to keep historic agricultural information and personal inputs, making sure convenient to get admission to and retrieval for evaluation and mannequin training.
5. Data Security: Incorporate strong safety measures to guard personal records and ensure the confidentiality and integrity of the system.

CONCLUSION

Crop suggestion systems using the SVM algorithm have the potential to revolutionize the way that farmers choose and develop crops. By presenting farmers with data-driven insights and recommendations, these systems can assist farmers in enhancing their yields, limiting their risk, higher efficiency, and making more sustainable choices. SVM algorithms are well-suited for crop suggestion due to the fact they are capable of dealing with high-dimensional datasets with a distinctly small quantity of observations, and they are sturdy to outliers and noise in the data. Additionally, SVM algorithms can analyze complicated relationships between the facets and the goal variable, which is essential for crop suggestion due to the fact the relationship between crop productiveness and the more than a few enter facets is regularly complex. Crop suggestion structures the use of the SVM algorithm can be using farmers of all sizes, in each developed and creating country. These structures can assist farmers in making higher choices about crop determination and enhance their universal profitability.

FUTURE WORK

In future work, crop recommendation Systems using the SVM algorithm is to develop models that are more personalized to the individual farmer's needs. This could be done by incorporating additional data into the models, such as the farmer's risk tolerance, financial constraints, and access to resources. Additionally, future work could focus on developing models that are more robust to changes in the climate and other environmental factors.

REFERENCES

Latu, . (2021). Sustainable Development: The Role of Gis and Visualisation. *The Electronic Journal on Information Systems in Developing Countries. EJISDC*, 38(5), 1–17.

Medar, et.al., (2020). A Survey on Data Mining Techniques for Crop Yield Prediction. *International Journal of Advance Research in Computer Science and Management Studies*, *2*(9).

Oikonomidis, A., Catal, C., & Kassahun, A. (2023). Deep learning for crop yield prediction: A systematic literature review. *New Zealand Journal of Crop and Horticultural Science*, 51(1), 1–26. 10.1080/01140671.2022.2032213

Palepu. (2021). An Analysis of Agricultural Soils by using Data Mining Techniques. *International Journal of Engineering Science and Computing,7*(10).

Pritam, B., & Nikola K. K., (2020). Spiking Neural Networks for Crop Yield Estimation Based on Spatio-temporal Analysis of Image Time Series. *IEEE Transactions On Geoscience And Remote Sensing*. IEEE.

Rani. (2020). The Impact of Data Analytics in Crop Management based on Weather Conditions. *International Journal of Engineering Technology Science and Research*, *4*(5).

Shinde. (2020). Web Based Recommendation System for farmers. *International Journal on Recent and Innovation Trends in Computing and Communication*, *3*(3).

Shreya, B., &Kalyani, B., (2020). Crop And Yield Prediction Model. *International Journal of Advance Scientific Research and Engineering Trends, 1.*

Tripathy, A. K. (2021). Data mining and wireless sensor network for agriculture pest/disease predictions. *Information and Communication Technologies (WICT), 2011 World Congress*. IEEE.

Vikas, K., & Vishal, D. (2021). KrishiMantra: Agricultural Recommendation System. *Proceedings of the 3rd ACM Symposium on Computing for Development*. ACM.

Chapter 5
Artificial Intelligence (AI)-Driven IoT (AIIoT)-Based Agriculture Automation

Kutubuddin Sayyad Liyakat Kazi
http://orcid.org/0000-0001-5623-9211
Baldev Ram Mirdha Institute of Technology, Solapur, India

ABSTRACT

The fusion of IoT and AI encompasses given rise to a new concept - artificial intelligence driven IoT (AIIoT). AIIoT is the intersection of AI and IoT, where AI algorithms are used to assess the data collected by IoT devices, improving their effectiveness and intelligence. AIIoT-based agriculture has become the outcome of the new opportunities for agricultural automation that this technology has made possible. Agriculture automation powered by AI and IoT is revolutionising the sector. It could lead to higher output, lower expenses, and better sustainability. The need for food will only grow as by 2050, it is expected that there will be 9.7 billion people on the planet. In addition to meeting this demand, AIIoT-based agricultural automation can help with resource scarcity and climate change issues.

INTRODUCTION

Global population growth is expected to accelerate, with estimates of 10 billion people within 2050. This places a great deal of pressure on the agriculture industry to raise yield per hectare and improve crop productivity. Agriculture is becoming less profitable due to a number of issues that farmers face, including small land holdings, a labor shortage, climate change, extreme weather, and a decline in soil fertility. Climate change along with persistent danger to agriculture has come from other environmental problems in recent years, making it extremely difficult to achieve increased productivity by Subeesh(2021).

Agriculture automation is the practice of using technology and machinery to improve and streamline processes in the agricultural industry. It involves the integration of various technologies such as sensors, robotics, and AI to perform tasks that were previously done manually by farmers. This trend has been gaining traction in recent years, and it is transforming the way we produce, process, and distribute food by Priya(2023)& Wale(2019).

DOI: 10.4018/979-8-3693-3583-3.ch005

The need for agriculture automation has become increasingly evident as the need for food increases along with the growth of the global population. United Nations projects that by 2050, there will be 9.7 billion people on the planet, meaning that farmers must produce 70% more food compared to what they do now in order to meet demand. This is a daunting task, especially when considering the challenges that farmers face, such as climate change, labor shortages, and limited resources.

One of the main benefits of agriculture automation is increased efficiency and productivity. With the use of advanced technologies, farmers can now perform tasks such as planting, watering, and harvesting crops at a much faster rate and with greater accuracy. This not only saves time and labor but also reduces the risk of human error, leading to higher yields and better quality produce. Automation also allows farmers to monitor and manage their crops in real-time, making it easier to detect and address any issues that may arise by Sultan Kazi(2023).

Another advantage of agriculture automation is cost savings. Automation technology may require a large initial investment, but over time, it can save farmers a substantial sum of money. With the use of autonomous vehicles and drones, farmers can reduce their reliance on manual labor, which can be costly and hard to find in some regions. Automation also reduces the need for fuel, water, and pesticides, which can result in substantial cost savings for farmers. Additionally, automation can help reduce waste by precisely applying resources only where needed, further cutting down on costs by Vahida (2023) & Karale (2023).

The use of automation in agriculture also has environmental benefits. By using sensors and data analysis, farmers can gather information about their crops' health and growth, allowing them to use resources such as water and fertilizers more efficiently. This can help reduce the environmental impact of agriculture, such as water pollution and soil degradation. Automation can also aid in the transition to more sustainable farming practices, such as precision farming, which involves using data to tailor farming techniques to specific areas and crops by Gouse (2018), & Madhupriya(2022) .

Moreover, agriculture automation can help address labor shortages in the industry by Kutubuddin K(2023). As the average age of farmers increases, there is a growing concern about who will take over the responsibility of feeding the world. Automation can attract young people to the industry by offering a more technologically advanced and less physically demanding work environment. It can also allow farmers to focus on more complex tasks, such as data analysis and decision making, rather than spending long hours doing manual labor by Vinay (2022).

However, there are some challenges to implementing agriculture automation. Small-scale farmers may find it challenging to compete with bigger, technologically advanced farms due to the upfront costs associated with technology and equipment purchases. There is also a concern about the potential job loss that may result from automation by Kazi Kutubuddin (2022). However, many experts argue that automation can create new job opportunities in areas such as data analysis and maintenance of equipment by by Kazi Sul (2023a, b).

In all, agriculture automation is transforming the way we produce food and has the potential to address many of the challenges faced by the agricultural industry. By increasing efficiency, reducing costs, and promoting sustainable practices, it can assist farmers in reducing their environmental impact while satisfying the world's expanding food demand. As technology develops, we should anticipate more agricultural innovations that will influence how food is produced in the future by Priya (2023).

The farming sector has come to understand the value of precision farming over the years. By precisely measuring inputs and minimizing the overuse of potentially harmful pesticides and other inputs, precision agriculture has an environmentally friendly option which can boost productivity. Digital technology-driven agriculture offers a multitude of approaches for automating and improving agricultural productivity and output, despite obstacles brought on by climate change as well as other factors. Real-time analysis made possible by digitization in agriculture facilitates better managing land, irrigation planning, spraying, and additionally land oversight by Kazi K(2022a).

Utilizing cutting-edge digital technology will enable the agriculture sector to reap numerous additional benefits, including decreased input costs and waste, adoption of sustainable practices, and increased productivity to fulfil the world's expanding food demand. The use of digital technology in agriculture is becoming more and more popular around the world because of its powerful monitoring systems that are constantly updated and remarkably simple field management capabilities by Kazi K (2024).

Agriculture-related tasks that once required a few days filled with human sweat and animal labor were reduced by only a few hours of work thanks to mechanization. This can be viewed as the initial automation stage that changed the tasks associated with agriculture in developing nations such as India. In India, agricultural mechanization is still in its infancy and is expanding at an average of 7.5% annually. As digital technologies advance, this mechanization will become more sophisticated and rapid. Prolonged labor migration constitutes one of the main problems of the modern era. The proportion of workers in Indian agriculture that were employed in the sector fell coming from 59.1 in 1991 to 54.6% in 2011 along with is projected to reach 40.6% in 2020, according to research on the workforce involved in this industry by Subeesh(2021).

These were the principal justifications for increasing mechanization and decreasing manual labor. Reducing the tediousness of agricultural tasks encourages women workers to take the lead and contribute significantly to agricultural activities. The degrees of mechanization in maize crop agriculture are displayed in Figure 1. Mechanization of agriculture was essential in developing nations like India to boost economic growth and break through all these barriers.

Figure 1. Levels of mechanization in maize crop farming operations

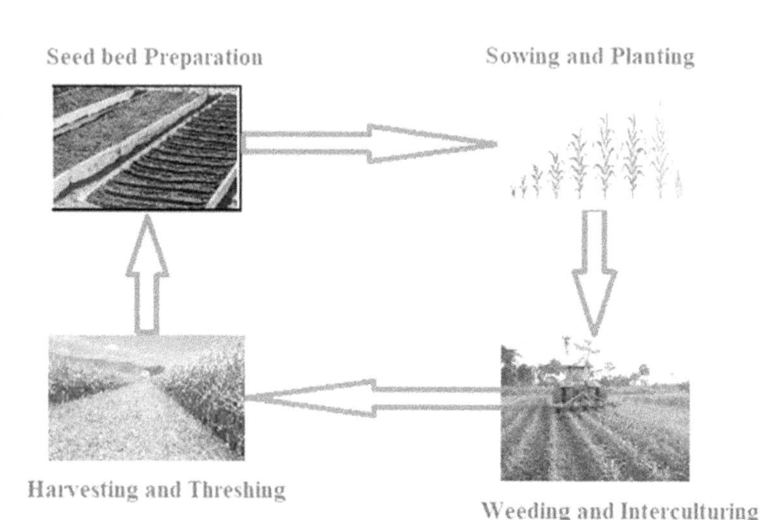

Two possible ways to address the food crisis are to increase the amount of land used for large-scale farming and to apply best practices and technology support to increase productivity. The only option in developing nations with densely populated areas—where expanding land area is practically unfeasible—is to become more intelligent through the application of revolutionary innovations including AI-Artificial Intelligence & IoT-Internet of Things by Umar(2022), Sayyad Liyakat(2023), Kazi K S(2023),Sultanabanu(2023) & Sultana(2023). AI and IoT have been identified as the two main technologies that could completely transform how modern agriculture is conducted thanks to recent developments in ICT (information and communication technology) and related research. Improved conclusions may be formed from field data by utilising technological advances including artificial intelligence and internet of things, which also make it possible to plan farming practices systematically with the least amount of manual labour by Kutubuddin Kazi(2022).

IOT (INTERNET OF THINGS)

IoT has taken the world by storm, revolutionizing the way we live, work, and interact with technology. From smart homes and cities to connected cars and wearables, the possibilities of IoT seem endless. The concept of IoT is simple – it is the interconnection of everyday devices, objects, and machines through the internet, allowing them to communicate and exchange data. This has opened up a whole new world of opportunities and is transforming the way we live our lives by K S L(2022), Kutubu(2022), KS(2022)(2023).

The revolution of IoT began with the rapid development of technology, particularly in the areas of wireless connectivity, sensors, and data analytics. These advancements have made it possible for devices to be connected to the internet and collect and transmit data in real-time(Figure 2). This data can then be used to automate processes, improve efficiency, and provide valuable insights by Sultanabanu Liyakat(2023), S S L(2023), KSSL(2023), SSLK(2023).

Figure 2. IoT structure

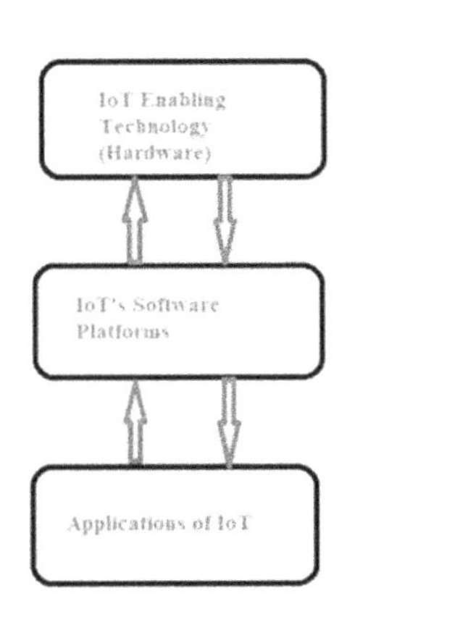

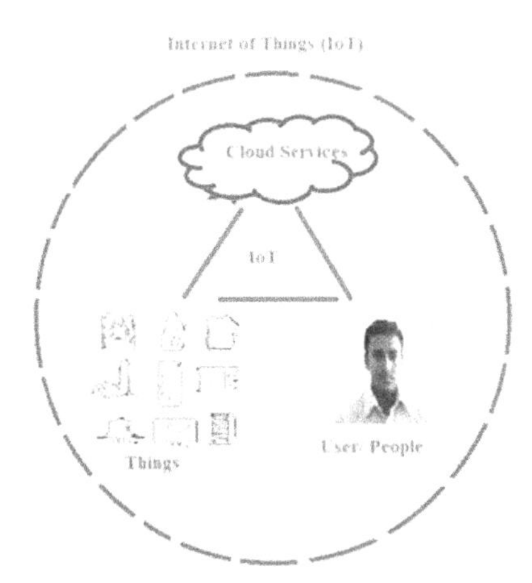

One of the most significant impacts of the IoT revolution is in our homes. With the rise of smart home devices such as voice assistants, smart thermostats, and security systems, our homes have become more interconnected and intelligent. We can now control and monitor our homes remotely, making our lives more convenient and comfortable. For example, we can turn off lights, adjust the temperature, and even make a cup of coffee, all with a simple voice command.

The IoT revolution has also brought significant changes in the healthcare industry. With the use of connected devices, doctors can monitor patients remotely, reducing the need for frequent hospital visits. Wearable devices such as smartwatches and fitness trackers can collect real-time data on our health and fitness, allowing us to make better decisions about our well-being. This has not only improved the quality of healthcare but also made it more accessible and cost-effective by S Liyakat(2023).

Another area where IoT is making a significant impact is in transportation. With the development of connected cars, we can now have real-time information about traffic, weather conditions, and even our vehicle's performance. This not only makes our driving experience safer and more convenient but also helps in reducing traffic congestion and carbon emissions. In the future, we can expect to see more self-driving cars, which will revolutionize the way we travel.

The rise of smart cities is also a result of the IoT revolution. By using connected sensors and data analytics, cities can optimize resources and improve the quality of life for its residents. For example, smart streetlights can automatically adjust their brightness based on the time of day, reducing energy consumption. Smart waste management systems can detect when bins are full, optimizing garbage collection routes and reducing costs.

The IoT revolution is not only impacting our daily lives but also transforming industries such as manufacturing and agriculture. By using connected sensors and data analytics, factories can monitor their equipment and predict when maintenance is needed, reducing downtime and costs. In agriculture, connected devices can monitor soil conditions, weather patterns, and crop health, allowing farmers to make better decisions and increase productivity.

Figure 3 shows the application scenario in IoT for Year 2023.

Figure 3. IoT application scenario in 2023

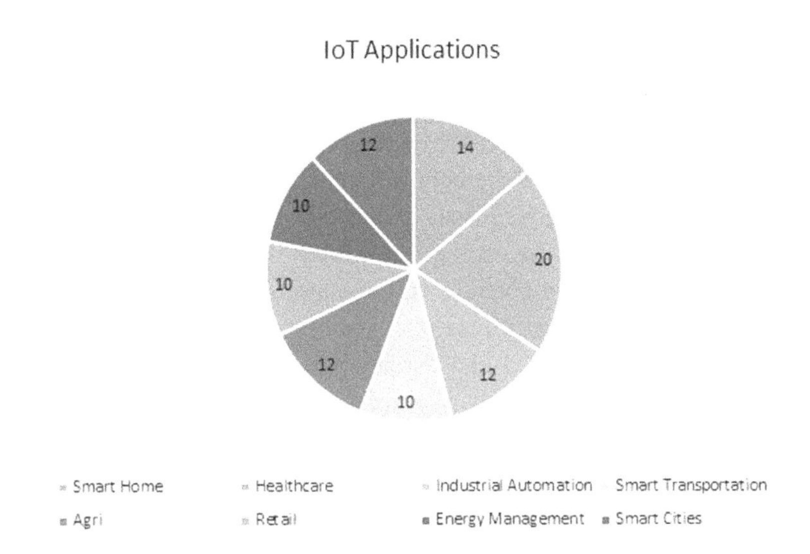

However, with all the benefits that IoT brings, there are also concerns about privacy and security. Cyber attacks and information breaches are more likely when there are more devices online. Thus, it is essential to have appropriate security measures established to safeguard our private data and stop illegal access to our gadgets. For IoT security we recommends KK approach, and if decisions are made automatically by system, we recommends KSK approach.

In all, the IoT revolution has brought about significant changes in our lives, making it more convenient, efficient, and connected. It has opened up a whole new world of opportunities and has the potential to transform industries and societies in ways we never thought possible. As the technology continues to evolve, we can expect to see even more exciting advancements and innovations in the world of IoT. It is up to us to embrace this revolution and use it to create a better and smarter future.

AI (ARTIFICIAL INTELLIGENCE)

Artificial Intelligence, commonly referred to as AI, has been a rapidly developing technology in recent years. It refers to the capacity of a computer or other machine to carry out operations like learning, problem-solving, and decision-making that typically call for human intelligence. This technology has shown immense potential in various industries and has the power to transform the way we live and work by Dixit(2014),Dixit(2015).

Although AI have been approximately for many years, real progress has only recently been achieved. This is due to the advancements in computing power, data storage, and algorithms. With the increasing availability of big data and the rise of machine learning, AI has become more sophisticated and is now capable of performing complex tasks that were previously thought to be exclusive to humans.

Automation is one of the most important areas where AI technology is being used. It can automate a number of processes, increasing their accuracy and efficiency. AI-powered robots, for instance, can execute repetitive tasks in the manufacturing sector with accuracy and consistency, increasing productivity and lowering costs. AI can help physicians diagnose conditions and analyse medical images in the healthcare industry, enabling more rapid and precise treatment by SSL(2023).

AI technology has also made its way into the world of business. Companies are using AI to improve their customer service by implementing chatbots, which can interact with customers in real-time and provide them with quick and accurate responses. AI-powered virtual assistants are also being used to streamline administrative tasks, allowing employees to focus on more critical tasks by Kazi K(2022).

In the education sector, AI has the potential to revolutionize the way students learn. Students can have individualised learning experiences that are tailored to their unique strengths and weaknesses with the use of AI. AI-powered tutors can also assist students in their studies, providing them with instant feedback and personalized learning materials by Kuubuddin SL(2023).

Transportation is another industry in which AI technology shows enormous promise. AI-powered self-driving car technology possesses the potential to improve road safety and lower the prevalence of accidents brought on by human error. AI can also improve traffic flow, cutting down on delays and traffic.

However, with all its potential, AI technology also raises concerns about its impact on society. Many fear that AI will lead to job losses, as machines and robots take over human tasks. There are also concerns about the ethical implications of AI, as it is programmed by humans and can potentially perpetuate biases and discrimination by Kutub(2022), Kazi(2017), Kazi(2018).

To address these concerns, it is essential to have proper regulations in place to govern the development and use of AI technology. Companies must also ensure that their AI systems are unbiased and transparent, with clear guidelines on how they make decisions. Figure 4 shows the scenario of application of AI in 2023.

Figure 4. AI applications scenario in 2023

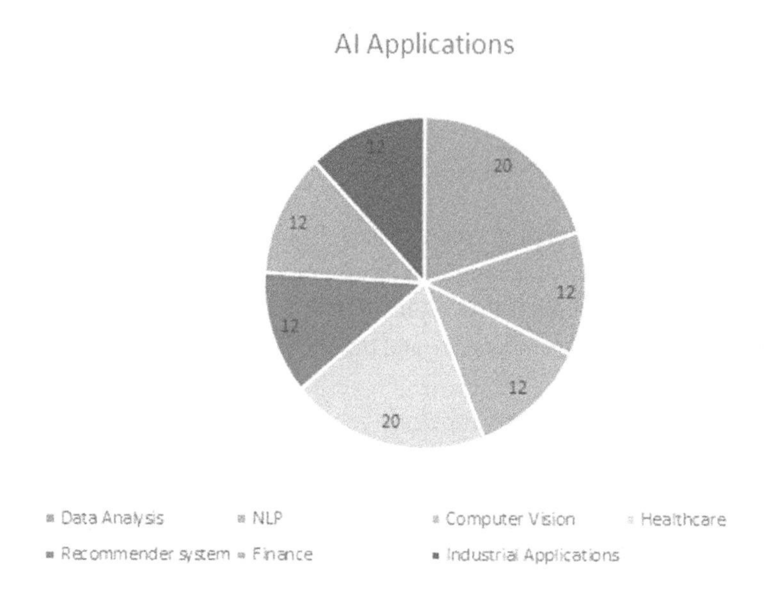

In all, AI technology has the potential to bring about significant changes in our lives, from improving efficiency and productivity to creating new job opportunities. However, it is crucial to use this technology ethically and responsibly to ensure its benefits are enjoyed by all. With proper regulations and ethical guidelines, AI has the potential to bring about a brighter and more advanced future for humanity by Mulani(2019), Mulani(2023).

LITERATURE SURVEY

Priya et al.(2023), explains Heart Healthcare system utilizes ML. Recently, cardiovascular diseases (CVD) have surpassed all other causes of death in both developed and developing countries. Lowering the death rate can be achieved by early identification of cardiac illnesses and ongoing expert therapeutic supervision. However, because of the increased intelligence, effort, and expertise required, precise identification of cardiac problems under any situation and 24-an hour physician consultation remain unfeasible. This study presented the fundamental idea for a machine learning (ML)-based system that predicts the likelihood of developing heart disease by employing ML techniques to detect heart disease in the near future. There are fewer synthesised research articles in this field, despite the growing number of empirical studies on the subject, especially from developing nations. Predictive analytics is becoming an increasingly vital tool for human protection and heart welfare services in an era when data availability is rapidly growing. This cutting-edge technology helps heart-care agencies make more informed choices regarding how they can best assist their clients by using data gathered from past events to forecast future patterns and outcomes. Predictive analytics, like every other data-driven technology, must be utilised properly to ensure morally and practically sound business practices. The included studies in this study concentrate on using machine learning algorithms to forecast the heart healthcare system (HHS). For registration as well as notification, they used the K-means Elbow technique; for HHS, they used a decision tree; and for immunisation reminders, they used MySQL.

Sunita et al.(2023), explain ML and IoT for Food and Fruit quality in food safety. This article describes a machine learning as well as IoT-based approach for monitoring perishable goods. The proposed system entails using Internet of Things (IoT) devices to upload images taken with high-resolution cameras to a cloud server. K-means clustering is used to segment these images before they are uploaded onto a cloud server. After principal component analysis is used to extract attributes from the images, trained ML method have been utilised to classify the pictures. This suggested method uses machine learning, image processing, and the Internet of Things for tracking perishable food.

Mulani et al. (2019), utilizes Ml for Hand Gesture recognition system. Human-computer interaction that is intuitive and natural can be achieved through hand gestures. To maximise recognition of hand gestures for mouse operation, our new approach combines established methods of Viola-Jones-Haar-like feature-based recognition of objects and skin color-based ROI segmentation. A mouse operation consists of moving the cursor along with clicking with either the left or right mouse button. This paper first defines a Region of Interest (ROI) using colour as a robust feature. Then, using the AdaBoost learning algorithm and Haar-like features, hand postures are detected within this ROI. By combining a series of poor classifiers, the AdaBoost learning algorithm dramatically improves performance and creates an accurate cascaded classifier.

Pradeepa et al.(2022), explains the use of ML & IoT for students Health perdition. As more students live independently and are spread out over wide geographic areas, it is now necessary to keep an eye on their health status. This study proposes an IoT-based approach to student health management that continuously monitors students' vital signs and uses cutting-edge medical technology to identify biological and behavioural changes. In this concept, the IoT module collects vital data, and NN models have been employed to evaluate the data and identify potential risks to children's physiological and behavioural changes. The results of the experiment indicate that the proposed model is a reliable and accurate way to assess the pupils' states. After evaluating the suggested model, the SVM achieved an optimal performance of 99.1%, which is satisfactory for our goals. The outcomes also outperformed algorithms for random forests, decision trees, and multilevel perceptron neural systems.

Kazi (2018), explains the use of ML in Aquatic study. By using colour matching, the colour correction over the whole sequence is lessened as well as the colour differences between neighbouring images. In aquatic image applications, we apply linear correction and gamma correction, respectively, for the brightness and chrominance elements in the original images (marine). The problem of colour consistency in 360-degree panoramic photos is addressed by colour correspondence and colour difference distribution techniques. This article integrates the stitching approach into a panoramic imaging system to generate high-resolution and high-quality panoramic photographs to cellphones.

Nikita(2020), explain use of IoT for ITS. Within intelligent transportation systems (ITS), vehicle tracking data is essential for fleet managers who wish to track vehicles for logistics and transportation as well as for drivers and passengers who want to know where they are to cross or obtain location-relevant data. For the purpose of conducting rescue operations, the ability to identify a car that is in danger and locate illegal cars or vehicles carrying hazardous chemicals is essential to government authorities. Information technology and communication are necessary for ITS to operate. Some of these tools, like loop detectors, are well known to individuals in transportation sector. However, an assortment of less well-known systems and technologies are necessary for ITS to function. While communication and control technologies form the foundation of ITS, human factors are equally important and can present difficulties. This paper outlines the most significant technological solutions for ITS and clarifies how human factors specialists should be included from the start in the network and instrument design of ITS. The study used the example of an Indian transportation scenario centre to discuss how to find IoT solutions for ITS and evaluate them inside the corporate design.

Ravi et al(2022), explain use of IoT on LOVE. A system is needed to demonstrate the love and closeness to one another as well as their remembrance. Even today, the only ways to express closeness are still by phone or mail. We will therefore demonstrate our love and remembrance through the use of contemporary methodology and the IoT. There has never been a system or concept like this before. We present this system using the Internet of Things and sensors.

Sunil Kumar et al.(2022), explains Deep Learning for diseases identification in Plants. Plant diseases are the focus of this research because they seriously jeopardise the food production and livelihoods of small-scale farmers. In traditional farming, each row is visually inspected by trained personnel to detect plant diseases. Because it is labor-intensive and time-consuming, this task is flawed by nature because it is done by humans. The goal of this research is to combine image recognition and deep learning techniques (Faster R-CNN+ResNet50) to assess real-time photos in order to develop an automated identification algorithm for three of the most common diseases of maize plants: Northern Leaf Blight, Cercospora Spot, and Frequent Rust. The proposed system achieved a 93.5% reliability rate in effectively identifying three diseases of maize.

The system that Wale et al. (2019) have presented for use in agriculture is an excellent one; it determines whether the motor should be turned on or off based on the amount of moisture that is present in the soil. Additionally, the user is provided with information regarding the water level in the water tank through the usage of the IoT in addition to the LCD display. There is information provided by the system on the temperature and humidity of the atmosphere. We are able to take action in accordance with this. In addition, the presence of animals can be identified thanks to the PIR sensor. A low cost, a low power consumption, and a shorter amount of time dedicated to analysis were among the positives.

UNIFIED DESIGN OF AI-POWERED AGRICULTURAL IOT (AIIOT) ECOSYSTEM

A system of agricultural automation based on IoT integrates various technologies to accomplish its intended purpose. The IoT sensors are deployed on the intended site at the foundation of the entire system (Figure 5). This could include farm machinery, sensors placed in greenhouses, cattle ear tags, soil monitoring fields, and so forth by Sunil(2022). An Internet of Things device utilised in the farm site consists of several parts. Among the most widely used devices are the RaspberryPi, Arduino, as well as Beagle Bone by Gund(2023). These devices typically have a small amount of RAM for processing, but they can effectively communicate with the outside world through the use of a communication module by K Kazi(2017). They have the ability to send and receive signals from outside sources.

The sensors record any modifications in the surroundings in real time, and an IoT gateway will transmit the data to a distant server or cloud. The management of data is the responsibility of the cloud or remote server. Typically, data are kept in databases. Conventional relational databases shouldn't be used for storing this kind of data because it is unstructured and large in volume.

The information is extracted from a database and exported as CSV, Excel, images, or as any other format that the analysing programme can handle in AI-enabled data pipeline. The primary task of data pre-processing is cleaning the data—that is, eliminating outliers, normalising, and other tasks. The data are separated into train, validation, and test sets for supervised algorithms. Numerous algorithms may be utilised to the data, depending on the kind of processing that needs to be carried out and the data itself. Using the training dataset, a model is built and trained. Validation and testing data are then used to assess the model's accuracy and performance. To verify model stability, K-Fold Cross-validation techniques are frequently employed. The precision and recall values obtained from confusion matrices are utilised to gauge a classification model's quality.

The model can be used to generate results over unknown data in the cloud once it has undergone thorough validation. The AI model is able to generate intelligent forecasts and choices according to the information that is received, which can be shared with the farmers or used to start any farm operations. With a smartphone, farmers can quickly obtain recommendations, simplified versions of prediction results, and the necessary actions. Email and SMS notifications are also possible. In the recent past, farmer-friendly web and mobile applications have also become popular means of informing farmers. Additionally, the farmer has control over farm operations via mobile applications. They have the ability to send control signals, which the actuators can then use to carry out the required tasks.

Figure 5. AIIoT in agriculture system

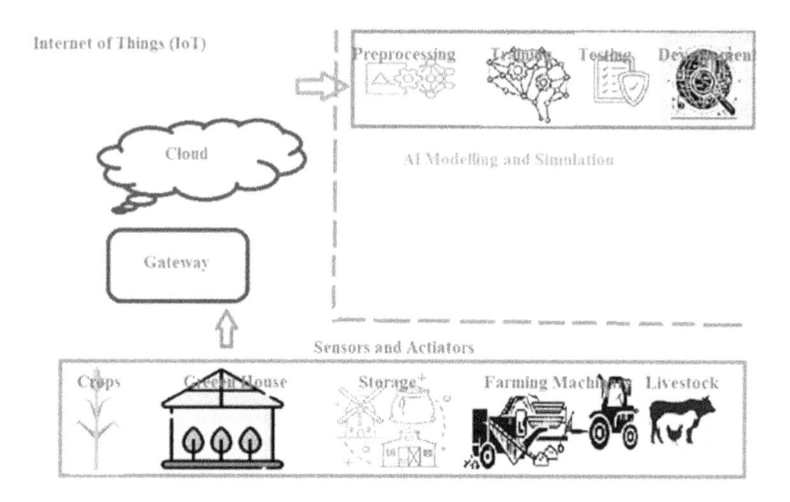

The integration of AI and IoT has given rise to a new concept - Artificial Intelligence driven IoT (AIIoT). AIIoT is the convergence of AI and IoT, where AI algorithms are used to analyze the data collected by IoT devices, making them smarter and more efficient by Kazi k(2024). This technology has opened up new avenues for automation in agriculture, leading to what is known as AIIoT-based agriculture automation.

AIIoT-based agriculture automation involves the use of AI-powered sensors, drones, robots, and other IoT devices to collect and analyze data from farms. This data is then used to make informed decisions and automate various farming processes, resulting in increased productivity, reduced costs, and improved sustainability.

Precision farming is one of the main advantages of AI/IoT-based agricultural automation. Farmers can gather information on the condition of plants and their development rate, as well as soil temperature, moisture content, and nutrient levels, with the use of sensors and drones. The precise amount of water, fertiliser, and pesticides required for each crop is then determined by AI algorithms analysing this data, doing away with guesswork and cutting down on waste. This lessens farming's negative environmental effects while also increasing crop yields.

AIIoT-based agriculture automation also enables farmers to monitor their crops remotely. With the help of IoT devices, farmers can keep track of their crops' growth and health, even when they are not physically present on the farm. This allows them to detect any issues early on and take timely action, preventing crop losses.

In addition to precision farming and remote monitoring, AIIoT-based agriculture automation also includes automated irrigation systems, robotic harvesting, and autonomous tractors. These technologies not only save time and labor costs but also make farming safer and more efficient.

Moreover, AIIoT-based agriculture automation also has the potential to revolutionize livestock farming. With the help of sensors and AI algorithms, farmers can monitor the health and behavior of their livestock, detect any signs of illness, and take necessary measures. This can prevent the spread of diseases and improve animal welfare.

Apart from improving productivity and reducing costs, AIIoT-based agriculture automation can also contribute to sustainability. By optimizing the use of resources and reducing wastage, it can help in conserving water, reducing the use of chemicals, and minimizing the carbon footprint of farming. This is crucial in today's world, where sustainability has become a key concern.

However, like any other technology, AIIoT-based agriculture automation also has its challenges. The initial investment cost can be high, making it inaccessible for small-scale farmers. Moreover, there is a lack of awareness and technical expertise among farmers, which may hinder the adoption of this technology. Governments and organizations need to provide support and training to farmers to overcome these challenges and reap the benefits of AIIoT-based agriculture automation.

IOT AND AI APPLICATIONS FOR AUTOMATED AGRICULTURE

By introducing smarter applications, an AI-powered IoT ecosystem has the potential to introduce significant improvements in control and precision in farming practices. The potential of these technologies in agriculture is limitless because they can automate labor-intensive tasks with little need for human intervention. The following provides a summary of the IoT and AI applications in agricultural engineering (Figure 6).

Figure 6. Applications of AIIoT in automated agriculture

1. Smart Cropping

Smart cropping is a modern and innovative approach to farming that utilizes technology to optimize crop production and increase efficiency. This technique involves the use of advanced tools and precision farming methods to monitor and manage crops, resulting in higher yields and reduced environmental impacts by Wale(2019).

In traditional farming practices, farmers rely on their experience and intuition to determine the best time to plant, irrigate, and harvest their crops. However, with the introduction of smart cropping, farmers can now take advantage of technology to make informed decisions and improve their farming practices.

A crucial element of smart cropping involves the application of precision agriculture. Utilising sensors, drones, and GPS technology, precision agriculture gathers information on plant health, soil conditions, and moisture content. Farmers gain important insights into the particular requirements of their crops by using this data to create intricate maps of the farm.

For example, by using soil sensors, farmers can determine the nutrient levels and pH balance of their soil, allowing them to make necessary adjustments to optimize plant growth. This not only results in healthier crops but also reduces the use of fertilizers and other chemicals, leading to a more sustainable farming approach.

In addition to precise data collection, smart cropping also utilizes automated systems to perform tasks such as planting and irrigation. This not only reduces the physical labor required by farmers but also ensures that crops are planted and watered at the most optimal times.

Another advantage of smart cropping is the ability to detect and address potential issues in the early stages. For instance, drones equipped with thermal cameras can detect variations in plant temperature, indicating the presence of pests or diseases. This allows farmers to take timely action and prevent the spread of the problem, saving both time and resources.

Smart cropping also incorporates the use of smart machines, such as robots and autonomous tractors, which can perform tasks with high precision and efficiency. These machines are equipped with sensors and cameras that can analyze and map the field, identify weeds, and apply herbicides only where necessary. This method reduces the use of chemicals, resulting in cost savings for farmers and a healthier environment.

Furthermore, smart cropping also promotes sustainable farming practices. By using precision agriculture, farmers can reduce their water and energy usage, resulting in a more efficient use of resources. It also allows for a more targeted approach to farming, leading to a reduction in waste and a more sustainable use of land.

One of the most significant benefits of smart cropping is its potential to increase food production. With a growing global population, the demand for food is also increasing. By utilizing technology and optimizing crop production, smart cropping can help meet this demand and ensure food security for future generations.

In all, smart cropping is a game-changer in the field of agriculture. By leveraging technology, farmers can make informed decisions, increase efficiency, and reduce their environmental impact. This approach not only benefits farmers but also contributes to a more sustainable and secure food system for all. As technology continues to advance, the potential for smart cropping to revolutionize farming and address the challenges faced by the agriculture industry is limitless. It is undoubtedly the future of farming.

2. Smart Greenhouse

The world is facing a growing population and a changing climate, leading to challenges in food production. Traditional farming methods are not sustainable in the long run, and the need for innovative solutions to enhance agricultural production is becoming more pressing. This is where the concept of a 'Smart Greenhouse' comes into play.

A Smart Greenhouse is a modern take on the traditional greenhouse, which uses advanced technology to create a controlled environment for the cultivation of plants. It is equipped with sensors, actuators, and other technologies that monitor and control the internal environment of the greenhouse, including temperature, humidity, light, and irrigation. This allows for optimal growing conditions for plants, regardless of the external weather conditions.

One of the biggest advantages of a Smart Greenhouse is its ability to optimize the use of resources. The sensors in the greenhouse collect data on the internal environment and transmit it to a computer system, which then analyzes the data and adjusts the settings accordingly. This results in efficient use of resources such as water, energy, and fertilizers, reducing the environmental impact of traditional farming methods. For example, the use of drip irrigation systems in a Smart Greenhouse can reduce water consumption by up to 70% compared to traditional irrigation methods by Magdum(2024).

Another benefit of a Smart Greenhouse is its ability to protect crops from pests and diseases. The controlled environment and strict monitoring of the greenhouse make it difficult for pests and diseases to thrive, reducing the need for harmful pesticides and herbicides. This not only benefits the environment but also produces healthier and safer crops for consumption.

Furthermore, the use of technology in a Smart Greenhouse allows for year-round production of crops. With the ability to adjust temperature and light levels, it is possible to grow plants even during the off-season. This not only increases the availability of fresh produce but also provides a more stable income for farmers.

One of the most significant advantages of a Smart Greenhouse is its ability to adapt to the changing climate. With the rise of extreme weather events such as droughts and floods, traditional agriculture is becoming increasingly vulnerable. Smart Greenhouses, on the other hand, can withstand these challenges by adjusting the internal environment to ensure the optimal growth of plants.

The concept of Smart Greenhouses is still relatively new, but it is gaining popularity around the world. In countries like Japan and the Netherlands, where land and resources are limited, Smart Greenhouses are being used to increase food production and reduce the environmental impact of agriculture. In developing countries, Smart Greenhouses are being used to improve food security and provide a sustainable source of income for small-scale farmers.

In all, a Smart Greenhouse is a game-changer in the world of agriculture. It combines the benefits of technology with sustainable farming practices to create a more efficient and environmentally friendly way of growing crops. With the world's population projected to reach 9.7 billion by 2050, the need for innovative solutions like Smart Greenhouses is crucial to ensure food security for all. It is, without a doubt, a revolution in agriculture that has the potential to shape the future of food production.

3. Smart Storage

The world is facing a growing population and a changing climate, leading to challenges in food production. Traditional farming methods are not sustainable in the long run, and the need for innovative solutions to enhance agricultural production is becoming more pressing. This is where the concept of a 'Smart Greenhouse' comes into play.

A Smart Greenhouse is a modern take on the traditional greenhouse, which uses advanced technology to create a controlled environment for the cultivation of plants. It is equipped with sensors, actuators, and other technologies that monitor and control the internal environment of the greenhouse, including temperature, humidity, light, and irrigation. This allows for optimal growing conditions for plants, regardless of the external weather conditions.

One of the biggest advantages of a Smart Greenhouse is its ability to optimize the use of resources. The sensors in the greenhouse collect data on the internal environment and transmit it to a computer system, which then analyzes the data and adjusts the settings accordingly. This results in efficient use of resources such as water, energy, and fertilizers, reducing the environmental impact of traditional farming methods. For example, the use of drip irrigation systems in a Smart Greenhouse can reduce water consumption by up to 70% compared to traditional irrigation methods.

Another benefit of a Smart Greenhouse is its ability to protect crops from pests and diseases. The controlled environment and strict monitoring of the greenhouse make it difficult for pests and diseases to thrive, reducing the need for harmful pesticides and herbicides. This not only benefits the environment but also produces healthier and safer crops for consumption.

Furthermore, the use of technology in a Smart Greenhouse allows for year-round production of crops. With the ability to adjust temperature and light levels, it is possible to grow plants even during the off-season. This not only increases the availability of fresh produce but also provides a more stable income for farmers.

One of the most significant advantages of a Smart Greenhouse is its ability to adapt to the changing climate. With the rise of extreme weather events such as droughts and floods, traditional agriculture is becoming increasingly vulnerable. Smart Greenhouses, on the other hand, can withstand these challenges by adjusting the internal environment to ensure the optimal growth of plants.

The concept of Smart Greenhouses is still relatively new, but it is gaining popularity around the world. In countries like Japan and the Netherlands, where land and resources are limited, Smart Greenhouses are being used to increase food production and reduce the environmental impact of agriculture. In developing countries, Smart Greenhouses are being used to improve food security and provide a sustainable source of income for small-scale farmers.

In all, a Smart Greenhouse is a game-changer in the world of agriculture. It combines the benefits of technology with sustainable farming practices to create a more efficient and environmentally friendly way of growing crops. With the world's population projected to reach 9.7 billion by 2050, the need for innovative solutions like Smart Greenhouses is crucial to ensure food security for all. It is, without a doubt, a revolution in agriculture that has the potential to shape the future of food production.

4. Smart Farming Machinery

Farming has been the backbone of human civilization since time immemorial. It has been the primary source of food and livelihood for people all around the world. As technology advances, the agricultural sector has also been revolutionized with the introduction of smart farm machinery. Smart farm machinery refers to the use of advanced technology and data-driven systems in farming operations. It includes a wide range of equipment such as tractors, harvesters, crop monitoring systems, drones, and many more. These machines are equipped with sensors, cameras, and GPS technology, which collect and analyze data to make farming more efficient and productive.

One of the primary benefits of smart farm machinery is its ability to automate various farming tasks. With the use of GPS technology, tractors can now be operated with minimal human intervention. They can precisely plow, sow, and harvest crops, saving time and labor costs. This also reduces the chances of human error, leading to better crop yields.

Real-time data collection and analysis is another benefit of smart farm equipment. These devices have sensors and cameras built in to track temperature, moisture content in the soil, and other crucial factors, giving farmers insightful information. They can use this information to make more informed decisions about pest management, fertiliser application, and irrigation, which will increase crop yields and make better use of available resources by Neeraja(2024).

Drones have also become an essential part of smart farming. They can be used to monitor crop growth, identify areas of stress, and detect pests and diseases. With the help of infrared and thermal imaging, they can provide detailed and accurate information about the health of crops. This allows farmers to take timely action and prevent potential crop losses.

One of the significant challenges faced by farmers is the unpredictable weather conditions. However, with the help of smart farm machinery, they can now have a better understanding of weather patterns. This information can help them plan their farming activities accordingly and minimize the risk of crop damage due to adverse weather conditions.

Apart from increasing productivity, smart farm machinery also has a positive impact on the environment. By using data-driven systems, farmers can reduce the use of fertilizers, pesticides, and water, which not only saves costs but also minimizes the environmental impact of farming. This can lead to sustainable farming practices that benefit both the farmers and the environment.

The use of smart farm machinery is not limited to large-scale commercial farms. Even small-scale farmers can benefit from this technology. With the increasing availability of affordable and user-friendly equipment, small farmers can also boost their productivity and profitability.

However, the adoption of smart farm machinery comes with its own set of challenges. The initial cost of these machines can be a barrier for small farmers, and they may require training to operate them effectively. Moreover, the use of technology also raises concerns about data privacy and security. These challenges need to be addressed to ensure the widespread adoption of smart farm machinery.

In summary, smart farm machinery has transformed the way farming is done. It boosted farming's productivity, sustainability, and efficiency. Given the constant demand for food, it has become crucial to adopt technology in agriculture to meet the world's food needs. The use of smart farm machinery is a step in the right direction towards achieving this goal.

5. Smart Livestock

Smart livestock management is a modern approach to the traditional methods of raising and caring for livestock. It involves the use of technology, data, and innovative practices to improve the overall health, productivity, and efficiency of the animals. This method is gaining popularity among farmers and ranchers around the world as it offers numerous benefits and advantages.

One of the key elements of smart livestock management is the use of technology. Advancements in technology have made it possible for farmers to monitor their livestock in real-time. This includes using sensors and GPS tracking devices to keep track of the animals' location, movement, and health. Farmers can also use drones to survey their pastures and identify potential hazards such as predators or damaged fences. This technology not only saves time and labor but also allows farmers to detect any issues with their livestock before they become serious problems.

Another aspect of smart livestock management is the use of data. Farmers can learn a great deal about the productivity, health, and behaviour of their livestock by gathering and evaluating data. This information can help them make informed decisions about their herd's diet, breeding, and healthcare.

For example, data on milk production can help dairy farmers adjust their feeding and milking schedules to optimize their cows' milk yield. Similarly, data on weight gain can help ranchers determine the best time to sell their animals for maximum profit.

In addition to technology and data, smart livestock management also involves implementing innovative practices. This includes using alternative feed sources, such as grass and legumes, to reduce the reliance on expensive and environmentally damaging feed crops. Farmers can also use rotational grazing techniques to improve the quality of their pastures and prevent overgrazing. This not only benefits the animals but also helps to maintain the health of the land.

One of the primary benefits of smart livestock management is improved animal welfare. With real-time monitoring and data analysis, farmers can quickly identify and address any health issues or injuries in their livestock. This results in better care and treatment for the animals, leading to healthier and more productive livestock.

Moreover, smart livestock management also has a positive impact on the environment. With the use of alternative feed sources and sustainable grazing practices, farmers can reduce their carbon footprint and contribute towards mitigating climate change. This method also promotes biodiversity by ensuring the preservation of natural habitats for wild animals and plants.

Furthermore, smart livestock management can help to increase profitability for farmers. By using technology and data to optimize their operations, farmers can reduce costs and improve productivity. This can result in higher profits and a more sustainable business model.

In all, smart livestock management is a modern and effective approach to raising and caring for livestock. It combines the use of technology, data, and innovative practices to improve animal welfare, promote sustainability, and increase profitability. By adopting this method, farmers can ensure the long-term success of their operations while also contributing towards a healthier and more sustainable agriculture industry.

The validation of work is shown in figure 7 below where we can represent the system response in terms of accuracy with ANN and DT approach

Figure 7. Validation of work with respect to ANN and DT in terms of accuracy

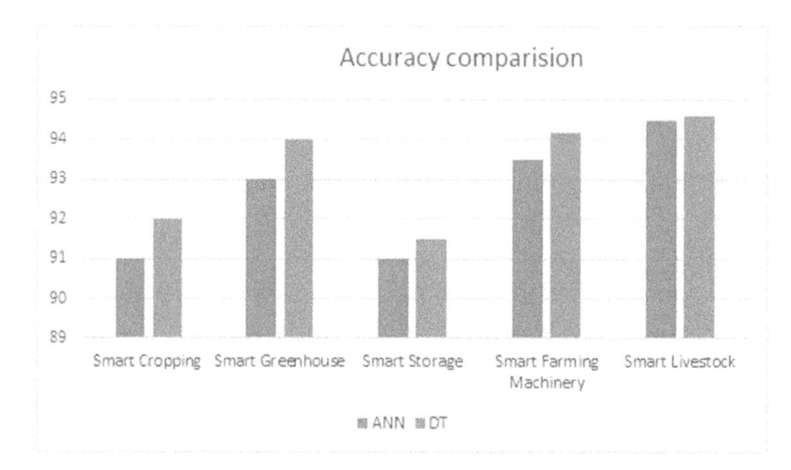

CHALLENGES AND OPPORTUNITIES

Even though technology has the potential to completely transform the agricultural industry, one of the biggest problems facing the ecosystem is farmers' lack of technical expertise in operating technology-driven equipment. Developing the systems with the farmers in mind is the best way to address this. When it comes to digital products, designers must concentrate on the user interface. One potential solution is to offer solutions in the local languages. When it comes to implementing cutting edge technology, small-scale farmers are primarily concerned about the cost and quality of the gadgets and sensors by Kadam(2022).

A system's dependability is essential for IoT solutions. Decisions are made with the use of decision support systems have a direct impact on agricultural practices, so any risks to their functionality or component failures will result in problems with dependability. With regard to the information management as well as security associated with broadly IoT applications, getting a network about tiny devices that's have a broad coverage, IoT systems possess very little capacity for the purpose of their data storage and processing capabilities. In order to maximise the potential of these networks with limited resources, appropriate data management techniques must be implemented.

Figure 8. Security concern in IoT

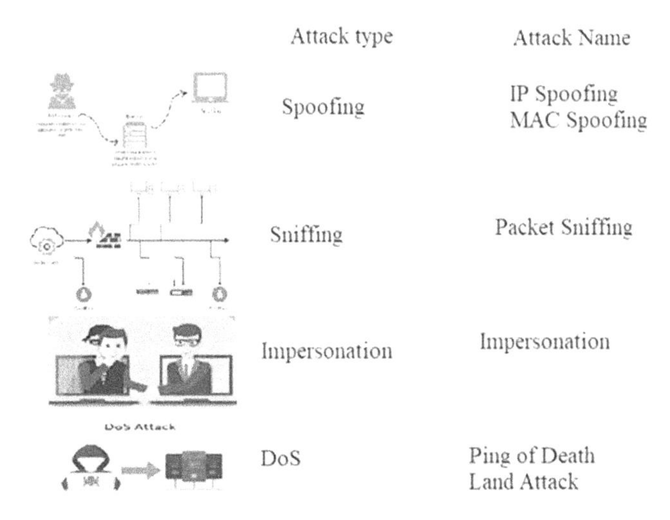

Because IoT devices are diverse, interoperability is essential, and proper device synchronization is required for optimal performance by Halli(2022). This is a challenging task due to the numerous vendors and devices involved. Since the amount of data coming from IoT devices is growing daily, horizontal scaling will eventually be necessary. Because the data from the devices can be unstructured, semi-structured, or structured, heterogeneity in the data is another significant challenge that researchers have tackled by Liyakat (2023),(2024).

Every type of data ought to be able to be easily handled and processed by the system. This challenge can be met by creating cloud-based architecture for Internet of Things applications, since cloud services offer massive processing power, enormous storage capacity, and high scalability. The quantity and quality of data determine the decision-making quality of AI algorithms as well. Finding an enormous amount of high-quality information is an important anxiety when developing models using AI by Kazi K (2024a, 2024b).

Figure 8 illustrates some of the main obstacles to the widespread use of IoT as well as analytics solutions, and it is evident that security is a top priority for all of them.

AI and IoT systems are expanding and improving opportunities for more value development and capturing. Despite all of the difficulties, they are anticipated to help automate and improve agriculture in the future. These technologies have the potential to revolutionize agricultural practices. In the coming years, the development of 5G infrastructure will become essential to expanding the potential of IoT. The capacity of 5G networks is a hundred times greater than that of 4G networks, which can significantly boost internet speed.

The evolution of 5G, where response times can be faster than in the blink of an eye, will address the communication delay issue that limits current IoT systems. In the future, sensors as well as embedded technologies are going to stay widely accessible and affordable, which will greatly increase the potential of the Internet of Things. The development of sophisticated algorithms and artificial intelligence powers the creation of smarter applications.

CONCLUSION

The IoT and AI are two of the newest and most talked-about technologies. IoT, with its network of connected devices, has made our lives more convenient, and AI, with its capacity for learning and decision-making, has revolutionized a number of industries. The agriculture sector could undergo a radical change if these two potent technologies are integrated. The integration of AI and IoT has given rise to a new concept - Artificial Intelligence driven IoT (AIIoT). The intersection of AI and IoT is known as AIIoT, wherein AI algorithms are employed to evaluate the data gathered by IoT devices, thereby enhancing their intelligence and efficacy. AIIoT-based agriculture is the result of new automation opportunities in agriculture made possible by this technologization-based agriculture automation is a game-changer for the agriculture industry. It has the potential to increase productivity, reduce costs, and improve sustainability. AIIoT-based agriculture automation can help meet this demand while also addressing the challenges of climate change and resource scarcity. It is time for farmers to embrace this technology and take their farming practices to the next level.

FUTURE SCOPE

Main Limitations of IoT system is security. Security concern will harm the system performance. Also the decisions are made automatically by the system whenever the data is available. Hence in Future we will implement the system by using KK approach to solve security issue and the decisions are made automatically by system by means of KSK approach which is AIIoT based decision support system.

REFERENCES

Drixit, A. (2014). A Review paper on Iris Recognition. *Journal GSD International society for green. Sustainable Engineering and Management*, 1(14), 71–81.

Akansha, K. (2022). Email Security. *Journal of Image Processing and Intelligent remote sensing,2*(6).

Dixit, J. (2015). Iris Recognition by Daugman's Algorithm – an Efficient Approach. *Journal of applied Research and Social Sciences*.

Mulani, A. (2019). Effect of Rotation and Projection on Real time Hand Gesture Recognition system for Human Computer Interaction. *Journal of The Gujrat Research Society*, 21(16), 3710–3718.

Gund, V. D. (2023). PIR Sensor-Based Arduino Home Security System. *Journal of Instrumentation and Innovation Sciences*, 8(3), 33–37.

Halli Umar, M. (2022). Nanotechnology in IoT Security. *Journal of Nanoscience. Nanoengineering & Applications*, 12(3), 11–16.

Dixit, J. (2015). Iris Recognition by Daugman's Method. *International Journal of Latest Technology in Engineering, Management &. Applied Sciences (Basel, Switzerland)*, 4(6), 90–93.

Karale Aishwarya A. (2023). Smart Billing Cart Using RFID, YOLO and Deep Learning for Mall Administration. *International Journal of Instrumentation and Innovation Sciences,8*(2).

Kazi, K. (2017). Lassar Methodology for Network Intrusion Detection. *Scholarly Research Journal for Humanity science and English Language*, *4*(24), 6853 - 6861.

Kazi, K. (2022). Hybrid optimum model development to determine the Break. *Journal of Multimedia Technology & Recent Advancements*, 9(2), 24–32.

Kazi K. (2022a). Model for Agricultural Information system to improve crop yield using IoT. *Journal of open Source development*, *9*(2), 16 – 24.

Kazi, K. (2024). AI-Driven IoT (AIIoT) in Healthcare Monitoring. In Nguyen, T., & Vo, N. (Eds.), *Using Traditional Design Methods to Enhance AI-Driven Decision Making* (pp. 77–101). IGI Global. 10.4018/979-8-3693-0639-0.ch003

Kazi, K. S. (2017). Significance And Usage Of Face Recognition System. *Scholarly Journal For Humanity Science and English Language*, 4(20), 4764–4772.

Kazi, K. S. (2022). IoT-Based Healthcare Monitoring for COVID-19 Home Quarantined Patients. *Recent Trends in Sensor Research & Technology*, 9(3), 26–32.

Kazi, K. S. (2023). Detection of Malicious Nodes in IoT Networks based on Throughput and ML. *Journal of Electrical and Power System Engineering*, 9(1), 22–29.

Kazi, K. S. (2024b). Computer-Aided Diagnosis in Ophthalmology: A Technical Review of Deep Learning Applications. In Garcia, M., & de Almeida, R. (Eds.), *Transformative Approaches to Patient Literacy and Healthcare Innovation* (pp. 112–135). IGI Global.10.4018/979-8-3693-3661-8.ch006

Kazi, K. S. L. (2018). Significance of Projection and Rotation of Image in Color Matching for High-Quality Panoramic Images used for Aquatic study. *International Journal of Aquatic Science*, 09(02), 130–145.

Kazi K S L. (2022). IoT-based weather Prototype using WeMos. *Journal of Control and Instrumentation Engineering*, *9*(1), 10 - 22.

Kazi, S. (2023). Fruit Grading, Disease Detection, and an Image Processing Strategy. *Journal of Image Processing and Artificial Intelligence*, 9(2), 17–34.

Kazi, S. S. L. (2023). IoT in Electrical Vehicle: A Study. *Journal of Control and Instrumentation Engineering*, 9(3), 15–21.

Kazi, S. S. L. (2023). IoT Changing the Electronics Manufacturing Industry. *Journal of Analog and Digital Communications*, 8(3), 13–17.

Kazi, S. S. L. (2023). Electronics with Artificial Intelligence Creating a Smarter Future: A Review. *Journal of Communication Engineering and Its Innovations*, 9(3), 38–42.

Kazi, V. (2023). Deep Learning, YOLO and RFID based smart Billing Handcart. *Journal of Communication Engineering & Systems*, 13(1), 1–8.

Kazi Kutubuddin, S. L. (2022). Predict the Severity of Diabetes cases, using K-Means and Decision Tree Approach. *Journal of Advances in Shell Programming*, 9(2), 24–31.

Kazi Kutubuddin, S. L. (2022). A novel Design of IoT based 'Love Representation and Remembrance' System to Loved One's. *Gradiva Review Journal*, 8(12), 377–383.

Kazi Kutubuddin S. L. (2022). Business Mode and Product Life Cycle to Improve Marketing in Healthcare Units. *E-Commerce for future & Trends*, *9*(3), 1-9.

Kosgiker, G. M. (2018). Machine Learning- Based System, Food Quality Inspection and Grading in Food industry. *International Journal of Food and Nutritional Sciences*, 11(10), 723–730.

KS. (2023). IoT based Healthcare system for Home Quarantine People. *Journal of Instrumentation and Innovation sciences,8.*

KSSL. (2023). IoT in the Electric Power Industry. *Journal of Controller and Converters*, 8(3), 1–7.

Kutub, K. (2022). Reverse Engineering's Neural Network Approach to human brain. *Journal of Communication Engineering & Systems*, 12(2), 17–24.

Kutubu, K. (2022). Detection of Malicious Nodes in IoT Networks based on packet loss using ML. *Journal of Mobile Computing, Communication & mobile. Networks*, 9(3), 9–16.

Kutubuddin, K. (2022). Big data and HR Analytics in Talent Management: A Study. *Recent Trends in Parallel Computing*, 9(3), 16–26.

Liyakat, K. K. S. (2023). *Detecting Malicious Nodes in IoT Networks Using Machine Learning and Artificial Neural Networks. 2023 International Conference on Emerging Smart Computing and Informatics (ESCI)*, Pune, India.10.1109/ESCI56872.2023.10099544

Liyakat, K. K. S. (2023). Machine Learning Approach Using Artificial Neural Networks to Detect Malicious Nodes in IoT Networks. In Shukla, P. K., Mittal, H., & Engelbrecht, A. (Eds.), *Computer Vision and Robotics. CVR 2023. Algorithms for Intelligent Systems*. Springer. 10.1007/978-981-99-4577-1_3

Liyakat, K. K. S. (2024). Machine Learning Approach Using Artificial Neural Networks to Detect Malicious Nodes in IoT Networks. In Udgata, S. K., Sethi, S., & Gao, X. Z. (Eds.), *Intelligent Systems. ICMIB 2023. Lecture Notes in Networks and Systems* (Vol. 728). Springer. https://link.springer.com/chapter/10.1007/978-981-99-3932-9_12, 10.1007/978-981-99-3932-9_12

Liyakat, K. S. (2023). Integrating IoT and Mechanical Systems in Mechanical Engineering Applications. *Journal of Mechanisms and Robotics*, 8(3), 1–6.

Liyakat, S. S. (2023). IoT Based Arduino-Powered Weather Monitoring System. *Journal of Telecommunication Study*, 8(3), 25–31. 10.46610/JTC.2023.v08i03.005

Mulani, A. O., & Patil, R. M. (2023). Discriminative Appearance Model For Robust Online Multiple Target Tracking. *Telematique*, 22(1), 24–43.

Neeraja, P., Kumar, R. G., Kumar, M. S., Liyakat, K. K. S., & Vani, M. S. (2024). *DL-Based Somnolence Detection for Improved Driver Safety and Alertness Monitoring. 2024 IEEE International Conference on Computing, Power and Communication Technologies (IC2PCT)*. Greater Noida. https://ieeexplore.ieee.org/document/10486714, 10.1109/IC2PCT60090.2024.10486714

Nerkar, P., & Shinde, S. (2023). Monitoring Fresh Fruit and Food Using Iot and Machine Learning to Improve Food Safety and Quality. *Tuijin Jishu/Journal of Propulsion Technology*, *44*(3), 2927 – 2931.

Nerkar, P. M., & Dhaware, B. U. (2023). Predictive Data Analytics Framework Based on Heart Healthcare System (HHS) Using Machine Learning. *Journal of Advanced Zoology,44*(2).

Nikita, K., & Supriya, J. (2020). Design of Vehicle system using CAN Protocol. *International Journal for Research in Applied Science and Engineering Technology*, 8(V), 1978–1983. 10.22214/ijraset.2020.5321

Prashant, K. Magadum (2024). Machine Learning for Predicting Wind Turbine Output Power in Wind Energy Conversion Systems, *Grenze International Journal of Engineering and Technology*, *10*.https://thegrenze.com/index.php?display=page&view=journalabstract&absid=2514&id=8

Sayyad, L. (2023), System for Love Healthcare for Loved Ones based on IoT. *Research Exploration: Transcendence of Research Methods and Methodology, 2*.

SSLK. (2023). IoT in Electrical Vehicle: A Study. *Journal of Control and Instrumentation Engineering*, 9(3), 15–21.

Subeesh, A., & Mehta, C. R. (2021). Automation and digitization of agriculture using artificial intelligence and internet of things. *Artificial Intelligence in Agriculture*, 5, 278–291. 10.1016/j.aiia.2021.11.004

Sul, K. (2023a). IoT Based Arduino-Powered Weather Monitoring System. *Journal of Telecommunication Study*, 8(3), 25–31.

Sul, K. (2023b). ArduinoBased Weather Monitoring System. *Journal of Switching Hub*, 8(3), 24–29.

Sultanabanu, K. (2023). Arduino Based Weather Monitoring System. *Journal of Switching Hub*, 8(3), 24–29.

Wale Anjali, D. (2019). Rokade Dipali. Smart Agriculture System using IoT. *International Journal of Innovative Research In Technology*, 5(10), 493–497.

Chapter 6
Artificial Intelligence in the Agri-Business Sector:
Prioritizing the Barriers Through Application of Analytical Hierarchy Process (AHP)

Sarita Kumari Singh
KIIT University, India

Puspalata Mahapatra
KIIT University, India

ABSTRACT

The agri-business sector stands at the nexus of global food production, supply chain management, and rural development, yet it grapples with multifaceted challenges. In response, artificial intelligence (AI) emerges as a transformative force; however, the adoption of AI in agriculture faces significant barriers, particularly in countries like India. This study systematically identifies and prioritizes these barriers using the Analytical Hierarchy Process (AHP) methodology. The results highlight the paramount importance of technological infrastructure, data accessibility, and skill development. Ethical considerations around safety and transparency, economic constraints, and social-cultural acceptance also emerge as critical factors. The study offers insights into the relative significance of each barrier, facilitating informed decision-making and targeted interventions. Ultimately, by addressing these barriers, stakeholders can unlock new opportunities for growth, sustainability, and food security, ensuring prosperity for agricultural communities in the digital age.

INTRODUCTION

The agri-business sector plays a vital role in global food production, supply chain management, and rural development (Varchenko, 2019; Kireyenka, 2021; Tsoulfas et al., 2021). However, it faces numerous challenges, including climate change, resource scarcity, labor shortages, market volatility, and evolving consumer preferences (Ray et al., 2021; Migunov et al., 2023). In this context, AI offers promising

DOI: 10.4018/979-8-3693-3583-3.ch006

solutions by enabling data-driven decision-making, automation, predictive analytics, and optimization across the agricultural value chain (Smith, 2020; Ganeshkumar et al., 2021; Shadrin et al., 2020).

Artificial Intelligence (AI) has emerged as a transformative force across various industries, revolutionizing traditional practices and unlocking unprecedented opportunities for innovation and efficiency (Dwivedi et al., 2019). At its core, AI involves the development of computer systems capable of performing tasks that typically require human intelligence, such as learning from data, recognizing patterns, making predictions, and adapting to changing environment (Saba & Rehman, 2013). Machine learning, a subset of AI, focuses on building algorithms that can learn from data and improve performance over time without being explicitly programmed (Riihijärvi, & Mähönen, 2018; Jordan, & Mitchell, 2015) . Deep learning, a more advanced form of machine learning, employs neural networks with multiple layers to extract intricate patterns from vast datasets (Cockburn et al., 2018; Shrestha & Mahmood, 2019). In recent years, the agri-business sector has increasingly embraced AI technologies across the world to address challenges, enhance productivity, and ensure sustainable agricultural practices (Spanaki et al., 2021; Elbasi et al., 2023).

In agri-business, AI applications span a wide range of domains, including precision agriculture, crop management, livestock monitoring, supply chain optimization, market forecasting, and risk management (Sharma et al., 2021; Elbasi et al., 2023). Precision agriculture, in particular, has emerged as a prominent area where AI technologies are revolutionizing traditional farming practices (Shafi, 2019; Zhang et al., 2002). By leveraging data from sensors, satellites, drones, and other sources, AI-driven systems enable farmers to monitor and manage crops with unprecedented precision, optimizing resource usage, minimizing environmental impact, and maximizing yields (Nyéki & Neményi, 2022).

In the realm of agri-business, AI holds immense promise for transformation by enabling data-driven decision-making, automation, and optimization across the agricultural value chain. By harnessing the power of AI technologies, farmers and agri-businesses can overcome challenges, enhance productivity, and ensure sustainable agricultural practices for future generations (Sáiz-Rubio et al., 2020). However, India has yet to fully embrace AI in agriculture due to a myriad of challenges that must be surmounted to unlock its vast potential. These hurdles encompass concerns regarding data reliability and accessibility, the compatibility of different systems, affordability of adoption, as well as anxieties surrounding privacy, security, and regulatory adherence (Sundari, 2018; Hota & Verma, 2022). Furthermore, the digital gap between large and small-scale farmers presents a significant barrier (Dhillon & Moncur, 2023).

In addition to these practical obstacles, there are profound ethical considerations associated with the integration of AI in agriculture. These include the risk of displacing rural livelihoods, the consolidation of power within agri-tech corporations, and the unforeseen consequences of algorithmic decision-making (Dara et al., 2022; Sparrow & Howard, 2020). Effectively tackling these complex challenges is crucial for fully realizing the benefits of AI in propelling agricultural practices forward in India. Thus, the focus of this study is on prioritizing and ranking the barriers hindering the adoption of AI in agriculture.

LITERATURE REVIEW

In the contemporary agricultural landscape, the integration of artificial intelligence (AI) stands as a transformative force, fundamentally reshaping traditional farming practices (Fountas, 2020; Sharma, 2021, Phadnis, 2023). While conventional technologies have facilitated the transition from manual to digital processes in agriculture, AI introduces a paradigm shift, offering unparalleled capabilities to revolutionize agricultural operations (Gupta, 2023). Scholars such as Benos et al. (2021), Liakos et al. (2018), Elbasi et al. (2023) and Sharma et al., (2021) emphasize the transformative potential of AI in agriculture, highlighting its predictability, self-learning, human emulation, automation, and augmentation capabilities. These characteristics enable AI to revolutionize agricultural production by optimizing resource allocation, enhancing crop monitoring, and mitigating environmental impact (Elbasi et al., 2023).

Recent technological advancements, empowered by self-learning algorithms, big data availability, and enhanced computational power, have accelerated the proliferation of AI-driven systems in agriculture (Adli et al., 2023; Elbasi et al., 2023; Liu, 2020). As AI technologies become more pervasive and performant, they offer cost-effective solutions to address prevailing business challenges in agriculture (Smith, 2020; Liu, 2020; Peters et al., 2020; Sarkar et al., 2022).

Traditionally, agricultural firms have relied on digital or industrialized technologies to modernize their operations. However, AI distinguishes itself by offering unique capabilities that transcend traditional approaches. These advancements in AI are timely, contextualized within the growing use of AI across various industries (Ammulu, 2020).

AI has the potential to revolutionize agricultural production in institutional settings by enhancing the monitoring and management of environmental conditions and crop yields. It can also help reduce waste by optimizing the use of fertilizers and water (Nabavi-Pelesaraei et al., 2018). In addition, AI may integrate with the global positioning system on land and agricultural machinery to do various agricultural tasks such as weed management, application of plant nutrients, mapping soil moisture, detecting fruits, and eliminating waste (Adli et al., 2023). Furthermore, there are possibilities for combining technology, as artificial intelligence (AI) systems can be employed to oversee and regulate high-temperature greenhouse conditions in newly implemented geothermal heating and cooling solutions designed for agricultural applications (Riahi et al., 2020). Table 1 provides a comprehensive overview of various applications of Artificial Intelligence (AI) in the agri-business sector.

Table 1. Application of AI in agribusiness

Application Area	Description	Authors
Crop Monitoring and Management	Utilizing drones, satellite imagery, and IoT sensors to monitor crop health, growth, and pest/disease detection.	Elbasi et al. (2023), Gao et al. (2020), Filho et al. (2019)
Precision Agriculture	Implementing AI algorithms to optimize resource allocation, such as water, fertilizers, and pesticides, based on real-time data and analysis.	Jiang et al. (2018), Kumar et al. (2023), Elbasi et al. (2023)
Predictive Analytics	Using AI to forecast crop yields, market demand, and weather patterns to make data-driven decisions for planting, harvesting, and selling crops.	Kumar et al (2022), Phadnis (2023)
Supply Chain Optimization	Employing AI algorithms to streamline logistics, inventory management, and distribution processes to reduce costs and improve efficiency.	Li (2023), Gao et al. (2020)
Livestock Monitoring and Management	Utilizing AI-powered sensors and image recognition to monitor animal health, behavior, and productivity, enabling early detection of diseases and optimization of breeding programs.	Achour et al. (2020), Suresh & Sarath (2019).
Agricultural Robotics	Deploying AI-driven robots and automated machinery for tasks such as planting, weeding, harvesting, and sorting crops, reducing labor costs and improving productivity.	Marinoudi et al. (2019), Emmi et al. (2023)
Soil Health Assessment	Using AI-based analysis of soil samples to assess fertility levels, nutrient content, and contamination, guiding precision farming practices.	Reshma et al. (2020), Singh et al. (2023), Ashoka et al. (2023)

continued on following page

Table 1. Continued

Application Area	Description	Authors
Market Forecasting	Applying AI models to analyze market trends, consumer behavior, and commodity prices, assisting farmers in making informed decisions about crop selection and sales strategies.	Cavazza et al. (2023), Zhang et al. (2020)
Agri-Financing and Insurance	Utilizing AI algorithms for risk assessment, credit scoring, and insurance underwriting, providing financial services tailored to the needs of farmers and agribusinesses.	Cole et al. (2017), Śmietanka et al. (2020), Emmi et al. (2023)
Smart Irrigation Systems	Integrating AI with IoT sensors and weather data to optimize irrigation scheduling, conserving water resources and maximizing crop yield.	Kumar et al. (2023), Kuo et al. (2000)
Disease and Pest Management	Employing AI models to identify and predict outbreaks of diseases and pests, enabling timely interventions and minimizing crop losses.	David (2023), Savary et al. (2017), Zhang et al. (2020)
Food Quality and Safety	Using AI-powered systems for real-time monitoring and inspection of food products, ensuring compliance with quality standards and regulations.	Qian et al. (2022), Tutul et al. (2023), David (2023)
Climate Resilience	Leveraging AI to analyze climate data and develop strategies for mitigating the impact of climate change on agriculture, such as drought-resistant crop varieties and adaptive farming practices.	Gupta et al. (2023), Reshma et al. (2020), Singh et al. (2023)
Agri-Tech Innovation	Supporting research and development of AI-based solutions for various agricultural challenges, fostering innovation and sustainability in the agribusiness sector.	Linaza et al. (2021), Leong et al. (2023), Gupta et al. (2023)

Source: Author's Self Compiled

While the future of AI in agriculture holds immense promise, there are still barriers and restrictions to its adoption, particularly in countries like India (Tzachor, 2021). These barriers include lack of necessary resources to implement AI technologies, limited access to technical training, high upfront costs, reluctance to use advanced technology, lack of practical experience, job displacement concerns, ethical concerns, lack of trust in technology, and a language barrier compounded by high illiteracy rates and a digital divide (Cubric, 2020; Kumar et al., 2021). Additionally, regulatory challenges such as lack of enforcement of data regulations, privacy and transparency concerns, risk aversion, and resistance to change further hinder AI adoption in the agricultural sector (Adli et al., 2023). Table 2 summarizes key findings from various studies regarding the challenges and barriers to the adoption of AI technologies in agriculture, shedding light on crucial considerations for stakeholders in the industry.

Table 2. Barriers to AI integration in agribusiness

Author(s)	Findings
Smith et al. (2020), Ryan et al. (2023), Valdez et al. (2023), Ammulu (2020)	Lack of awareness and understanding of AI technologies among farmers.
Mishra et al. (2023), Ryan et al. (2023)	Resistance to change among traditional farming communities due to fear of job displacement and loss of control.
Elbasi et al. (2023), Mishra et al. (2023), Hota et al. (2022)	Limited technical expertise and skills among farmers and agribusiness professionals to effectively use AI tools.
Maraveas (2022), Elbasi et al. (2023)	Uncertainty about the return on investment (ROI) of AI technologies in agribusiness, leading to reluctance to adopt.
Taneja et al.(2023), Leong et al. (2023), Smith et al. (2020)	Lack of standardization in data formats and data collection methods, making it difficult to aggregate and analyze data across different systems.

continued on following page

Table 2. Continued

Author(s)	Findings
Mehrabi et al. (2020), Dhillon & Moncur (2023), Elbasi et al. (2023)	Challenges in accessing AI technologies for small-scale and subsistence farmers in developing countries due to affordability and lack of infrastructure.
Cubric (2020), Javaid et al. (2023), Bhat & Huang (2021).	Concerns about the reliability and accuracy of AI predictions in variable and unpredictable agricultural environments.
Jha et al. (2019), Eli-Chukwu (2019), Elbasi et al. (2023), Subeesh & Mehta (2021)	Limited availability of AI solutions tailored to the specific needs and conditions of different agricultural regions and crops.
Zheng et al. (2022), Kumar & Prakash (2020), Xie & Huang (2021)	Lack of government support and incentives for farmers to adopt AI technologies, including funding and training programs.
Leong et al. (2023), Elbasi et al. (2023), Mishra et al. (2023)	Limited integration of AI technologies into existing farm management systems and workflows, leading to inefficiencies and resistance.
Okengwu et al. (2023), Elbasi et al. (2023), Kaur et al. (2022), Liakos et al. (2018)	Concerns about data privacy and ownership rights, particularly regarding the sharing of sensitive agricultural data with AI providers.
Peters et al. (2020), Liakos et al. (2018), Elbasi et al. (2023), Veni & Rani (2023)	Challenges in effectively leveraging AI technologies to address complex and multifaceted issues such as climate change and sustainable agriculture.
Gardeazabal et al. (2023), Dal Mas et al. (2023), Subeesh & Mehta (2021)	Lack of collaboration and knowledge-sharing among stakeholders in the agri-food value chain, hindering the adoption and diffusion of AI innovations.
Sharma & Garg (2021), Elbasi et al. (2023), Kaur et al. (2022)	Challenges in designing and implementing AI solutions that are user-friendly and accessible to farmers with diverse educational backgrounds and technological literacy levels.
Yigezu et al. (2018), Okengwu et al. (2023)	Concerns about the potential negative impacts of AI adoption on small-scale farmers' livelihoods and socio-economic well-being.
Joshi et al. (2022), Elbasi et al. (2023), Dal Mas et al. (2023)	Limited access to high-quality training datasets for AI models, particularly in regions with sparse agricultural data collection infrastructure.
Maraveas (2022), Peters et al. (2020), Okengwu et al. (2023)	Regulatory hurdles and policy gaps regarding the use of AI in agriculture, leading to uncertainty and delays in adoption.
Holzinger et al. (2022), Joshi et al. (2022), Veni & Rani (2023)	Challenges in ensuring the reliability and robustness of AI algorithms in diverse and dynamic agricultural environments.
Songol et al (2021), Holzinger et al. (2022), Elbasi et al. (2023)	Limited scalability of AI solutions for smallholder farmers due to resource constraints and lack of tailored support services.
Dara et al. (2022), Leong et al. (2023), Elbasi et al. (2023), Mishra et al. (2023)	Challenges in addressing ethical considerations related to AI usage in agriculture, including fairness, transparency, and accountability.
Lin (1991), Ryan et al. (2023),	Lack of awareness and education programs on AI technologies for farmers and agribusiness stakeholders, hindering adoption.
Ryan et al. (2023), Bayan (2018), Elbasi et al. (2023)	Limited interoperability between AI systems and existing farm management software, leading to compatibility issues and data silos.
Cole & Xiong (2017), Adli et al. (2023)	Challenges in predicting and mitigating potential risks and negative consequences associated with AI adoption in agribusiness.

Source: Author's Self Compiled

Despite these challenges, the benefits of AI in agriculture are substantial. By collecting and analyzing large amounts of data, AI can help farmers make data-driven decisions, optimize resource usage, and reduce environmental impact (Elbasi et al., 2023). For example, the integration of AI in agriculture could lead to a 60% decrease in pesticide usage and a 50% reduction in water usage, according to the World Economic Forum. Moreover, research efforts should prioritize developing inclusive strategies that ensure equitable access to AI technologies and empower marginalized farming communities to leverage the benefits of digital transformation (Sanginga et al., 2004).

METHODOLOGY

This study employs a mixed-method approach, combining qualitative and quantitative techniques to systematically investigate and prioritize the barriers to the adoption of Artificial Intelligence (AI) in agriculture. An extensive review of existing literature on AI adoption in the agricultural sector has been done to identify the various barriers and challenges hindering AI implementation. The literature review serves to establish a theoretical foundation for understanding the key issues, identify gaps in existing research, and compile a preliminary list of potential barriers to AI adoption.

Following the literature review, a qualitative thematic analysis is conducted on the identified barriers. This involves categorizing the barriers into distinct themes based on their nature and impact. The thematic analysis includes grouping similar barriers into broader categories, and cross-referencing with existing studies to ensure the reliability of themes. The primary aim of this phase is to provide a comprehensive understanding of the multifaceted challenges faced by stakeholders in adopting AI technologies in agriculture.

To refine and validate the identified barriers, a panel of experts is engaged. This panel comprises three IT heads from agricultural AI providers, four senior research scholars specializing in agricultural technology, and two managers from the Food Corporation of India. These experts contribute to the quantitative analysis phase using the Analytical Hierarchy Process (AHP). AHP, a widely recognized decision-making technique, is selected for this study due to its effectiveness in handling complex decisions and its prior usage in similar studies (Saaty, 1980; Podvezko, 2009, Singh et al., 2023).

AHP provides a structured framework for organizing and analyzing the relative importance of various factors, making it suitable for prioritizing barriers to AI adoption in agriculture. It allows for hierarchical structuring of decision criteria, pairwise comparisons to determine relative importance, and synthesis of results to produce a comprehensive ranking of barriers. Furthermore, AHP offers a systematic approach to consensus-building among experts, ensuring robust and reliable results (Yang et al., 2019).

Experts perform pairwise comparisons of barriers to determine their relative importance, filling out comparison matrices. These comparisons are then analyzed to calculate priority weights using the eigenvector method. The consistency of these comparisons is ensured using a consistency check. The resulting priority weights are synthesized to produce a comprehensive ranking of barriers.

KEY CHALLENGES IN THE DEPLOYMENT OF ARTIFICIAL INTELLIGENCE IN INDIAN BANKING SECTOR

After Delphi analysis the determined factors were categorized under five broader thematic heads i.e. Technological, Ethical, Economic, Social & Cultural and Knowledge & Skill. AHP model was developed based on the identified goal, criteria and sub criteria.

1. Technological Barriers:

    - Shortage of infrastructure: Inadequate technological infrastructure, such as hardware, software, and network systems, hinders the effective implementation of AI in agriculture (Ammulu, 2020; Ryan et al., 2023).

- Insufficient support from technical people: Lack of expertise and technical support personnel poses challenges in deploying and maintaining AI systems in agricultural settings (Linanza et al, 2021).
- Lack of communication: Poor communication channels between stakeholders, such as farmers, technology providers, and policymakers, impede the dissemination of information and knowledge about AI adoption (Kutter et al., 2011; Ryan et al., 2023).
- Lack of contextual awareness: Limited understanding of the local agricultural context and its specific requirements leads to mismatches between AI solutions and farmers' needs (Bestelmeyer et al., 2020).
- Lack of standardization: Absence of standardized protocols and guidelines for AI implementation in agriculture results in interoperability issues and inefficiencies (Mentsiev & Gatina, 2021).
- Response time and accuracy level: AI systems' performance in terms of response time and accuracy may not meet the demands of real-time agricultural operations, affecting their usability and effectiveness (Liakos et al., 2018).
- Poor internet connectivity: Limited access to high-speed internet services in rural areas restricts farmers' ability to utilize cloud-based AI applications and data-driven insights (Salemink et al., 2017).
- Inadequate electric supply: Unreliable or insufficient electricity supply in rural regions disrupts the operation of AI-powered agricultural equipment and systems, hampering their adoption and functionality (Liu et al., 2021).
- Limited availability and quality of training datasets: It refers to the scarcity and inadequate quality of data needed to train AI algorithms effectively for agricultural applications, hindering the development and deployment of robust AI solutions in farming practices (Misra et al., 2020).

2. Ethical Barriers:

- AI safety-related issues: Concerns about the safety and reliability of AI technologies, including potential risks of malfunction, data breaches, and unintended consequences, create barriers to adoption (Aleshkovski, 2022; Skitsko et al., 2023).
- Lack of governance: Absence of regulatory frameworks and governance mechanisms for AI in agriculture undermines trust and confidence in its use, leading to reluctance among stakeholders (Hohma et al., 2023).
- Lack of enforcement of data regulation: Weak enforcement of data protection laws and regulations fails to safeguard farmers' data privacy and rights, raising ethical concerns (Kaur et al., 2022).
- Lack of privacy and transparency: Inadequate transparency in AI algorithms and data processing practices, coupled with privacy breaches, erodes trust and raises ethical dilemmas among farmers and other stakeholders (Schmidt et al., 2020).

3. Economic Barriers:

- Less ROI: Uncertainty about the return on investment (ROI) from AI adoption, coupled with high upfront costs and long payback periods, deters farmers from investing in AI technologies (Tzachor, 2021).
- Less public investment: Limited government funding and support for AI research, development, and implementation in agriculture constrain innovation and adoption efforts (Lin et al., 2022).
- Implementation costs: High costs associated with implementing AI systems, including hardware, software, training, and infrastructure upgrades, pose financial barriers to adoption (Cubric, 2020).
- Maintenance fees: Ongoing expenses related to AI system maintenance, updates, and technical support add to the overall cost burden for farmers, especially those with limited financial resources (Bharti et al., 2018).
- Affordability issues: Limited affordability of AI technologies and services, particularly for smallholder farmers and resource-constrained agricultural enterprises, hinders widespread adoption and diffusion (Lamsal et al., 2023).

4. Social and Cultural Barriers:

- Unwillingness of farmers: Resistance or reluctance among farmers to embrace new technologies, including AI, due to inertia, skepticism, or fear of change, slows down adoption efforts (Qazi et al., 2022).
- Risk aversion and risk resistance: Farmers' cautious approach towards adopting AI technologies, driven by concerns about potential risks, uncertainties, and negative impacts on traditional farming practices and livelihoods (Marra et al., 2003).
- Lack of trust amongst service providers: Mistrust or skepticism towards technology providers, consultants, and extension services, stemming from past experiences, communication gaps, or perceived biases, undermines collaboration and partnership for AI adoption (Tzachor, 2021).

5. Knowledge and Skill Barriers:

- Lack of formal education: Limited access to formal education and training programs on AI technologies and applications for farmers and agricultural stakeholders hampers their capacity-building and skill development (Phillips, 1994).
- Language barrier: Language barriers, including limited availability of AI-related resources and training materials in local languages, impede knowledge dissemination and understanding among non-English speaking farmers and communities (Tzachor, 2021).
- Lack of skill for using AI: Insufficient skills and competencies among farmers and agricultural workers in using AI tools, platforms, and data analytics techniques hinder effective utilization and adoption (Kumar et al., 2020).
- High illiteracy to use digital devices: High literacy and digital literacy requirements for operating AI-enabled devices and platforms exclude illiterate or digitally inexperienced farmers from benefiting from AI technologies, exacerbating digital divides and inequalities in access to agricultural innovation (Abdulai, 2022).

RESULTS

The result presents a detailed analysis of the barriers to AI adoption in agriculture, assessed using the Analytic Hierarchy Process (AHP). Expert input was crucial in filling out the comparison matrix, allowing for precise ranking and weighting of the identified barriers. Table 3 illustrates the priority weights assigned to various barriers in the AHP decision tree. The table provides a hierarchical breakdown, showing the percentage weights both between and within categories, along with the overall ranking of each barrier.

Table 3. Priority weights in the AHP decision tree

	Percentage weight between the categories	Percentage weight within the categories	Ranking	Percentage weight among the factors	Ranking
Technology Barriers	**0.392**		**1**		
Shortage of infrastructure		0.25	1	0.098	1
Insufficient support from technical people		0.16	3	0.063	6
Lack of communication		0.08	5	0.031	13
Lack of contextual awareness		0.07	6	0.027	14
Lack of standardization		0.10	4	0.039	10
Response time and accuracy level		0.04	7	0.016	21
Poor internet connectivity		0.03	8	0.012	22
Limited availability and quality of training datasets		0.21	2	0.082	3
Inadequate electric supply		0.02	9	0.008	24
Ethical Barriers	**0.066**		**5**		
AI safety related issues		0.32	2	0.021	17
Lack of governance		0.16	3	0.011	23
Lack of enforcement of data regulation		0.10	4	0.007	25
Lack of privacy and transparency		0.40	1	0.026	15
Economic Barriers	**0.198**		**3**		
Less ROI		0.30	2	0.059	7
Less public investment		0.09	5	0.018	20
Implementation costs		0.37	1	0.073	4
Maintenance fees		0.11	3	0.022	16
Affordability issues		0.10	4	0.020	18
Social & Cultural Barriers	**0.106**		**4**		
Unwillingness of farmers		0.31	2	0.033	11
Risk aversion and risk resistance		0.19	3	0.020	19
Lack of trust amongst service providers		0.49	1	0.052	8
Knowledge & Skill Barriers	**0.238**		**2**		
Lack of formal education		0.27	2	0.064	5

continued on following page

Table 3. Continued

	Percentage weight between the categories	Percentage weight within the categories	Ranking	Percentage weight among the factors	Ranking
Language barrier		0.19	3	0.045	9
Lack of skill for using AI		0.39	1	0.093	2
High illiteracy to use digital device		0.13	4	0.031	12

Source: Author's Calculation

Interpretation

The Analytic Hierarchy Process (AHP) results, delineated in Table 2, furnish insightful perspectives into the relative importance of various barriers hindering the implementation of AI in agriculture. This table categorizes the barriers into five main domains: Technology, Ethical, Economic, Social & Cultural and Knowledge and skill facilitating an assessment of their inter-category and intra-category rankings. These weighted rankings offer a comprehensive understanding of the significance of each barrier and its sub-components within the agricultural AI implementation landscape.

In Column 1, the barriers are listed, encompassing factors under each domain. Column 2, labeled "Percentage weight between the categories," divulges the relative significance attributed to each domain. Remarkably, the "Technology Barriers" domain emerges as the most substantial with an overall weight of 39.2%, indicating the need to give paramount importance to these criteria. Following this, "Economic Barriers" and "Knowledge & Skill Barriers" domains hold weights of 19.8% and 23.8%, respectively, signifying their notable roles. The "Ethical Barriers" and "Social & Cultural Barriers" domains account for 6.6% and 10.6% of the total weight, respectively.

Column 3, titled "Percentage weight within the categories," delineates the relative importance of specific factors within their respective domains. Notably, within the Technology domain, "Shortage of infrastructure" and "Limited availability and quality of training datasets" emerge as the most influential factors. These factors are pivotal due to their direct impact on the technological foundation required for successful AI integration in agricultural practices. The shortage of infrastructure, encompassing hardware, software, and network systems, poses fundamental obstacles to deploying and maintaining AI systems effectively. Similarly, the limited availability and quality of training datasets hinder the development and training of AI algorithms, crucial for generating accurate insights and recommendations for farmers. Within the Economic domain, "Implementation costs" garners the highest weight; these costs encompass a range of expenses, including hardware, software, training, and infrastructure upgrades, which pose significant financial barriers to adoption, particularly for farmers with limited resources. In the Ethical domain, "Lack of privacy and transparency" is the most weighted factor; it directly impacts trust and confidence in AI systems among farmers and stakeholders. Addressing data privacy and transparency issues is essential for building trust and addressing ethical dilemmas associated with AI adoption in agriculture. Meanwhile, in the Social & Cultural domain, "Lack of trust amongst service providers" is identified as the most influential factor. Trust amongst service providers directly affects the willingness of farmers to engage with AI technologies and service providers. Building trust is essential for fostering collaboration and partnerships, ensuring effective deployment and utilization of AI solutions in agricultural practices.

Column 4 establishes a hierarchical ranking of factors within their respective domains, offering insights into the relative importance of each barrier. Column 5, denoted "Percentage weight among all the factors," provides a comprehensive outlook by indicating the percentage weight each factor contributes to the total number of factors examined. Lastly, Column 6 extends the classification to encompass all factors, offering a thorough assessment of the significance of each barrier across the entire spectrum.

This structured analysis enables a comprehension of the barriers obstructing the integration of AI in agriculture, facilitating informed decision-making and targeted interventions. By employing such comprehensive methodologies for prioritization, stakeholders can navigate the challenges effectively and foster the adoption of AI technologies in agricultural practices.

Figure 1. The arrangement of all factors in accordance with their percentage weight among the factors

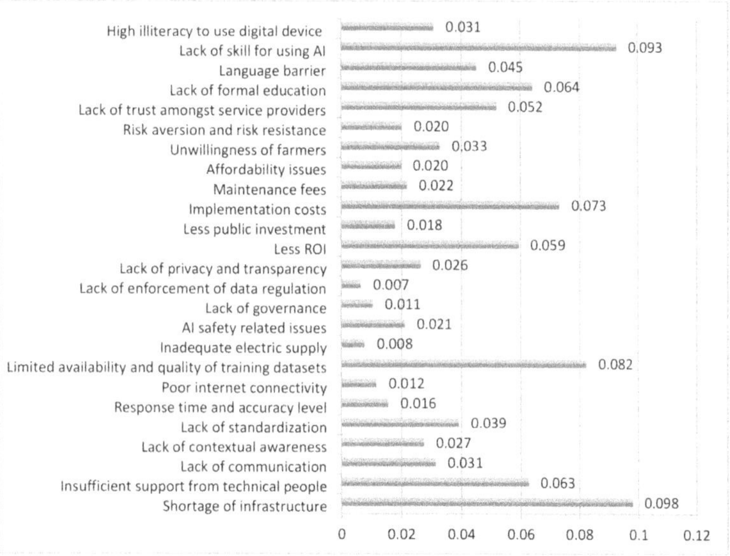

(Author's self-complied)

Figure 1 shows "Shortage of infrastructure" emerges as the most critical factor with a weight of 0.098, indicating its substantial impact on hindering AI implementation. Following closely is "Limited availability and quality of training datasets" with a weight of 0.082, ranking third overall. This factor highlights the scarcity and inadequacy of data needed to train AI algorithms effectively for agricultural applications, underscoring its crucial role in hindering the development and deployment of robust AI solutions in farming practices. "Lack of skill for using AI" obtains the second-highest weight of 0.093, ranking second overall. This factor points to the insufficient skills and competencies among farmers and agricultural workers in utilizing AI tools and platforms, posing significant challenges to their effective adoption and utilization. Other notable factors include "Implementation costs" (weight: 0.073, rank: 4), "Lack of formal education" (weight: 0.064, rank: 5), and "AI safety-related issues" (weight: 0.021, rank: 17), highlighting the key barriers faced in AI adoption within the agri-business sector.

CONCLUSION

In conclusion, this study sheds light on the pivotal role of Artificial Intelligence (AI) in revolutionizing the agri-business sector, offering solutions to address its myriad challenges and enhance productivity, sustainability, and resilience. Through a comprehensive exploration of AI's applications, benefits, and implications, the study underscores the transformative potential of AI in shaping the future of agriculture. However, despite its immense promise, the widespread adoption of AI in agriculture faces significant barriers and challenges, ranging from technological and economic constraints to ethical and social considerations.

The study systematically identifies and prioritizes these barriers, providing valuable insights for policymakers, industry stakeholders, and technology providers. From technological infrastructure limitations to ethical concerns surrounding AI safety and data governance, addressing these barriers requires a collaborative and multi-faceted approach that fosters inclusivity, transparency, and trust. Moreover, the study highlights the importance of tailored strategies that consider the unique needs and contexts of different agricultural communities, including smallholder farmers and rural populations. Initiatives aimed at promoting digital literacy, providing financial assistance, and fostering partnerships between stakeholders can help bridge the gap and ensure equitable access to AI-driven solutions.

In essence, while the challenges may be formidable, the potential benefits of AI in agri-business are vast. By overcoming these barriers and embracing AI-driven innovations, the agricultural sector can unlock new opportunities for growth, sustainability, and food security, paving the way for a more resilient and efficient food system for generations to come.

IMPLICATIONS OF THE STUDY

The insights gleaned from this study hold profound implications for stakeholders within the agri-business sector. Farmers, agricultural enterprises, and policymakers can leverage these findings to navigate the complexities of AI adoption and address barriers hindering its implementation. By prioritizing investments in technological infrastructure, such as reliable internet connectivity and consistent electricity supply, stakeholders can ensure that farmers have access to the necessary tools and resource. Enhancing data accessibility and developing high-quality, comprehensive training datasets are crucial for training effective AI models that address specific agricultural challenges. Capacity-building initiatives, including training programs that enhance digital literacy and technical skills among farmers and agricultural workers, are essential. These programs should be designed to be accessible and relevant to local contexts, including language considerations.

Moreover, fostering collaboration, trust, and knowledge sharing within the agricultural community is essential for promoting social and cultural acceptance of AI-driven innovations. Managers should advocate for the establishment and enforcement of comprehensive regulatory measures that address data privacy, security, and ethical concerns, thereby fostering trust among stakeholders. Policymakers are urged to develop supportive regulations and policies that facilitate ethical AI adoption, safeguard data privacy, and promote inclusive growth in the agri-business sector. Demonstrating the return on investment (ROI) of AI technologies through pilot projects and case studies can encourage wider adoption among farmers.

Furthermore, standardizing protocols and guidelines for AI implementation in agriculture is vital to ensure interoperability and efficiency. Developing industry-wide standards can help streamline the integration of AI technologies, making them more accessible and user-friendly for farmers. By addressing both managerial and technological challenges, stakeholders can harness the transformative potential of AI to enhance productivity, sustainability, and resilience in agriculture. Ultimately, by embracing AI technologies and overcoming the identified challenges, stakeholders can unlock new opportunities for growth and development, ensuring food security and prosperity for agricultural communities.

REFERENCES

Abdulai, A. (2022). *Toward digitalization futures in smallholder farming systems in Sub-Sahara Africa: A social practice proposal., 6*. Frontiers. .10.3389/fsufs.2022.866331

Achour, B., Belkadi, M., Filali, I., Laghrouche, M., & Lahdir, M. (2020). Image analysis for individual identification and feeding behaviour monitoring of dairy cows based on Convolutional Neural Networks (CNN). *Biosystems Engineering*, 198, 31–49. 10.1016/j.biosystemseng.2020.07.019

Adli, H., Remli, M., Wong, K., Ismail, N., González-Briones, A., Corchado, J., & Mohamad, M. (2023). Recent Advancements and Challenges of AIoT Application in Smart Agriculture: A Review. *Sensors (Basel)*, 23(7), 3752. 10.3390/s2307375237050812

Ajaykumar, K., & Madhavi, S. (2022). Review on Crop Yield Prediction with Deep Learning and Machine Learning Algorithms. *2022 4th International Conference on Inventive Research in Computing Applications (ICIRCA)*, 903-909. 10.1109/ICIRCA54612.2022.9985016

Aleshkovski, I. (2022). *Social Risks and Negative Consequences of Diffusion of Artificial Intelligence Technologies*. ISTORIYA., 10.18254/S207987840019849-2

Ammulu, D. (2020). The Impact of Artificial Intelligence in Agriculture. *International Journal of Advanced Research in Science. Tongxin Jishu*. 10.48175/IJARSCT-739

Ashoka, P., Avinash, G., Apoorva, M., Raj, P., Sekhar, M., Singh, S., Kumar, R., & Singh, B. (2023). *Efficient Detection of Soil Nutrient Deficiencies through Intelligent Approaches*. BIONATURE., 10.56557/bn/2023/v43i21877

Bayan, B. (2018). Factors influencing extent of adoption of artificial insemination (AI) technology among cattle farmers in Assam. *Indian Journal of Economics and Development*, 14(3), 528–534. 10.5958/2322-0430.2018.00166.X

Benos, L., Tagarakis, A., Dolias, G., Berruto, R., Kateris, D., & Bochtis, D. (2021). Machine Learning in Agriculture: A Comprehensive Updated Review. *Sensors (Basel)*, 21(11), 3758. 10.3390/s2111375834071553

Bestelmeyer, B., Marcillo, G., McCord, S., Mirsky, S., Moglen, G., Neven, L., Peters, D., Sohoulande, C., & Wakie, T. (2020). Scaling Up Agricultural Research With Artificial Intelligence. *IT Professional*, 22(3), 33–38. 10.1109/MITP.2020.2986062

Bharti, V., Bhan, S., Meetali, , & Deepshikha, . (2018). Impact of artificial intelligence for agricultural sustainability. *Journal of Soil and Water Conservation*, 17(4), 393–399. 10.5958/2455-7145.2018.00060.7

Bhat, S. A., & Huang, N. F. (2021). Big data and ai revolution in precision agriculture: Survey and challenges. *IEEE Access : Practical Innovations, Open Solutions*, 9, 110209–110222. 10.1109/ACCESS.2021.3102227

Cavazza, A., Mas, F., Campra, M., & Brescia, V. (2023). Artificial intelligence and new business models in agriculture: The "ZERO" case study. *Management Decision*. 10.1108/MD-06-2023-0980

Cockburn, I., Henderson, R., & Stern, S. (2018). *The Impact of Artificial Intelligence on Innovation. IRPN: Innovation & Cyberlaw & Policy*. Topic. 10.3386/w24449

Cole, S., & Xiong, W. (2017). Agricultural Insurance and Economic Development. *Annual Review of Economics*, 9(1), 235–262. 10.1146/annurev-economics-080315-015225

Cubric, M. (2020). Drivers, barriers and social considerations for AI adoption in business and management: A tertiary study. *Technology in Society*, 62, 101257. 10.1016/j.techsoc.2020.101257

Dal Mas, F., Massaro, M., Ndou, V., & Raguseo, E. (2023). Blockchain technologies for sustainability in the agrifood sector: A literature review of academic research and business perspectives. *Technological Forecasting and Social Change*, 187, 122155. 10.1016/j.techfore.2022.122155

Dara, R., Fard, S., & Kaur, J. (2022). Recommendations for ethical and responsible use of artificial intelligence in digital agriculture. *Frontiers in Artificial Intelligence*, 5, 884192. 10.3389/frai.2022.88419235968036

David, D. (2023). Weather Based Prediction Models for Disease and Pest Using Machine Learning: A Review. *Asian Journal of Agricultural Extension. Economia e Sociologia*, 41(11), 334–345. 10.9734/ajaees/2023/v41i112290

Dhillon, R., & Moncur, Q. (2023). Small-Scale Farming: A Review of Challenges and Potential Opportunities Offered by Technological Advancements. *Sustainability (Basel)*, 15(21), 15478. 10.3390/su152115478

Dwivedi, Y., Hughes, L., Ismagilova, E., Aarts, G., Coombs, C., Crick, T., & Duan, Y. (2019). Artificial Intelligence (AI): Multidisciplinary perspectives on emerging challenges, opportunities, and agenda for research, practice and policy. *International Journal of Information Management*. 10.1016/j.ijinfomgt.2019.08.002

Elbasi, E., Mostafa, N., AlArnaout, Z., Zreikat, A., Cina, E., Varghese, G., Shdefat, A., Topcu, A., Abdelbaki, W., Mathew, S., & Zaki, C. (2023). Artificial Intelligence Technology in the Agricultural Sector: A Systematic Literature Review. *IEEE Access : Practical Innovations, Open Solutions*, 11, 171–202. 10.1109/ACCESS.2022.3232485

Eli-Chukwu, N. (2019). Applications of Artificial Intelligence in Agriculture: A Review. *Engineering, Technology &. Applied Scientific Research*, 9(4), 4377–4383. 10.48084/etasr.2756

Emmi, L., Fernández, R., & Guerrero, J. (2023). Editorial: Robotics for smart farms. *Frontiers in Robotics and AI*, 9, 1113440. 10.3389/frobt.2022.111344036686213

Filho, F., Heldens, W., Kong, Z., & Lange, E. (2019). Drones: Innovative Technology for Use in Precision Pest Management. *Journal of Economic Entomology*, 113(1), 1–25. 10.1093/jee/toz26831811713

Fountas, S., Espejo-García, B., Kasimati, A., Mylonas, N., & Darra, N. (2020). The Future of Digital Agriculture: Technologies and Opportunities. *IT Professional*, 22(1), 24–28. 10.1109/MITP.2019.2963412

Ganeshkumar, C., Jena, S., Sivakumar, A., & Nambirajan, T. (2021). Artificial intelligence in agricultural value chain: Review and future directions. *Journal of Agribusiness in Developing and Emerging Economies*. 10.1108/JADEE-07-2020-0140

Gao, D., Sun, Q., Hu, B., & Zhang, S. (2020). A Framework for Agricultural Pest and Disease Monitoring Based on Internet-of-Things and Unmanned Aerial Vehicles. *Sensors (Basel)*, 20(5), 1487. 10.3390/s2005148732182732

Gardeazabal, A., Lunt, T., Jahn, M. M., Verhulst, N., Hellin, J., & Govaerts, B. (2023). Knowledge management for innovation in agri-food systems: A conceptual framework. *Knowledge Management Research and Practice*, 21(2), 303–315. 10.1080/14778238.2021.1884010

Gupta, S. (2023). *Artificial Intelligence in Smart Agriculture: Applications and Challenges*. CURRENT APPLIED SCIENCE AND TECHNOLOGY. 10.55003/cast.2023.254427

Gupta, S., Singh, N., & Kashyap, S. (2023). Management of agriculture through artificial intelligence in adverse climatic conditions. *Environment Conservation Journal*, 24(2), 408–412. Advance online publication. 10.36953/ECJ.23602638

Hohma, E., & Lütge, C. (2023). From Trustworthy Principles to a Trustworthy Development Process: The Need and Elements of Trusted Development of AI Systems. *AI*, 4(4), 904–926. Advance online publication. 10.3390/ai4040046

Holzinger, A., Saranti, A., Angerschmid, A., Retzlaff, C., Gronauer, A., Pejaković, V., Medel-Jiménez, F., Krexner, T., Gollob, C., & Stampfer, K. (2022). Digital Transformation in Smart Farm and Forest Operations Needs Human-Centered AI: Challenges and Future Directions. *Sensors (Basel)*, 22(8), 3043. Advance online publication. 10.3390/s2208304335459028

Hota, J., & Verma, V. (2022). Challenges to Adoption of Digital Agriculture in India. *2022 International Conference on Maintenance and Intelligent Asset Management (ICMIAM)*, (pp. 1-6). IEEE. 10.1109/ICMIAM56779.2022.10147002

Javaid, M., Haleem, A., Khan, I. H., & Suman, R. (2023). Understanding the potential applications of Artificial Intelligence in Agriculture Sector. *Advanced Agrochem*, 2(1), 15–30. 10.1016/j.aac.2022.10.001

Jha, K., Doshi, A., Patel, P., & Shah, M. (2019). A comprehensive review on automation in agriculture using artificial intelligence. *Artificial Intelligence in Agriculture*, 2, 1–12. 10.1016/j.aiia.2019.05.004

Jiang, Y., Hao, K., Cai, X., & Ding, Y. (2018). An improved reinforcement-immune algorithm for agricultural resource allocation optimization. *Journal of Computational Science*, 27, 320–328. 10.1016/j.jocs.2018.06.011

Jordan, M., & Mitchell, T. (2015). Machine learning: Trends, perspectives, and prospects. *Science*, 349(6245), 255–260. 10.1126/science.aaa841526185243

Joshi, A., Guevara, D., & Earles, M. (2022). Standardizing and Centralizing Datasets for Efficient Training of Agricultural Deep Learning Models. *Plant Phenomics (Washington, D.C.)*, 5, 0084. 10.34133/plantphenomics.008437680999

Kireyenka, N. (2021). *Models of agrarian business development in international practice., 59*, 22-40. .10.29235/1817-7204-2021-59-1-22-40

Kumar, G., Ramachandran, K., Sharma, S., Ramesh, R., Qureshi, K., & Ganesh, K. (2023). AI-Assisted Resource Allocation for Improved Business Efficiency and Profitability. *2023 3rd International Conference on Advance Computing and Innovative Technologies in Engineering (ICACITE)*, 54-58. IEEE. 10.1109/ICACITE57410.2023.10182679

Kumar, J., Chawla, R., Katiyar, D., Chouriya, A., Nath, D., Sahoo, S., Ali, A., & Singh, B. (2023). Optimizing Irrigation and Nutrient Management in Agriculture through Artificial Intelligence Implementation. *International Journal of Environment and Climate Change*. .10.9734/ijecc/2023/v13i103077

Kumar, S., Raut, R., Nayal, K., Kraus, S., Yadav, V., & Narkhede, B. (2021). To identify industry 4.0 and circular economy adoption barriers in the agriculture supply chain by using ISM-ANP. *Journal of Cleaner Production*, 293, 126023. 10.1016/j.jclepro.2021.126023

Kumar, T., & Prakash, N. (2020). ADOPTION OF AI IN AGRICULTURE: THE GAME-CHANGER FOR INDIAN FARMERS. *Proceedings of the International Conferences on ICT, Society and Human Beings (ICT 2020), Connected Smart Cities (CSC 2020) and Web Based Communities and Social Media (WBC 2020)*. ICT. 10.33965/ict_csc_wbc_2020_202008C025

Kuo, S., Merkley, G., & Liu, C. (2000). Decision support for irrigation project planning using a genetic algorithm. *Agricultural Water Management*, 45(3), 243–266. 10.1016/S0378-3774(00)00081-0

Kutter, T., Tiemann, S., Siebert, R., & Fountas, S. (2011). The role of communication and co-operation in the adoption of precision farming. *Precision Agriculture*, 12(1), 2–17. 10.1007/s11119-009-9150-0

Lamsal, R., Karthikeyan, P., Otero, P., & Ariza, A. (2023). Design and Implementation of Internet of Things (IoT) Platform Targeted for Smallholder Farmers: From Nepal Perspective. *Agriculture*, 13(10), 1900. Advance online publication. 10.3390/agriculture13101900

Leong, Y., Lim, E., Subri, N., & Jalil, N. (2023). Transforming Agriculture: Navigating the Challenges and Embracing the Opportunities of Artificial Intelligence of Things. *2023 IEEE International Conference on Agrosystem Engineering, Technology & Applications (AGRETA)*, (pp. 142-147). IEEE. 10.1109/AGRETA57740.2023.10262747

Li, X. (2023). Application of Intelligent Algorithm in the Research of Logistics Distribution Positioning System. *2023 Asia-Europe Conference on Electronics, Data Processing and Informatics (ACEDPI)*, (pp. 460-464). IEEE. 10.1109/ACEDPI58926.2023.00094

Liakos, K., Busato, P., Moshou, D., Pearson, S., & Bochtis, D. (2018). Machine Learning in Agriculture: A Review. *Sensors (Basel)*, 18(8), 2674. 10.3390/s1808267430110960

Lin, J. (1991). Education and Innovation Adoption in Agriculture: Evidence from Hybrid Rice in China. *American Journal of Agricultural Economics*, 73(3), 713–723. 10.2307/1242823

Lin, W., Liu, C., & Li, M. (2022). Research On Factors Influencing the Efficiency of Agricultural Science and Technology Innovation. *Proceedings of the International Conference on Information Economy, Data Modeling and Cloud Computing, ICIDC 2022*, Qingdao, China. 10.4108/eai.17-6-2022.2322808

Linaza, M., Posada, J., Bund, J., Eisert, P., Quartulli, M., Döllner, J., Pagani, A., Olaizola, I., Barriguinha, A., Moysiadis, T., & Lucat, L. (2021). Data-Driven Artificial Intelligence Applications for Sustainable Precision Agriculture. *Agronomy (Basel)*. 10.3390/agronomy11061227

Liu, Y., Ji, D., Zhang, L., An, J., & Sun, W. (2021). Rural Financial Development Impacts on Agricultural Technology Innovation: Evidence from China. *International Journal of Environmental Research and Public Health*, 18(3), 1110. 10.3390/ijerph1803111033513778

Maraveas, C. (2022). Incorporating Artificial Intelligence Technology in Smart Greenhouses: Current State of the Art. *Applied Sciences (Basel, Switzerland)*, 13(1), 14. 10.3390/app13010014

Marinoudi, V., Sørensen, C., Pearson, S., & Bochtis, D. (2019). Robotics and labour in agriculture. A context consideration. *Biosystems Engineering*, 184, 111–121. 10.1016/j.biosystemseng.2019.06.013

Marra, M., Pannell, D., & Ghadim, A. (2003). The economics of risk, uncertainty and learning in the adoption of new agricultural technologies: Where are we on the learning curve? *Agricultural Systems*, 75(2-3), 215–234. 10.1016/S0308-521X(02)00066-5

Mehrabi, Z., McDowell, M., Ricciardi, V., Levers, C., Martinez, J., Mehrabi, N., Wittman, H., Ramankutty, N., & Jarvis, A. (2020). The global divide in data-driven farming. *Nature Sustainability*, 4(2), 154–160. 10.1038/s41893-020-00631-0

Mentsiev, A., & Gatina, F. (2021). Data analysis and digitalisation in the agricultural industry. *IOP Conference Series. Earth and Environmental Science*, 677(3), 032101. 10.1088/1755-1315/677/3/032101

Migunov, R., Syutkina, A., Zaruk, N., Kolomeeva, E., & Arzamastseva, N. (2023). *Global Challenges and Barriers to Sustainable Economic Growth in the Agribusiness Sector*. WSEAS TRANSACTIONS ON BUSINESS AND ECONOMICS., 10.37394/23207.2023.20.85

Mishra, D., Muduli, K., Raut, R., Narkhede, B., Shee, H., & Jana, S. (2023). Challenges Facing Artificial Intelligence Adoption during COVID-19 Pandemic: An Investigation into the Agriculture and Agri-Food Supply Chain in India. *Sustainability (Basel)*, 15(8), 6377. 10.3390/su15086377

Misra, N., Dixit, Y., Al-Mallahi, A., Bhullar, M., Upadhyay, R., & Martynenko, A. (2020). IoT, Big Data, and Artificial Intelligence in Agriculture and Food Industry. *IEEE Internet of Things Journal*, 9(9), 6305–6324. 10.1109/JIOT.2020.2998584

Nabavi-Pelesaraei, A., Rafiee, S., Mohtasebi, S., Hosseinzadeh-Bandbafha, H., & Chau, K. (2018). Integration of artificial intelligence methods and life cycle assessment to predict energy output and environmental impacts of paddy production. *The Science of the Total Environment*, 631-632, 1279–1294. 10.1016/j.scitotenv.2018.03.08829727952

Nyéki, A., & Neményi, M. (2022). Crop Yield Prediction in Precision Agriculture. *Agronomy (Basel)*, 12(10), 2460. 10.3390/agronomy12102460

Okengwu, U., Onyejegbu, L., Oghenekaro, L., Musa, M., & Ugbari, A. (2023). *Environmental and ethical negative implications of AI in agriculture and proposed mitigation measures*. Scientia Africana. 10.4314/sa.v22i1.13

Peters, D., Rivers, A., Hatfield, J., Lemay, D., Liu, S., & Basso, B. (2020). Harnessing AI to Transform Agriculture and Inform Agricultural Research. *IT Professional*, 22(3), 16–21. 10.1109/MITP.2020.2986124

Phadnis, A. (2023). Implementation of Prediction of Crop Using SVM Algorithm. *International Journal for Research in Applied Science and Engineering Technology*, 11(5), 3812–3816. 10.22214/ijraset.2023.52265

Phillips, J. (1994). Farmer Education and Farmer Efficiency: A Meta-Analysis. *Economic Development and Cultural Change*, 43(1), 149–165. 10.1086/452139

Podvezko, V. (2009). Application of AHP technique. *Journal of Business Economics and Management*, 10(2), 181–189. 10.3846/1611-1699.2009.10.181-189

Qazi, S., Khawaja, B., & Farooq, Q. (2022). IoT-Equipped and AI-Enabled Next Generation Smart Agriculture: A Critical Review, Current Challenges and Future Trends. *IEEE Access : Practical Innovations, Open Solutions*, 10, 21219–21235. 10.1109/ACCESS.2022.3152544

Qian, C., Murphy, S., Orsi, R., & Wiedmann, M. (2022). How Can AI Help Improve Food Safety? *Annual Review of Food Science and Technology*. 10.1146/annurev-food-060721-01381536542755

Ray, P., Duraipandian, R., Kiranmai, G., Rao, R., & Jose, M. (2021). An Exploratory Study of Risks and Food Insecurity in the Agri Supply Chain. *Management Science*, 8(S1-Feb), 1–12. 10.34293/management.v8iS1-Feb.3752

Reshma, R., Sathiyavathi, V., Sindhu, T., Selvakumar, K., & Sairamesh, L. (2020). IoT based Classification Techniques for Soil Content Analysis and Crop Yield Prediction. *2020 Fourth International Conference on I-SMAC (IoT in Social, Mobile, Analytics and Cloud) (I-SMAC)*, (pp. 156-160). IEEE. 10.1109/I-SMAC49090.2020.9243600

Riahi, J., Vergura, S., Mezghani, D., & Mami, A. (2020). Intelligent Control of the Microclimate of an Agricultural Greenhouse Powered by a Supporting PV System. *Applied Sciences (Basel, Switzerland)*, 10(4), 1350. 10.3390/app10041350

Riihijärvi, J., & Mähönen, P. (2018). Machine Learning for Performance Prediction in Mobile Cellular Networks. *IEEE Computational Intelligence Magazine*, 13(1), 51–60. 10.1109/MCI.2017.2773824

Ryan, M., Isakhanyan, G., & Tekinerdogan, B. (2023). An interdisciplinary approach to artificial intelligence in agriculture. *NJAS: Impact in Agricultural and Life Sciences*, 95(1), 2168568. 10.1080/27685241.2023.2168568

Saaty, T. L., & Özdemir, M. S. (2014). How many judges should there be in a group? *Annals of Data Science*, 1(3-4), 359–368. 10.1007/s40745-014-0026-4

Saba, T., & Rehman, A. (2013). Effects of artificially intelligent tools on pattern recognition. *International Journal of Machine Learning and Cybernetics*, 4(2), 155–162. 10.1007/s13042-012-0082-z

Sáiz-Rubio, V., & Rovira-Más, F. (2020). From Smart Farming towards Agriculture 5.0: A Review on Crop Data Management. *Agronomy (Basel)*, 10(2), 207. 10.3390/agronomy10020207

Salemink, K., Strijker, D., & Bosworth, G. (2017). Rural development in the digital age: A systematic literature review on unequal ICT availability, adoption, and use in rural areas. *Journal of Rural Studies*, 54, 360–371. 10.1016/j.jrurstud.2015.09.001

Sanginga, P., Best, R., Chitsike, C., Delve, R., Kaaria, S., & Kirkby, R. (2004). Enabling rural innovation in Africa: An approach for integrating farmer participatory research and market orientation for building the assets of rural poor. *Uganda Journal of Agricultural Sciences*, 9, 934–949. 10.4314/UJAS.V9I1

Sarkar, M., Masud, S., Hossen, M., & Goh, M. (2022). A Comprehensive Study on the Emerging Effect of Artificial Intelligence in Agriculture Automation. *2022 IEEE 18th International Colloquium on Signal Processing & Applications (CSPA)*, (pp. 419-424). IEEE. 10.1109/CSPA55076.2022.9781883

Savary, S., Nelson, A., Djurle, A., Esker, P., Sparks, A., Amorim, L., Filho, A., Caffi, T., Castilla, N., Garrett, K., McRoberts, N., Rossi, V., Yuen, J., & Willocquet, L. (2017). Concepts, approaches, and avenues for modelling crop health and crop losses. *European Journal of Agronomy*. 10.1016/j.eja.2018.04.003

Schmidt, P., Biessmann, F., & Teubner, T. (2020). Transparency and trust in artificial intelligence systems. *Journal of Decision Systems*, 29(4), 260–278. 10.1080/12460125.2020.1819094

Shadrin, D., Menshchikov, A., Somov, A., Bornemann, G., Hauslage, J., & Fedorov, M. (2020). Enabling Precision Agriculture Through Embedded Sensing With Artificial Intelligence. *IEEE Transactions on Instrumentation and Measurement*, 69(7), 4103–4113. 10.1109/TIM.2019.2947125

Shafi, U., Mumtaz, R., García-Nieto, J., Hassan, S., Zaidi, S., & Iqbal, N. (2019). Precision Agriculture Techniques and Practices: From Considerations to Applications. *Sensors (Basel)*, 19(17), 3796. 10.3390/s1917379631480709

Sharma, A., Jain, A., Gupta, P., & Chowdary, V. (2021). Machine Learning Applications for Precision Agriculture: A Comprehensive Review. *IEEE Access : Practical Innovations, Open Solutions*, 9, 4843–4873. 10.1109/ACCESS.2020.3048415

Sharma, L., & Garg, P. K. (Eds.). (2021). *Artificial intelligence: technologies, applications, and challenges*. 10.1201/9781003140351

Sharma, R. (2021). Artificial Intelligence in Agriculture: A Review. *2021 5th International Conference on Intelligent Computing and Control Systems (ICICCS)*, (pp. 937-942). IEEE. 10.1109/ICICCS51141.2021.9432187

Shrestha, A., & Mahmood, A. (2019). Review of Deep Learning Algorithms and Architectures. *IEEE Access : Practical Innovations, Open Solutions*, 7, 53040–53065. 10.1109/ACCESS.2019.2912200

Singh, H., Halder, N., Singh, B., Singh, J., Sharma, S., & Shacham-Diamand, Y. (2023). Smart Farming Revolution: Portable and Real-Time Soil Nitrogen and Phosphorus Monitoring for Sustainable Agriculture. *Sensors (Basel)*, 23(13), 5914. 10.3390/s2313591437447764

Singh, S. K., Parida, J. K., & Pattnaik, P. K. (2023). Ranking of Attributes for Commercial Banks Using Multi Criteria Decision Making. [EEL]. *European Economic Letters*, 13(5), 142–155.

Skitsko, O., Skladannyi, P., Shyrshov, R., Humeniuk, M., & Vorokhob, M. (2023). THREATS AND RISKS OF THE USE OF ARTIFICIAL INTELLIGENCE. *Cybersecurity: Education, Science, Technique*. .10.28925/2663-4023.2023.22.618

Śmietanka, M., Koshiyama, A., & Treleaven, P. (2020). Algorithms in future insurance markets. *International Journal of Data Science and Big Data Analytics*. .10.2139/ssrn.3641518

Smith, M. (2020). Getting value from artificial intelligence in agriculture. *Animal Production Science*, 60(1), 46–54. 10.1071/AN18522

Songol, M., Awuor, F., & Maake, B. (2021). Adoption of artificial intelligence in agriculture in the developing nations: A review. *Journal of Language. Technology & Entrepreneurship in Africa*, 12(2), 208–229.

Spanaki, K., Sivarajah, U., Fakhimi, M., Despoudi, S., & Irani, Z. (2021). Disruptive technologies in agricultural operations: A systematic review of AI-driven AgriTech research. *Annals of Operations Research*, 308(1-2), 491–524. 10.1007/s10479-020-03922-z

Sparrow, R., & Howard, M. (2020). Robots in agriculture: Prospects, impacts, ethics, and policy. *Precision Agriculture*, 22(3), 818–833. 10.1007/s11119-020-09757-9

Subeesh, A., & Mehta, C. R. (2021). Automation and digitization of agriculture using artificial intelligence and internet of things. *Artificial Intelligence in Agriculture*, 5, 278–291. 10.1016/j.aiia.2021.11.004

Sundari, T. (2018). Digital Transformation of Indian Agriculture. *Contemporary Social Science*, 27(4), 65–71. 10.29070/27/58309

Suresh, A., & Sarath, T. (2019). An IoT Solution for Cattle Health Monitoring. *IOP Conference Series. Materials Science and Engineering*, 561(1), 012106. Advance online publication. 10.1088/1757-899X/561/1/012106

Taneja, A., Nair, G., Joshi, M., Sharma, S., Sharma, S., Jambrak, A., Roselló-Soto, E., Barba, F., Castagnini, J., Leksawasdi, N., & Phimolsiripol, Y. (2023). Artificial Intelligence: Implications for the Agri-Food Sector. *Agronomy (Basel)*, 13(5), 1397. 10.3390/agronomy13051397

Tsoulfas, G., & Mouzakitis, Y. (2021). Framing the transition towards sustainable agri-food supply chains. *IOP Conference Series. Earth and Environmental Science*, 899(1), 012003. 10.1088/1755-1315/899/1/012003

Tutul, M., Alam, M., & Wadud, M. (2023). Smart Food Monitoring System Based on IoT and Machine Learning. *2023 International Conference on Next-Generation Computing, IoT and Machine Learning (NCIM)*, (pp. 1-6). IEEE. 10.1109/NCIM59001.2023.10212608

Tzachor, A. (2021). Barriers to AI adoption in Indian agriculture: An initial inquiry. [IJIDE]. *International Journal of Innovation in the Digital Economy*, 12(3), 30–44. 10.4018/IJIDE.2021070103

Valdez, A., Amba, O., Jr., R., Dimalna, H., Gomampong, A., & Manaol, N. (2023). Farmers Perceptions on Artificial Insemination (AI): A Mixed Method Design. *European Journal of Theoretical and Applied Sciences*. .10.59324/ejtas.2023.1(3).31

Varchenko, O. (2019). Theoretical aspects of functioning of agro-food chains and features of their development in Ukrainian. *Ekonomìka ta upravlìnnâ APK*. .10.33245/2310-9262-2019-148-1-6-20

Veni, A., & Rani, K. (2023). Improvement of Agriculture Productivity by using Artificial Intelligence and Block Chain Technology. *International Journal of Scientific Research in Science and Technology*, 445–456. 10.32628/IJSRST52310451

Xie, H., & Huang, Y. (2021). Influencing factors of farmers' adoption of pro-environmental agricultural technologies in China: Meta-analysis. *Land Use Policy*, 109, 105622. 10.1016/j.landusepol.2021.105622

Yang, Y. P., Tian, H. L., & Jiao, S. J. (2019). Product design evaluation method using consensus measurement, network analysis, and AHP. *Mathematical Problems in Engineering*, 2019, 1–9. 10.1155/2019/4042024

Yigezu, Y., Mugera, A., El-Shater, T., Aw-Hassan, A., Piggin, C., Haddad, A., Khalil, Y., & Loss, S. (2018). Enhancing adoption of agricultural technologies requiring high initial investment among smallholders. *Technological Forecasting and Social Change*, 134, 199–206. 10.1016/j.techfore.2018.06.006

Zhang, D., Chen, S., Liwen, L., & Xia, Q. (2020). Forecasting Agricultural Commodity Prices Using Model Selection Framework With Time Series Features and Forecast Horizons. *IEEE Access : Practical Innovations, Open Solutions*, 8, 28197–28209. 10.1109/ACCESS.2020.2971591

Zhang, N., Wang, M., & Wang, N. (2002). Precision agriculture—A worldwide overview. *Computers and Electronics in Agriculture*, 36(2-3), 113–132. 10.1016/S0168-1699(02)00096-0

Zheng, H., Ma, J., Yao, Z., & Hu, F. (2022). How Does Social Embeddedness Affect Farmers' Adoption Behavior of Low-Carbon Agricultural Technology? *Frontiers in Environmental Science*, 10, 10.3389/fenvs.2022.909803

Chapter 7
Barriers of Agrisupply Chain Management:
During Mental and Physical Stress During Farming in Tractor

Suchismita Satapathy
http://orcid.org/0000-0002-4805-1793
KIIT University, India

Hullash Chauhan
http://orcid.org/0000-0002-7636-3065
Bharati Vidya Pith, India

Meghana Mishra
http://orcid.org/0000-0002-6419-5420
KIIT University, India

Hemalata Jena
KIIT University, India

Rita K. Sahu
KIIT University, India

ABSTRACT

In the agriculture sector, farmers do supply chain activity on their fields continuously in difficult weather conditions without using any modern technology or tools. Their work is usually manual and repetitive in nature. So, farmers suffer from musculoskeletal disorders and body pain. The use of tractors has in some way resolved the heavy manual work of farmers. However, the noise and vibration of tractors during work also have a negative impact on their bodies. The discomfort and static posture of farmers during tractor driving on rough agricultural ground led to muscle pain and stiffness. Sometimes slipped discs occur due to jerks on rough ground. So, this study is aimed at finding the risk factors that lead to work-related musculoskeletal disorder (WMSD) in tractor operators or farmers using tractors for agricultural purposes. Hence, a study is conducted on farmers driving tractors for performing agricultural operations with ergonomic tools such as rapid upper limb assessments (RULA) and quick exposure checks (QEC).

DOI: 10.4018/979-8-3693-3583-3.ch007

INTRODUCTION

The agricultural sector is a traditional one, and farmers use conventional tools for farming. There are many different technologies or tools being developed to improve the agriculture sector in terms of workplace safety and productivity. Compared to manual labour in farming for preparing land, the use of a tractor has a lot of benefits in terms of reduction of labour, reduction of heavy manual work, and time-consuming operations in hot summers and heavy rain. Simultaneously, tractors can help with many farming jobs, like tilling, sowing, harrowing, pushing, pulling, loading, unloading, ploughing, etc. It could relieve farmers from heavy, hard, and repetitive work. When the farmers were habituated to work with the help of old tools, they either suffered from injuries and cuts or musculoskeletal disorder problems. This physical stress impacts their minds, and farmers suffer from mental illness. The use of tractors in farming has resolved the physical and mental problems of farmers, which occur due to manual labour, but the tractor diver or farmer who operates the tractor is in great danger. Tractors overturning and rolling over are hazards that occur due to uneven agricultural land and the heavy weight of tractors. Working at heights and on stairs increases the risk of falling for farmers. The noise generated during driving is very unpleasant and may be the cause of hearing loss for farmers using tractors. During farming operations, the farmer has to drive a tractor on uneven agricultural land, which causes back pain due to unrestricted movement, vibration, sustained posture, and poor posture. Kumar et al. (1999) explained the adverse effect of vibration generated during driving a tractor on the farmers' back bone. Kumar et al. (2005) explained that Indian tractors are the main cause of hearing loss in operators due to a lack of noise and vibration control systems. Toren et al. (2002) wrote that risk of low-back and hip symptoms among farmers might be due to driving tractor. Sometimes, tractor drivers face accidents due to unguarded power takeoffs. Hence, uncomfortable sitting posture for long hours in an adverse climate with heavy vibration offered by a tractor increases physical stress on the driver or farmer driving a tractor. It may cause psychological and mental stress among them, which affects their behavioural changes and reduces productivity in the workplace. Chauhan et al. (2020) have discussed muscle problems, body and back pain in agri work place are the main cause of mental pressure. Hence, more ergonomic research or analysis is essential to studying the posture problems of tractor-driving farmers and the impact of physical stress on mental health is not observed in any of the literature. So in this paper, mental stress of tractor farmers is measured with respect to physical stress /discomfort.

There are many ergonomic tools for analysis postural discomfort in the workplace. Tajvar et al. (2021) have identified few comfortable agri practices methods by tools like (RULA) and (QEC). Lin et al. (2020) have tried to reduce participants' ergonomic risk by considering work ablity index and demographic data by (QEC) tool. Mokhtarinia et al. (2020) have identified the Quick Exposure Check (QEC) to monitor the threat of musculoskeletal disorders in their place of work. Li and Buckle (1998) have used the Quick Exposure Check tool to measure workplace safety among health and safety professionals. Sukadarin et al. (2013) have quantified ergonomic risk factors by performing posture analysis for oil palm workers by QEC. Chiasson et al. (2012) have used Quick Exposure Check (QEC) as the other ergonomics method to assess the safety of manufacturing sectors. Maulik et al. (2012) have developed the Quick Exposure Check (QEC) for medical laboratory technicians and healthcare professionals to check work-place musculoskeletal symptoms. Hossain et al. (2018) have implemented the Quick Exposure Check (QEC) method for analyzing work-related musculoskeletal disorders (WMSD). From the study, he found that the most common work-related diseases affect mostly the lower back, neck, and upper and lower extremities. Yusuf et al. (2016) used Rapid Upper Limb Assessment (RULA) analysis to decrease the work risk level and personal disorders of strawberry farmers in Bali. After an extensive literature review about ergonomic tools, RULA and QEC are selected for analysis of the posture of tractor drivers as they help to identify pain in joints, nerves, and muscles, and, with the help of QEC, time of operations and level of exposure can be measured. Then the optimization method, the Sun Flower Algorithm, is used to find ergonomic risks for tractor drivers.

The Sun Flower Algorithm has many uses. Wang & Khodaei (2020) have developed a model of the Sunflower optimization (SFO) algorithm for getting better system effectiveness for systems with separate generation systems. Alshammari & Guesmi (2020) have proposed the Sunflower optimization (SFO) to replace disorderly sequences. Nguyen (2021) has studied a method for optimising the position and size of DG in the distribution system based on enhanced sunflower optimization (ESFO) to reduce the power loss of the system. By applying the crop model SUNFLO to the underlying optimization problem. Picheny et al. (2015) quantified the sunflower optimisation (SFO) phenotypic optimization.

A mathematical model of the job rotation problem has been put forth by Sana et al. (2019) incorporates ergonomic considerations for repetitive tasks and uncomfortable postures in high-variability work contexts. The rapid upper limb assessment (RULA) was developed by Menychtas et al. (2020) to check-employees' problematic workstation situations. Sung et al. (2015) used a regression technique and artificial neural networks to construct the maximal voluntary contraction (MVC) grip and key pinch asset prediction models. Sanchez et al. (2018) have designed a tool to determine stress in the workplace in its initial stages.Coskun et al. (2021) used electromyography (EMG) signal analysis and visual perception to give robots and prosthetic amputees information about grasping abilities. (EMG) is a diagnostic technique that evaluates the skeletal muscles' physical health. According to Wu et al. (2018), lifting the same amount of weight repeatedly at the same frequency can increase your risk of injury. The lifting index (LI) from NIOSH (the US National Institute for Occupational Safety and Health) was calculated and was regarded as an objective risk indicator. The gait pattern was first introduced by Youssef et al. in 2021 based on the patient's initial degree and angle of deviation. Artificial intelligence-based fuzzy-assistedPetrinet(AIFAS) technique has been put forth by Lin et al. (2021) for stress assessments on HR and BP monitoring. In order to distinguish the principal component analysis (PCA) from the combined findings of individual features retrieved by chaos analysis and STFT, Gupta, and Mittal (2019) have concentrated on the QRS complex using PCA. In their analysis, Gupta et al. (2020) took the ECG into account. The ECG signal is alsoimpacted by numerous disturbances while being recorded. The analogue filter will stop due to temperature and drift, and the digital filter won't work since the pass band and gain parameters were chosen incorrectly.“Rpeak” identification is crucial for detecting heart problems, as per Gupta & Mittal (2021), but it is difficult to detect because it contains several forms of sound. In order to analyse ECG signals, Gupta et al. (2021) suggested auto-regressive modelling that was tailored to ARTFA models by minimizing forward and back-ward prediction errors (least squares). A study by Chauhan et al. (2021) examined ergonomic tools like “Rapid Entire Body Assessment” (RULA) to examine farm workers' working postures. Raqadi et al. (2017) have examined how stress can evolve into such a significant issue in contemporary organisations, particularly in developing nations. Guan (2021) has examined how robotic welding technology is used in contemporary-structures.

The health and safety concerns that Indian construction site employees face have been noted by Satapathy (2021), with the help of Ergonmomic tool OWAS (Ovako Work Posture Assessment System).In order to assess the most effective scheduling of carer shifts, Nahand et al. (2019) created a new multi-objective integer mathematical model that takes into account “human errors” made by carers. The “minimum and maximum working hours in a week” and the “shift changes of each nurse are noted and considered for this study.. A regional medical cloud computing information platform is made using Zhang's Visual Studio (2021) and the platform's functionality is evaluated. Particle swarm optimisation (PSO) and improved ant colony optimisations for the process of voting in the cloud computing environment have been proposed by Sridev and Chakkravarth (2021). To provide a practical solution, Yadav & Agrawal (2021) suggested an assembly line by using an actual mathematical-model. The improvement of food placements in ABCs utilising principles from opposition-based learning (OBL) has been researched by Sharma & Abraham (2020). Aydemir (2020) has suggested a novel method to display the distribution of motivation in a participant. Grey wolf optimisation (GWO) has been employed by Alamiedy et al. (2020) to enhance multi-objective anomaly-based IDS models. Using a single channel and short-duration (EEG) signals of one second, Behnam and Pourghassem (2017) developed an offline seizure detection method based on time domain, novel frequency sequence features, and innovative feature reduction. The methods that take numerical association rule mining (ARM) into account as a multi-objective optimisation issue have been studied by Altay & Alatas (2020). The PSO-algorithms, when paired with the local search approach to convert any infeasible point into feasible point solutions, yield optimal and viable points, which Gul et al. (2021) explored. A genetic algorithm and tabu search (GBTC) for the internet of things have been suggested by Ahmad et al. (2020) for the design of clustering issues. Singh & Kumar (2020) have concentrated on how wireless sensor networks (WSN) can perform better by maximising the trajectory of moving elements. According to Satheshkumar and Mangai (2021), the EEF-MDRP offers a reliable and effective path between the originating and target vehicles with the best possible intermediate stages and a minimum amount of complexity. To prevent Cyberattacks during communications on a network, Cherkaoui et al. (2020) have presented a NDT (I.e., novel detection technique). The Hybrid Ant Colony Optimisation Routing Protocol (EHACORP), put forward by Ramamoorthy and Thangavelu in 2021, aims to increase the effectiveness of routing by usingthe shortest path. Anoop and Bipin (2021) have made use of the topology-based forms of mobile user activities and their optimisation. The multi-objective artificial bee colony techniques were utilised by Mahmoodabadi and Shahangian (2019) must store a non-dominated solution. The distance and levy-flight-based Crow Search Algorithms (DLCSA) were proposed by Bhullar et al. (2020) for the optimization of three fractional-order controllers. To determine the status of an unmanned aerial vehicle (UAV), Kumar & Nagabhooshanam (2021) presented the Extended Kalman Filter (EKF) using Modified Artificial Bee Colony (MABC) algorithms.

METHODOLOGY

A tractor driver working in agriculture fields was exposed to many body posture problems. The tractor driver or farmer has to plough in an agricultural field, which is rough ground. He has to work for 8 to 9 hours daily in a constant position holding the wheels of a tractor. During the ploughing work, the constant and repetitive movement of the hands generates pain in the shoulder, hand, wrist, neck, and back bone. Hence, ergonomic analysis is essential to finding muscular disorders and related problems. For this, the postures of tractor drivers engaged in farming were collected and recorded, and analysis was conducted by two of the widely used ergonomics tools such as "Rapid Upper Limb Assessments (RULA)" and "Quick Exposure Check (QEC)" for ergonomic evaluations. Working time and repetition of the work or frequency of work were recorded from the video, followed by a body-posture examination of the "neck, leg, trunk, and arm angle", correspondingly. Scores were calculated by the "Rapid Upper Limb Assessments (RULA) and Quick Exposure Check (QEC)" survey using "Ergo Fellow 3.0 Software".

Based on the final score, a recommendation was made for the particular work and posture. Further, the Sun Flower Optimization (SFO) method" was used to find the best fitness values for farmers, which will minimise mental workload by considering ergonomic discomfort. Hence, to measure mental stress, ergonomic data of 33 farmers and tractor drivers of Odisha, India, was noted (by RULA and QEC) using an ergonomic tool. They drove tractors from 5 to 15 years of age, 30 to 55 years of age, had no chronic disease, and mental stress was measured by taking their view with a yes/no type question.

Figure 1. Methodology of study

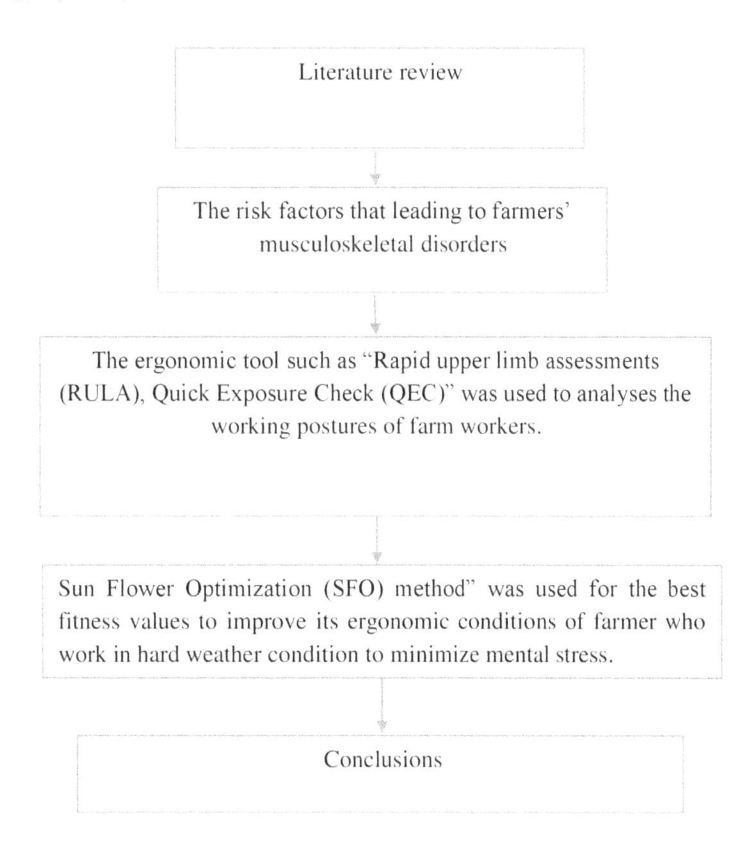

SUNFLOWER OPTIMIZATION (SFO) ALGORITHM

Sunflower conduct serves as the inspiration for the SFO concept. Sunflowers follow the same daily cycle; they rise and set with the sun like clock hands. To prepare for departure the following morning, they travelled in the other direction at night. Yang (2012) presented the SFO algorithm, which is based on blooming plant pollination and takes into account biological reproduction. His programme randomly selected and replicated the pollination along the shortest distance between flowers i and flowers (i+1)

pollen gamete. Additionally, a plant receives less heat the farther it is from the sun, so the same principle should be used to obtain.

Thus, the heat Q received by each plant McInnes (1999) were specified by

$$Q_i = \frac{P}{4\pi r^2 i} \quad (1)$$

Here, *P* is the power of the source and *ri* the Distance between the best current and the plant i the direction of the sunflowers towards the sun was:

$$\vec{s}_i = \frac{\dot{x} - x_i}{|\dot{x} - x_i| \vee}, \quad i = 1, 2 \ldots\ldots, n_p. \quad (2)$$

The methods of the sunflowers on the direction were evaluated by:

$$d_i = \lambda \times P_{i(}\|x_{i -} x_{i-1}\|) \times \|x_{i -} x_{i-1}\|, \quad (3)$$

Where the λ constant value that determines the "inertial" displacement of plants, Pi (|| xi - xi-1||) is the probability of pollination, that is, sunflower i is pollinated by its closest neighbour i - 1. Accordingly the process is repeated.

$$d_{max} = \frac{\vee x_{max - x_{min}} \vee}{2 \times N_{POP}} \quad (4)$$

Here, Xmax, and Xmin were the higher and inferior bounds values, and Npop was the number of plants in total population. The new plantation will be:

$$\overrightarrow{X_{i+1}} = \vec{x}_i + d_i \times \vec{s}_i \quad (5)$$

SUNFLOWER OPTIMIZATION ALGORITHMS

Step 1: Initially a uniform or random populations of flowers
Step 2: Novelty the sun (best solution s*) in the initial populations
Step 3: Place all plants towards the sun
Step 4: while (k < Max-Days)
Step 5: Estimate the positioning vector for individual plant
Step 6: Eradicate m (%) plant away from the sun
Step 7: Compute the steps for each plant
Step 8: Best b plant will pollinate around the sun
Step 9: Estimate the new entities
Step 10: If a new specific is a global best, update the sun
end while
got
Best solutions

RESULTS AND DISCUSSIONS

An image of a farmer is taken during ploughing in an agricultural field and Rapid upper limb assessments (RULA), ergonomic analysis is conducted. The farmer selected for analysis was 31 years old, a native of Odisha and a small-scale farmer who had 9 years of experience in tractor farming (Figure.2).

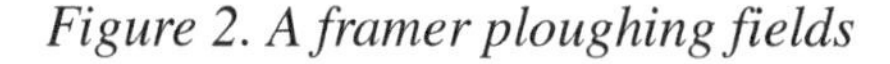
Figure 2. A framer ploughing fields

From Figure 2, it was found that a male framer is ploughing Agri fields in a tractor. Figures 3–10 describe the hand, wrist, neck, etc. poses in the Rula sheet. Figure 3 shows his upper arms were moved more than 20 degrees, supporting the weight of his arms. Figure 4 showed his wrists were twisted between 15 and 15 degrees and, additionally, his wrists were bent away from his midline. Figure 5 describes how his neck was bent more than 20 degrees and, additionally, twisted more than 180 degrees. Figure 6 shows the legs and feet were well supported and in an evenly balanced position. Figures 7 and 8 describe his lower arms moving between 0 and 60 degrees and, in addition, working across the middle line of his body or out to the side, his wrist twisted away from the handshake position. Figure 9 explains that his trunk, which was bent between 0 and 20 degrees and additionally twisted more than 180 degrees, Figure 10 elaborates on the fact that the upper arm, lower arm, and wrist muscles were found to be held for longer than 1 minute along with shock or force with rapid build-up; similarly, the neck, trunk, and legs muscles were found to be held for more than 4 times per minute with shock or force with rapid build-up. On the basis of this input, a RULA score of 7 was found. Hence, it was concluded that "investigation and changes are required immediately" from Figure 11.

Figure 3. RULA data for upper arm, upper arms were moved more than 20 degrees and in additional leaning or supporting the weight of arm

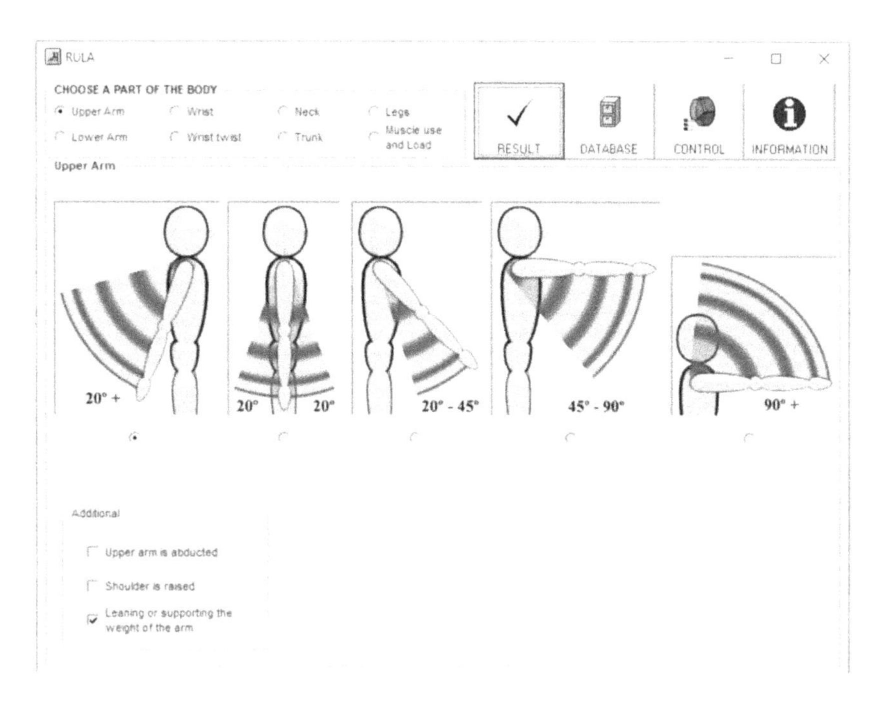

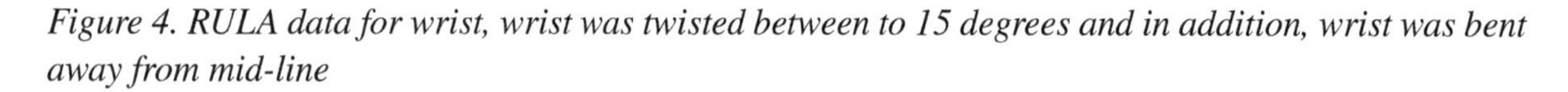

Figure 4. RULA data for wrist, wrist was twisted between to 15 degrees and in addition, wrist was bent away from mid-line

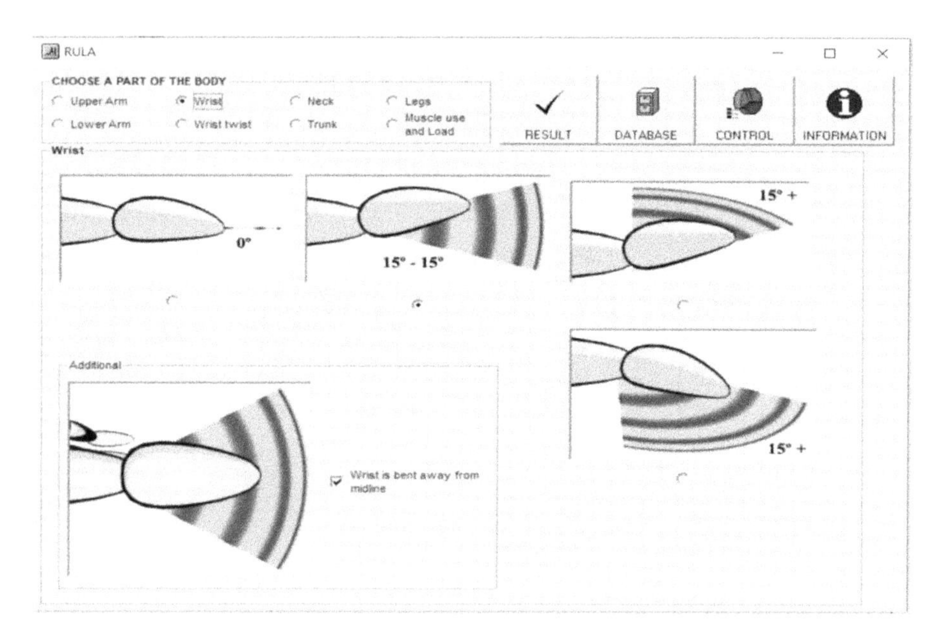

Figure 5. RULA data for neck, neck was bent more than 20 degrees and additionally, neck was twisted more than 180 degrees

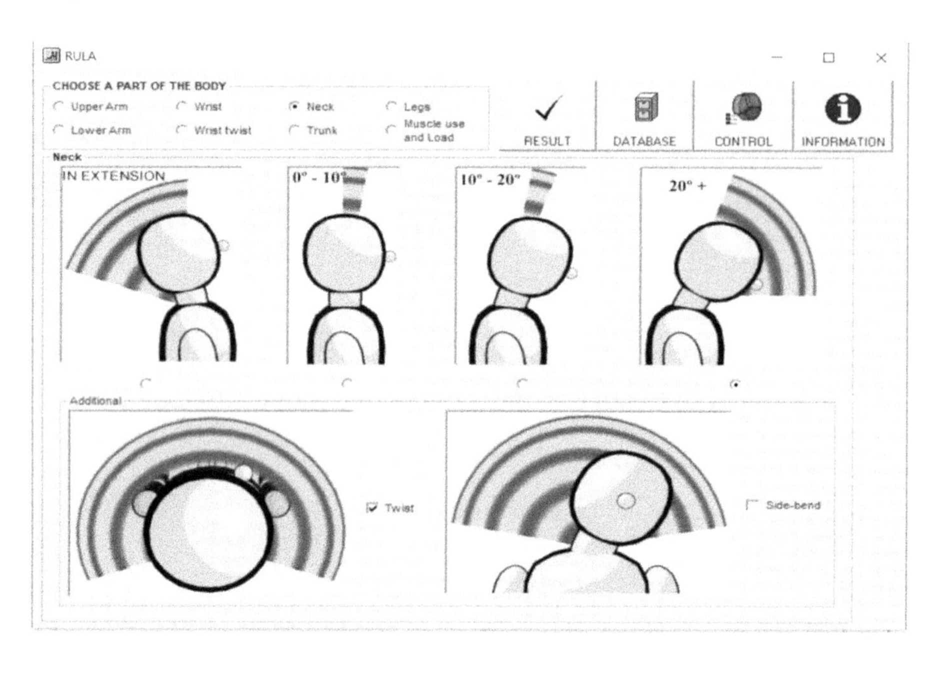

Figure 6. RULA data for legs, his legs and feet well supported and in an evenly-balanced poster

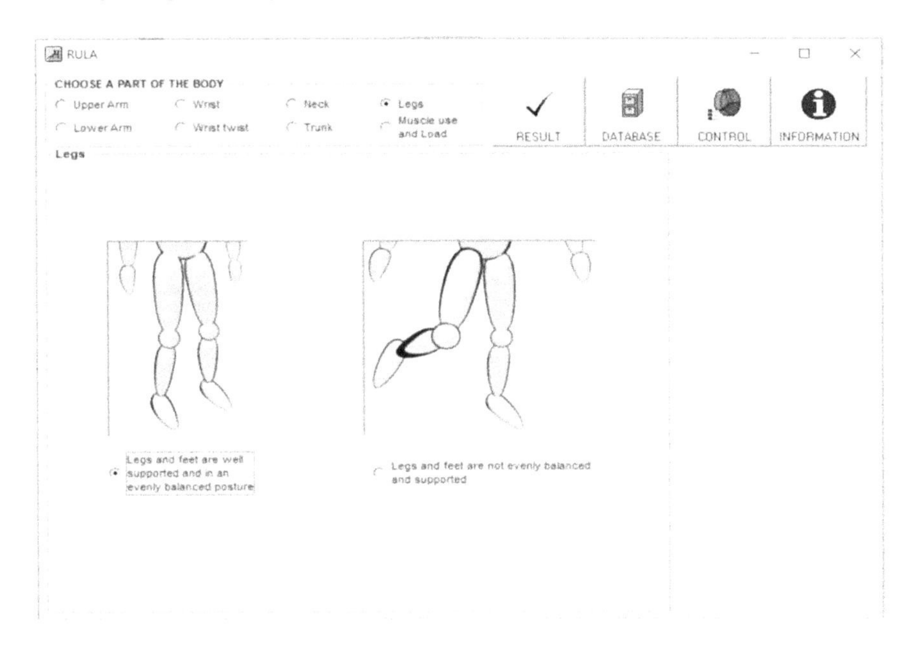

Figure 7. RULA data for lower arm, his lower arm moves between 0 degree to 60 degrees and in addition, works across the middle line of the body or out to the side

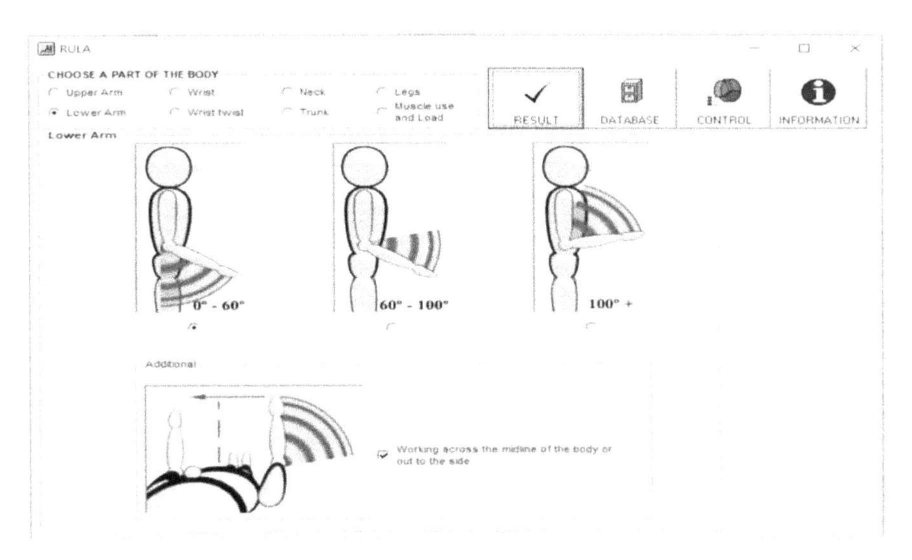

Figure 8. RULA data for wrist twist, his wrist twisted away from hand shake position

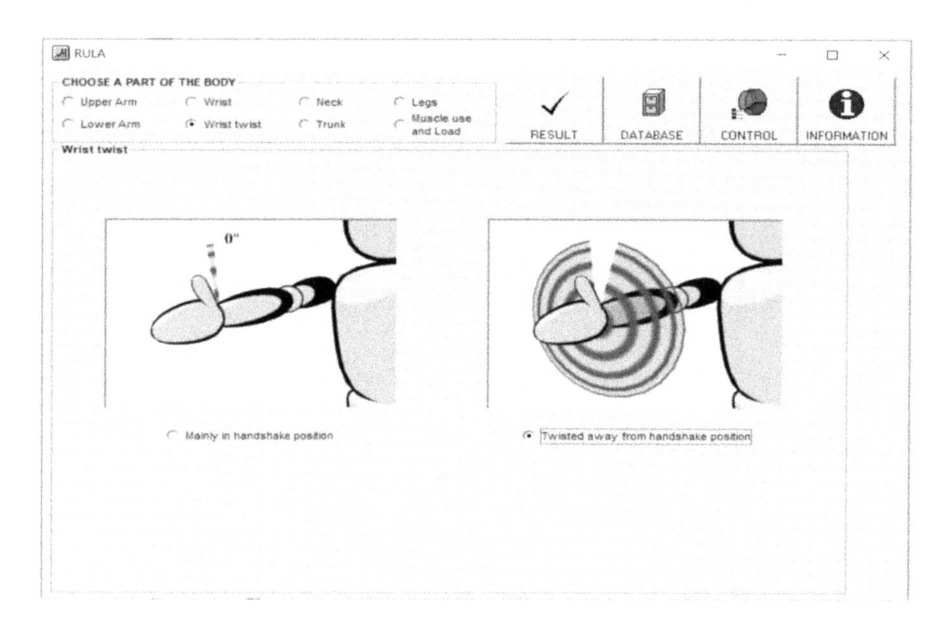

Figure 9. RULA data for trunk, His trunk was bent between 0 to 20 degrees and additionally twisted more than 180 degrees

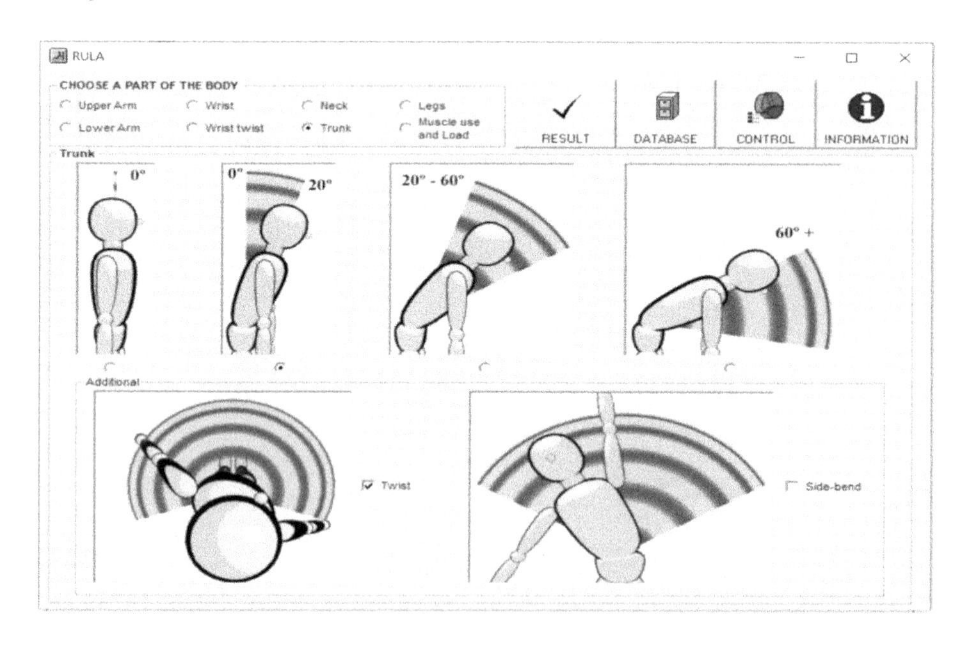

Figure 10. RULA data for muscle use and load

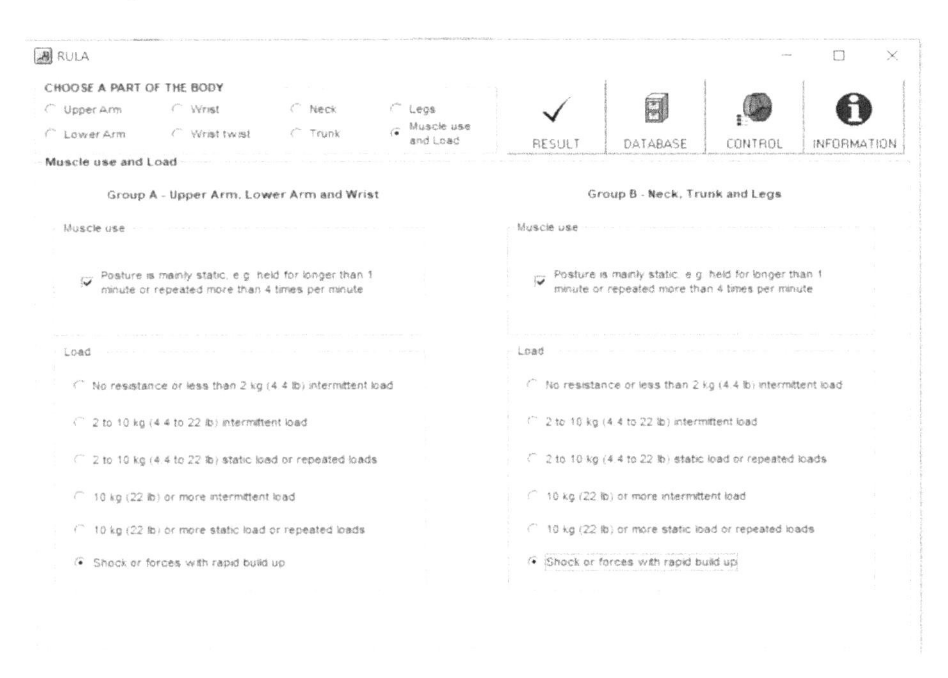

For Group A the upper arm, lower arm and wrist muscles were found to be held for longer than 1 minute along with shock or force with rapid build-up similarly for Group B her neck, trunk and legs muscles were found to be more than 4 times per minutes with shock or force with rapid build-up on the basis of these input.

Figure 11. RULA final score of 7 was found. Hence, it was concluded that "Investigation and changes are required immediately"

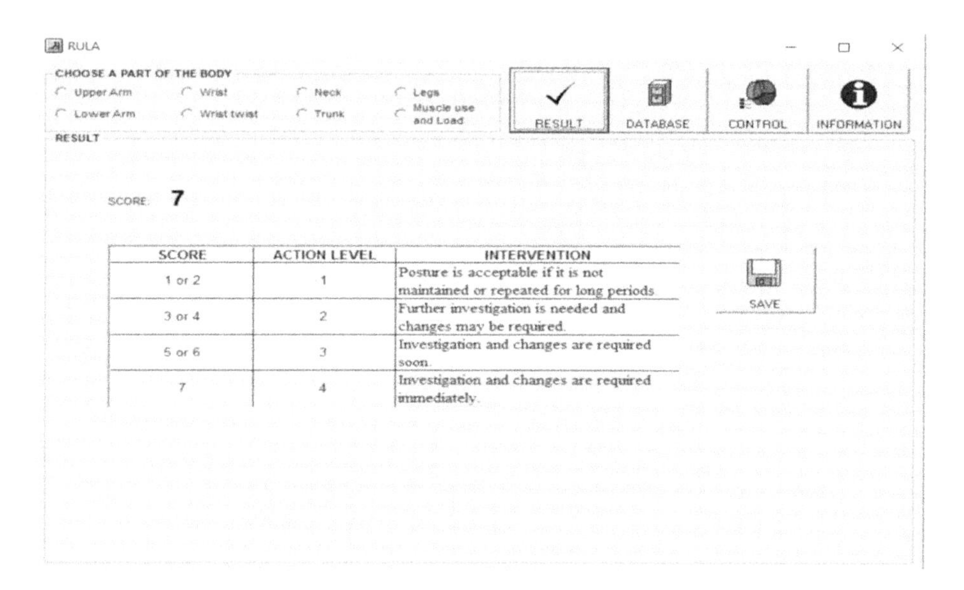

SCORE	ACTION LEVEL	INTERVENTION
1 or 2	1	Posture is acceptable if it is not maintained or repeated for long periods
3 or 4	2	Further investigation is needed and changes may be required.
5 or 6	3	Investigation and changes are required soon.
7	4	Investigation and changes are required immediately.

Further, QEC is used to check the work frequency of the tractor driver.
Quick exposure check for framer ploughing fields

Figure 12. Observer assessment

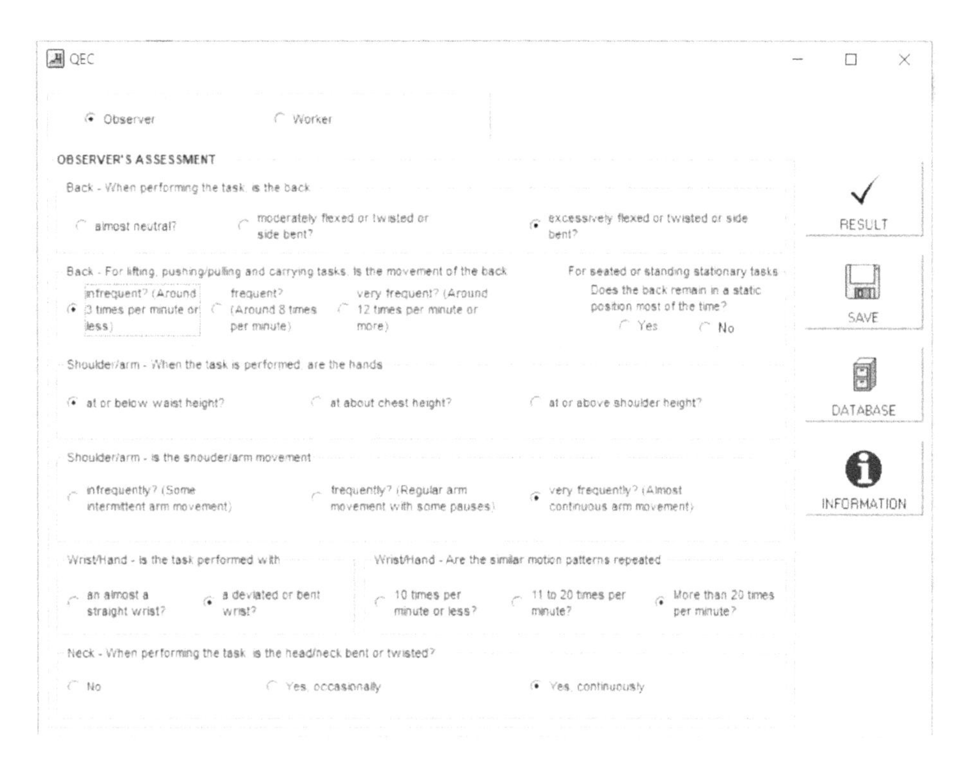

The farmer observes when performing the ploughing figure. 12, **his** back is excessively flexed, twisted, and bent while driving the tractor. He pushes and pulls the gear of a tractor frequently, around three times per minute or less. His shoulders and arms when performing the ploughing are below waist height, and his shoulder and arm movements when performing the ploughing move very frequently (almost continuous arm movement). His wrists and hands are deviated and bent while performing the task; his wrists and hands are moved in similar motion patterns repeated more than 20 times per minute; and his neck and head are bent or twisted when performing the task continuously.

Figure 13. Worker assessments

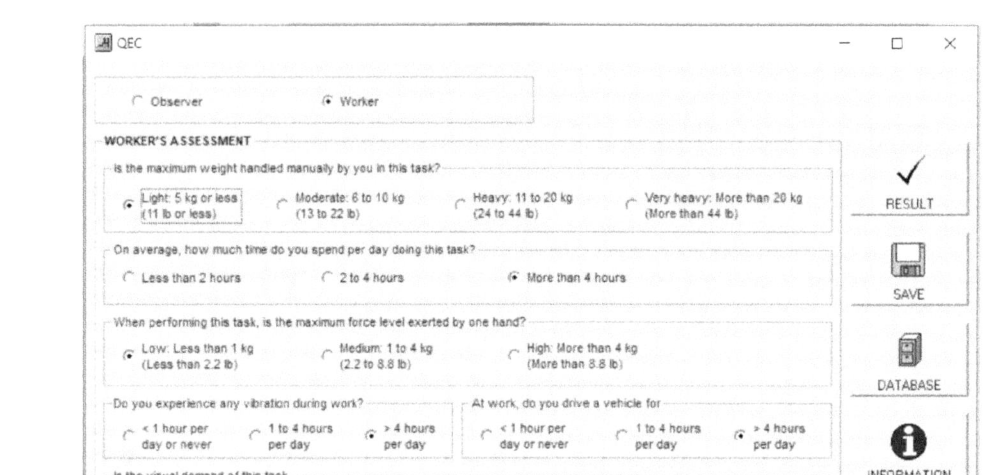

After observing the farmer (Figure 13), it was recorded that the farmer drives his tractor while working in his field to plough the crop, while doing the task manually is much lighter. On average, he does his task for more than 4 hours. When performing ploughing, there is less force exerted by one hand, and farmers experience more than 4 hours of vibration during their work. During ploughing, farmers drive a tractor for more than 4 hours per day. Due to the high visual demand of this task and the need to view some fine details, farmers often find difficulty while doing the ploughing and later find it highly stressful work.

Figure 14. Interpretation of the result of QEC

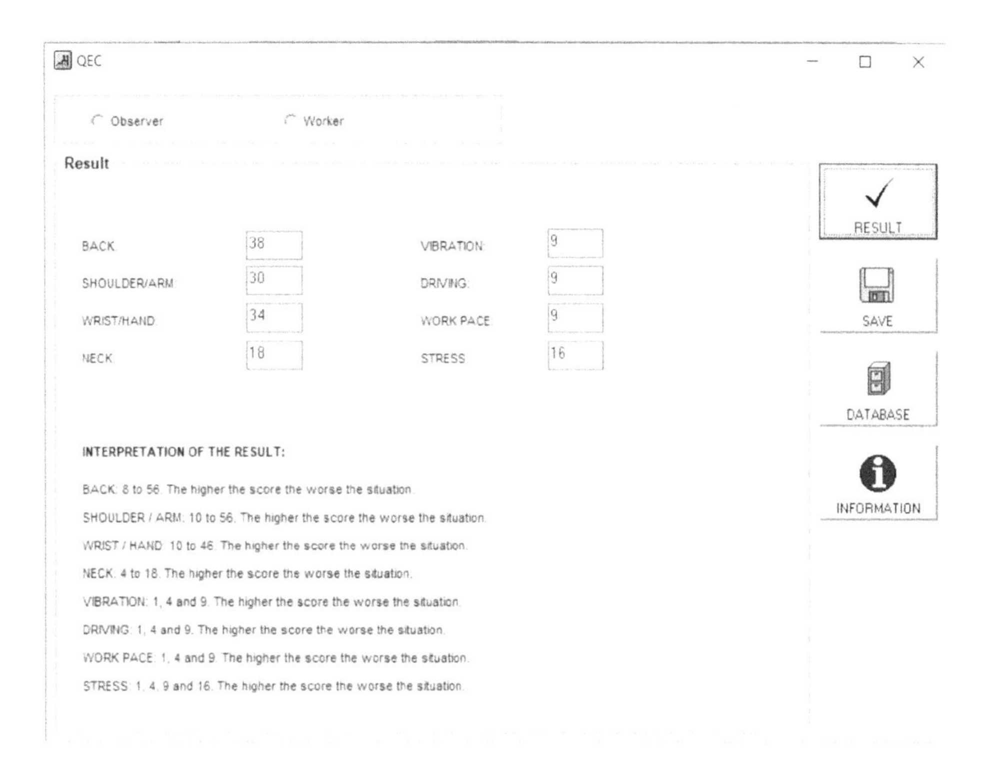

From the interpretation of the results of the QEC, the scores of farmers were between 8 and 56. The higher the score, the worse the situation. For tractor drivers, shoulder and arm lengths are found to be 10 to 56. Wrists and hands were found to be between 10 and 46. The neck was found to be between 4 and 18. Vibration during tasks was found to be 1, 4, and 9. Driving tractors is found to be 1, 4, and 9. During work, they were found to be 1, 4, and 9. While performing a task in adverse climatic conditions, stress was found to be 1, 4, and 16. Hence, it can be seen from the above discussion that the tractor operator is at high risk. Hence, further analysis is essential. So,SFO is implemented based on the collected data of 33 farmers engaged in driving tractors. Here, output is taken as the mental stress of tractor drivers (yes or no) (i.e., 1 or 0). Input of the tractor driver score (i.e., X1=back, X2=shoulder/arm, X3=wrist,X-4=neck,X5=vibration, X6=driving, X7=workplace condition,X8=physical stress, etc.) found by QEC. Then Minitab 17 is used to find output (y) and mental stress by linear regression analysis.

Mental stress(y)=-2.99+(-0.00038*back)+(0.0290*shoulder)+(0.038*wrist)+(0.123*neck)+(0.179*vibration)+(0.165*work place)+(-0.00251*Physical stress)

Then SFO is implemented to minimize mental stress.The MATLAB simulation worksheet of "Sun Flower Optimization" method used in this study was as illustrated in Figure 15. Further, based on the output obtained by the use of "Sun Flower Optimization" method as shown in Figure 16, the "Total number of function evaluation (Sun Flower) = 2000", and after "Iteration 100 function = 0.0075", ""Best solution (Sun Flower) = [416.1103 76.28657 110 68 29.63888 126.3093]", "Best Fitness (Sun Flower) = 0.0074671" of tractor driver during different seasons and during different time intervals.

Figure 15. MATLAB simulation database of "sun flower optimization"

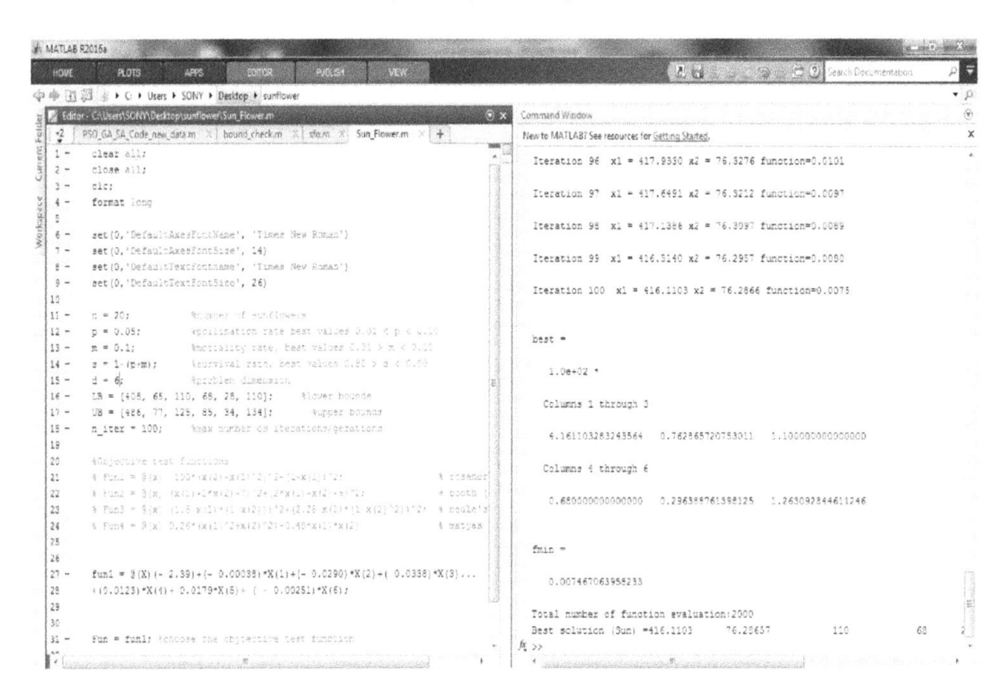

Total number of function evaluations (Sun Flower) = 2000", and after "Iteration 100 function = 0.0075", ""Best solution (Sun Flower) = [416.1103 76.28657 110 68 29.63888 126.3093]",

Figure 16. Result of "sun flower" "best fitness (sun flower) = 0.0074671"

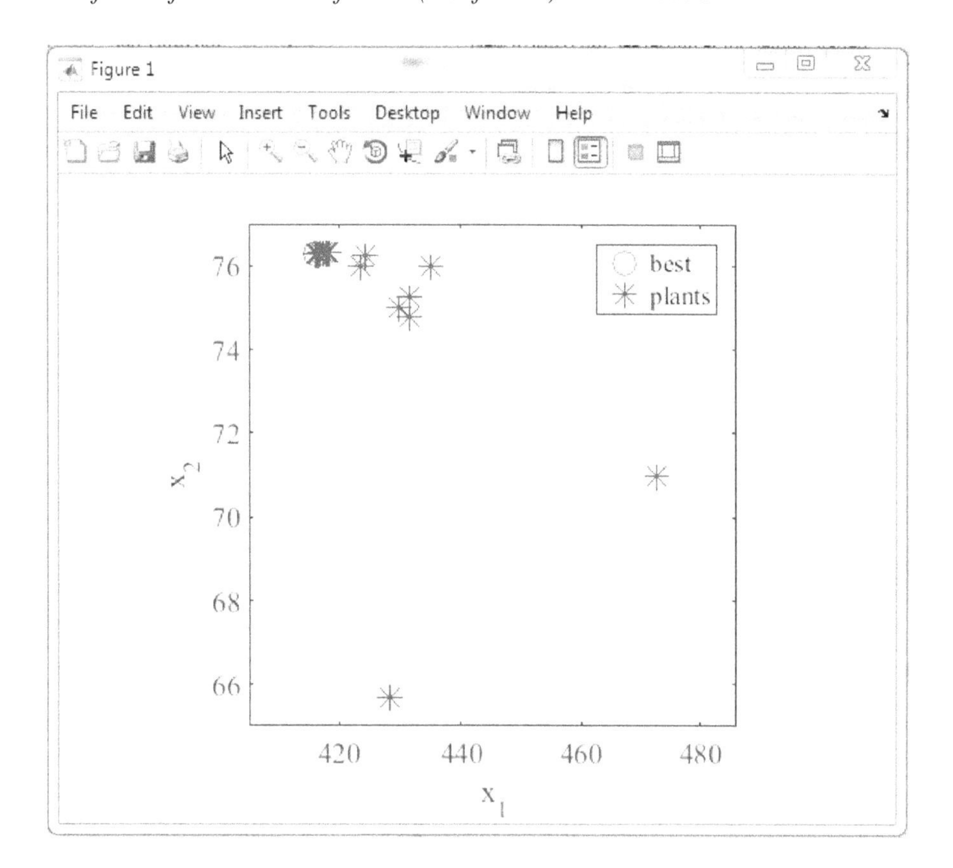

The best fitness value was 0. 0074671.It shows that the mental stress can be minimized by improving input parameters of tractor drivers/farmers .Hence, tractor models must be redesigned to reduce noise and vibration factors. The ergonomic design of tractors must be taken care of to provide maximum comfort to the drivers.

CONCLUSION

Tractor-driving farmers work in different types of agricultural fields (dry land, wet land, muddy land etc.). The force obtained due to driving in different ways directly affects their body posture and has an impact on their work efficiency and productivity in the agricultural sector. The risk factors leading to tractor farmers' musculoskeletal disorder (MSD) were identified. Ergonomic evaluation of working posture has been an initial step in reducing work-related body problems. Workplace discomfort increases the mental stress of farmers. Hence, proper ergonomic intervention by suitable farm-related and modern tools, technology, and equipment is needed to improve the workplace and provide maximum comfort, which will increase productivity and profit.

ACKNOWLEDGMENT

We want to sincerely thank each and every one of the farmers who took part in this poll.

Contributions of authors All the data for this study was gathered, examined, and analysed by Hullash Chauhan. Dr. S. Satapathy significantly assisted in drafting the manuscript.Ashok K. Sahoo reads and reviews the work.

Ethical endorsement All research were carried out in conformity with the ethical guidelines for human experimentation outlined in the Helsinki Declaration and its subsequent revisions, or with comparable standards. informed approval Each participant in this study gave informed consent after having the nature of the investigation explained to them.

REFERENCES

Ahmad, M., Hameed, A., Ullah, F., Wahid, I., Rehman, S. U., & Khattak, H. A. (2020). A bio-inspired clustering in mobile adhoc networks for internet of things based on honey bee and genetic algorithm. *Journal of Ambient Intelligence and Humanized Computing*, 11(11), 4347–4361. 10.1007/s12652-018-1141-4

Al-Raqadi, A. M., Rahim, A. A., Masrom, M., & Al-Riyami, B. S. N. (2017). Cooperation and direction as potential components for controlling stress on the perceptions of improving organisation's performance. *International Journal of System Assurance Engineering and Management*, 8(1), 327–341. 10.1007/s13198-015-0337-7

Alamiedy, T. A., Anbar, M., Alqattan, Z. N., & Alzubi, Q. M. (2020). Anomaly-based intrusion detection system using multi-objective grey wolf optimization algorithm. *Journal of Ambient Intelligence and Humanized Computing*, 11(9), 3735–3756. 10.1007/s12652-019-01569-8

Alshammari, B. M., & Guesmi, T. (2020). New chaotic sunflower optimization algorithm for optimal tuning of power system stabilizers. *Journal of Electrical Engineering & Technology*, 15(5), 1985–1997. 10.1007/s42835-020-00470-1

Anoop, V., & Bipin, P. R. (2021). Exploitation whale optimization based optimal offloading approach and topology optimization in a mobile ad hoc cloud environment. *Journal of Ambient Intelligence and Humanized Computing*, 1–20.

Aydemir, O. (2020). Detection of highly motivated time segments in brain computer interface signals. *Journal of the Institution of Electronics and Telecommunication Engineers*, 66(1), 3–13. 10.1080/03772063.2018.1476190

Behnam, M., & Pourghassem, H. (2017). Spectral correlation power-based seizure detection using statistical multi-level dimensionality reduction and PSO-PNN optimization algorithm. *Journal of the Institution of Electronics and Telecommunication Engineers*, 63(5), 736–753. 10.1080/03772063.2017.1308845

Chauhan, H., Satapathy, S. & Sahoo, A.K. (2021). Mental stress minimization in farmers: an approach using REBA, PSO, and SA. *Int J Syst Assur EngManag*. .10.1007/s13198-021-01167-y

Cherkaoui, B., Beni-hssane, A., & Erritali, M. (2020). Variable control chart for detecting black hole attack in vehicular ad-hoc networks. *Journal of Ambient Intelligence and Humanized Computing*, 11(11), 5129–5138. 10.1007/s12652-020-01825-2

Chiasson, M. È., Imbeau, D., Aubry, K., & Delisle, A. (2012). Comparing the results of eight methods used to evaluate risk factors associated with musculoskeletal disorders. *International Journal of Industrial Ergonomics*, 42(5), 478–488. 10.1016/j.ergon.2012.07.003

Coskun, M., Yildirim, O., & Demir, Y. (2021). *Efficient deep neural network model for classification of grasp types using sEMG signals*. J Ambient Intell Human Comput., 10.1007/s12652-021-03284-9

Guan, T. (2021). Research on the application of robot welding technology in modern architecture. *International Journal of System Assurance Engineering and Management*, 1-10.

Gul, F., Rahiman, W., Alhady, S. S., Ali, A., Mir, I., & Jalil, A. (2021). Meta-heuristic approach for solving multi-objective path planning for autonomous guided robot using PSO–GWO optimization algorithm with evolutionary programming. *Journal of Ambient Intelligence and Humanized Computing*, 12(7), 7873–7890. 10.1007/s12652-020-02514-w

Gupta, V., & Mittal, M. (2019). QRS complex detection using STFT, chaos analysis, and PCA in standard and real-time ECG databases. *Journal of The Institution of Engineers (India): Series B, 100*(5), 489-497.

Gupta, V., Mittal, M., & Mittal, V. (2020). R-peak detection-based chaos analysis of ECG signal. *Analog Integrated Circuits and Signal Processing*, 102(3), 479–490. 10.1007/s10470-019-01556-1

Hossain, M. D., Aftab, A., Al Imam, M. H., Mahmud, I., Chowdhury, I. A., Kabir, R. I., & Sarker, M. (2018). Prevalence of work related musculoskeletal disorders (WMSDs) and ergonomic risk assessment among readymade garment workers of Bangladesh: A cross sectional study. *PLoS One*, 13(7), e0200122. 10.1371/journal.pone.020012229979734

Kumar, A., Mathur, N. N., Varghese, M., Mohan, D., Singh, J. K., & Mahajan, P. (2005). Effect of tractor driving on hearing loss in farmers in India. *American Journal of Industrial Medicine*, 47(4), 341–348. 10.1002/ajim.2014315776468

Kumar, A., Varghese, M., Mohan, D., Mahajan, P., Gulati, P., & Kale, S. (1999). Effect of whole-body vibration on the low back: A study of tractor-driving farmers in north India. *Spine*, 24(23), 2506. 10.10 97/00007632-199912010-0001310626314

Kumar, N. R., & Nagabhooshanam, E. (2021). EKF with Artificial Bee Colony for Precise Positioning of UAV Using Global Positioning System. *Journal of the Institution of Electronics and Telecommunication Engineers*, 67(1), 60–73. 10.1080/03772063.2018.1528186

Li, G., & Buckle, P. (1998, October). A practical method for the assessment of work-related musculoskeletal risks-Quick Exposure Check (QEC). *Proceedings of the Human Factors and Ergonomics Society Annual Meeting*, 42(19), 1351–1355. 10.1177/154193129804201905

Lin, Q., Li, T., Shakeel, P. M., & Samuel, R. D. J. (2021). Advanced artificial intelligence in heart rate and blood pressure monitoring for stress management. *Journal of Ambient Intelligence and Humanized Computing*, 12(3), 3329–3340. 10.1007/s12652-020-02650-3

Mahmoodabadi, M. J., & Shahangian, M. M. (2019). A new multi-objective artificial bee colony algorithm for optimal adaptive robust controller design. *Journal of the Institution of Electronics and Telecommunication Engineers*, 1–14.

Maulik, S., De, A., & Iqbal, R. (2012, July). Work related musculoskeletal disorders among medical laboratory technicians. In *2012 Southeast Asian Network of Ergonomics Societies Conference (SEANES)* (pp. 1-6). IEEE. 10.1109/SEANES.2012.6299585

Menychtas, D., Glushkova, A., & Manitsaris, S. (2020). Analyzing the kinematic and kinetic contributions of the human upper body's joints for ergonomics assessment. *Journal of Ambient Intelligence and Humanized Computing*, 11(12), 6093–6105. 10.1007/s12652-020-01926-y

Nguyen, T. T. (2021). Enhanced sunflower optimization for placement distributed generation in distribution system. *Iranian Journal of Electrical and Computer Engineering*, 11(1), 107. 10.11591/ijece.v11i1.pp107-113

Picheny, V., Trépos, R., Poublan, B., & Casadebaig, P. (2015). Sunflower phenotype optimization under climatic uncertainties using crop models. *arXiv preprint arXiv:1509.05697.*

Ramamoorthy, R., & Thangavelu, M. (2021). An enhanced hybrid ant colony optimization routing protocol for vehicular ad-hoc networks. *Journal of Ambient Intelligence and Humanized Computing*, 1–32.

Sana, S. S., Ospina-Mateus, H., Arrieta, F. G., & Chedid, J. A. (2019). Application of genetic algorithm to job scheduling under ergonomic constraints in manufacturing industry. *Journal of Ambient Intelligence and Humanized Computing*, 10(5), 2063–2090. 10.1007/s12652-018-0814-3

Sanchez, W., Martinez, A., & Hernandez. (2018). A predictive model for stress recognition in desk jobs. *J Ambient Intell Human Comput*. .10.1007/s12652-018-1149-9

Satapathy, S. (2021). Work place discomfort and risk factors for construction site workers. *International Journal of System Assurance Engineering and Management*, 1-13.

Satheshkumar, K., & Mangai, S. (2021). EE-FMDRP: Energy efficient-fast message distribution routing protocol for vehicular ad-hoc networks. *Journal of Ambient Intelligence and Humanized Computing*, 12(3), 3877–3888. 10.1007/s12652-020-01730-8

Sharma, T. K., & Abraham, A. (2020). Artificial bee colony with enhanced food locations for solving mechanical engineering design problems. *Journal of Ambient Intelligence and Humanized Computing*, 11(1), 267–290. 10.1007/s12652-019-01265-7

Singh, S. K., & Kumar, P. (2020). A comprehensive survey on trajectory schemes for data collection using mobile elements in WSNs. *Journal of Ambient Intelligence and Humanized Computing*, 11(1), 291–312. 10.1007/s12652-019-01268-4

Sridevi, G., & Chakkravarthy, M. (2021). A meta-heuristic multiple ensemble load balancing framework for real-time multi-task cloud scheduling process. *International Journal of System Assurance Engineering and Management*, 12(6), 1459–1476. 10.1007/s13198-021-01244-2

Sung, P. C., Hsu, C. C., Lee, C. L., Chiu, Y. S. P., & Chen, H. L. (2015). Formulating grip strength and key pinch strength prediction models for Taiwanese: A comparison between stepwise regression and artificial neural networks. *Journal of Ambient Intelligence and Humanized Computing*, 6(1), 37–46. 10.1007/s12652-014-0245-8

Tajvar, A., Daneshmandi, H., Dortaj, E., Seif, M., Parsaei, H., Shakerian, M., & Choobineh, A. (2021). Common errors in selecting and implementing pen–paper observational methods by Iranian practitioners for assessing work-related musculoskeletal disorders risk: A systematic review. *International Journal of Occupational Safety and Ergonomics*, 1–7.33736566

Torén, A., Öberg, K., Lembke, B., Enlund, K., & Rask-Andersen, A. (2002). Tractor-driving hours and their relation to self-reported low-back and hip symptoms. *Applied Ergonomics*, 33(2), 139–146. 10.1016/S0003-6870(01)00061-812009120

Varol Altay, E., & Alatas, B. (2020). Performance analysis of multi-objective artificial intelligence optimization algorithms in numerical association rule mining. *Journal of Ambient Intelligence and Humanized Computing*, 11(8), 3449–3469. 10.1007/s12652-019-01540-7

Wang, Y., Li, A., & Khodaei, H. (2020). Optimal designing of a CCHP source system using balanced Sunflower optimization algorithm. *Energy Sources. Part A, Recovery, Utilization, and Environmental Effects*, 1–23. 10.1080/15567036.2020.1747575

Wu, H. C., Hong, W. H., & Chiu, M. C. (2018). Comparisons with subjective and objective indexes of lifting risk among different combinations of lifting weight and frequency. *Journal of Ambient Intelligence and Humanized Computing*, 1–5.

Yadav, A., & Agrawal, S. (2021). Mathematical model for robotic two-sided assembly line balancing problem with zoning constraints. *International Journal of System Assurance Engineering and Management,* 1-14.

Youssef, A. E., Kotb, Y., Fouad, H., & Mustafa, I. (2021). Overlapping gait pattern recognition using regression learning for elderly patient monitoring. *Journal of Ambient Intelligence and Humanized Computing*, 12(3), 3465–3477. 10.1007/s12652-020-02503-z

Yusuf, M., Adiputra, N., Sutjana, I. D. P., & Tirtayasa, K. (2016). The improvement of work posture using rapid upper limb assessment: Analysis to decrease subjective disorders of strawberry farmers in Bali. International Research Journal of Engineering. *IT and Scientific Research*, 2(9), 1–8.

Zhang, K. (2021). The design of regional medical cloud computing information platform based on deep learning. *International Journal of System Assurance Engineering and Management*, 1-8.

Chapter 8
Behind the Barriers:
Identifying Critical Credit Access Challenges in Agri-Business Sector of India

Puspalata Mahapatra
KIIT University, India

Lopamudra Lenka
KIIT University, India

ABSTRACT

Agricultural credit provision plays a vital role for farmers, producers, and business entrepreneurs giving them access to the funds for multiple purposes. The research culminates in the development of a conceptual model that outlines the multifaceted challenges confronted by the Indian agricultural industry while accessing the credit for their sustainable development. In the present study, thorough analysis has been done with the help of AHP technique to find out the identified key challenges which hinders the establishment and growth of agri-business activities in rural sector of India. This document will be helpful to the small agri-businessmen, rural agri-entrepreneurs, and producers while getting credit facility by the banks and other financial institutions. It will also be helpful to government and policy makers to overcome these challenges, so that proper coordination among financial markets, agri-entrepreneurs, and government will be established. It would further influence and encourage the investors to invest in agri projects on a massive scale in future.

INTRODUCTION

Rural Credit is considered as an instrumental support in achieving the goal of 'zero poverty', being the first Sustainable Development Goals (SDGs). However, formal financial institutions predominantly serve affluent, educated households from higher social strata, leaving out the weaker sections who are in greater need of credit, (Aditya, K. S.et al.2019).Agricultural credit is considered as a prime enabler for business expansion in rural sector. It supplies the financial means which help farmers and producers to invest in growth-focused activities by allowing them to boost in production, to introduce new technologies, expand new businesses and explore new markets. So far the sources of rural credit are concerned, it can be of two types, namely, institutional and non-institutional sources. The institutional sources include

DOI: 10.4018/979-8-3693-3583-3.ch008

NABARD, commercial banks, Cooperative Credits, RRBs (Regional Rural Banks), LDBs (Land Development Banks), Micro Finance Institutions etc. On the other hand the informal or non-institutional credit sources include the loans taken from the money lenders, village traders, commission agents, relatives, land lords etc. The provision of agricultural credit bring diversified scope of activities like capital for rural investment, crop diversification, commercialization of agriculture, infrastructure development, technological adoption, market facilitation, fostering innovation, research and development, trade and sustainable practices .The agricultural production process is inherently biological, leading to extended transition periods and creating a significant delay between investment and income generation. To boost production, farmers need to invest in modern inputs, which require provision of credit either from savings or through borrowing. Therefore, a greater flow of institutional credit is crucial for accelerating agricultural growth. However, developing the rural financing market is complex because agriculture is fragmented, geographically dispersed, heavily influenced by climate conditions, partially commercialized, and lacking in essential socio-economic and institutional infrastructure. Despite these challenges, agriculture in India has been given priority status for institutional financing, especially after the bank nationalization era, (Harisha, B. N. 2017).As the world evolves, agribusiness will continue to play a key role in addressing critical issues concerning to food production, distribution, consumption, economic growth and sustainability. Agribusiness is essential for feeding the world's expanding population, tackling food security challenges, and fostering economic growth of a nation. Agribusiness covers a diverse array of activities, including growing crops, raising livestock, processing agricultural products, distributing food, marketing, and providing support services. It combines cutting-edge technologies, inventive practices, and effective supply chain management to enhance productivity and profitability. Agribusinesses vary in size from small family farms to major multinational companies, each playing a distinct role in the agricultural industry. Agri-business is shaped by a variety of factors, such as climate conditions, consumers choices and preferences, producer ability and willingness to invest and government policies and regulations, technological adoption, international trade and agreements. To succeed, agribusinesses must effectively navigate these influences and should be financially well-equipped to attain sustainability and adapt to the dynamism of the world.

LITERATURE REVIEW

Agriculture is considered as the mainstay of life, where 47 per cent of the population is dependent on agriculture for livelihood (The Economic Survey of India 2021). To achieve a successful agricultural transformation, the focus has to be given for agricultural mechanization, skill development, competitive pricing models, adoption of innovative technology, climate smart practices, secured land ownership, agricultural diversification, public policy and investment. Hence, agricultural transformation will require setting up institutions that can provide support in various areas, such as: assisting small-scale farmers with resources and training, promoting regional development through strategic projects, rehabilitating degraded lands and replanting to increase agricultural productivity, facilitating marketing to ensure farmers can reach broader markets, encouraging specialization in specific crops for increased efficiency and bringing added values to the entire sector, Nyoni, J. (2021). However, a key factor behind agriculture's under performance has been credit inadequacy followed by low institutional and poor rate of investment. The transition from extensive farming practices to innovative farming requires huge investment, which creates demand for rural credit facilities. It further contributes to employment and income generation indirectly by providing raw materials and other intermediate inputs to various manufacturing and industrial sectors. These sectors included agribusinesses, food processing companies, and retail outlets like supermarkets (Adam, Bevan, Gollin & Mkenda, 2012). The growth in agricultural production is unlikely without substantial financial investments to drive innovation. In a market-based economy, funding for production usually comes from internal sources, namely business profits. However, most agricultural enterprises struggle with low profitability or even losses, limiting their capacity for self-funding. This financial constraint, in turn, hinders their ability to invest in and rejuvenate their operations. The low return on agricultural investments also tends to deter potential investors, especially international ones, who could contribute not only with capital but also with cutting-edge machinery, equipment, and technology, (Finagina O., 2021) However, enhancing crop production requires extensive investment in agricultural infrastructure, including irrigation, supply of inputs, farm mechanization, innovation, supply chain management and R& D activities in agricultural sector. Additionally, to ensure a successful harvest, farmers, peasants, and agricultural producers need to invest considerable funds annually and often, depending on financial institutions for loans to incur essential expenses before the harvest and allowing them to earn revenue through crop sales. As agriculture's contribution to the total GDP has caught the declining trend, it is predicted that the share of agricultural credit as a proportion of total credit will also decrease. However, it's crucial to ensure that the share of credit relative to agricultural GDP does not fall; ideally, it should increase and continue the upward trend observed recently. To achieve this, banks must conduct a thorough analysis to identify the risks involved in extending agricultural credit and then devise market-oriented solutions to mitigate them. This further requires banks to rethink their traditional view of agriculture and adopt innovative strategies to support the sector's evolving needs (Harisha, B. N. 2017). The formal banking system has been able to meet some of the needs for productive credit, but its performance has been hampered by a lack of professionalism and accountability. Moreover, banks have fallen short in addressing the "consumption" needs of the poor. When it comes to promoting equity, the banking system often mirrors the biases inherent in a deeply divided society. This has allowed predatory moneylenders to thrive in an environment where markets are fragmented and interconnected in ways that create a suffocating network of exploitation Devaraja, T. S. (2011).The major Challenges of agriculture credit in India include credit insufficiency, low santion of loans, less priorities to small and marginal farmers, low institutional support, red tapism and other challenges related to sustainability (Sahni, M.2020).Despite the expansion of the rural credit system, the amount of rural credit in India remains insufficient to meet the increasing demand driven by rising prices of agricultural inputs. The loan amounts approved for farmers by various credit agencies are often too low to cover the full range of agricultural operations. Due to the inadequacy of the sanctioned loan amounts, farmers sometimes use the funds for unproductive purposes, defeating the original intent of the credit. Rural credit agencies have largely failed to address the needs of small and marginal farmers. As a result, these needy farmers receive less attention from credit schemes, while relatively wealthier farmers get more focus due to their better creditworthiness. In India, institutional credit arrangements are inadequate compared to the growing needs. The existing cooperative credit institutions, such as Primary Agricultural Credit Societies, Land development banks, Commercial banks, and Regional rural banks, have not reached the majority of rural farmers. As per the United Nations Food and Agriculture Organization (UNFAO) differences of enthusiasm are found while going for agricultural finance due to the factors like agriculture is a politically and seasonally sensitive sector. Additionally, agricultural borrowers often engage in similar activities, creating covariant risks due to market and price fluctuations, yield uncertainties, and changes in domestic and international policies. State actions, such as waivers for overdue loans, can also influence lending risks. Externally funded credit schemes often suffer from low loan repayment discipline and high financial transaction costs followed by long distances to serve dispersed rural customers, poorly developed transportation and communication infrastructure, insufficient understanding of diverse farm households, expensive management and supervision of rural bank branches. Similarly, a few more factors related to such category of concerns are lack of required collateral, funding liabilities including reduced turnover.

RESEARCH METHODOLOGY AND STRUCTURE OF THE STUDY:

The present study is based on both primary and secondary data. Here the whole process begins with an extensive literature review and the synthesis of existing literature. From the widespread literature review, 31 factors were extracted which act as barriers for agri-business financing. These barriers or challenges are very crucial and faced by the farmers or agricultural entrepreneurs while getting the loan from the banks and other financial institutions. On the basis of the opinion of total respondents (220) which consists of farmers, agricultural businessman and agricultural entrepreneurs are taken into consideration. Set of Questionnaires were given to them to and they were asked to fill up using the 5 points Likert scale. On the basis of high mean and SD, 24 factors were selected and grouped under 6 heads on the basis of their similarity nature. Then the study proceeds to rank the key challenges to know which one is the topmost challenge and which one is the least important challenge. For prioritizing and ranking the challenging factors, AHP technique has been used here where the opinion of 20 experts who are working as senior level professional in the banking and on banking financial companies involved in disbursing the agri-loan have taken into consideration. The analytic hierarchy process (AHP) is a process that uses hierarchical decomposition to deal with complex information in multi-criterion decision making. The analytic hierarchy process (AHP) helps -to deal with complex information in multi-criterion decision making. By utilizing the findings of the survey research and statistical methods like AHP, the factors that acts as more challenging (on the basis of ranking) for getting the credit for agri-business have been taken into consideration and then some suggestions have been provided in this regard.

The structure of the study starts with an introduction in section (1), followed by literature in the next section. Methodology is explained in section (III) and finally analysis and its results in section (IV). Discussions and conclusions are given finally.

Figure 1. Structure of the study

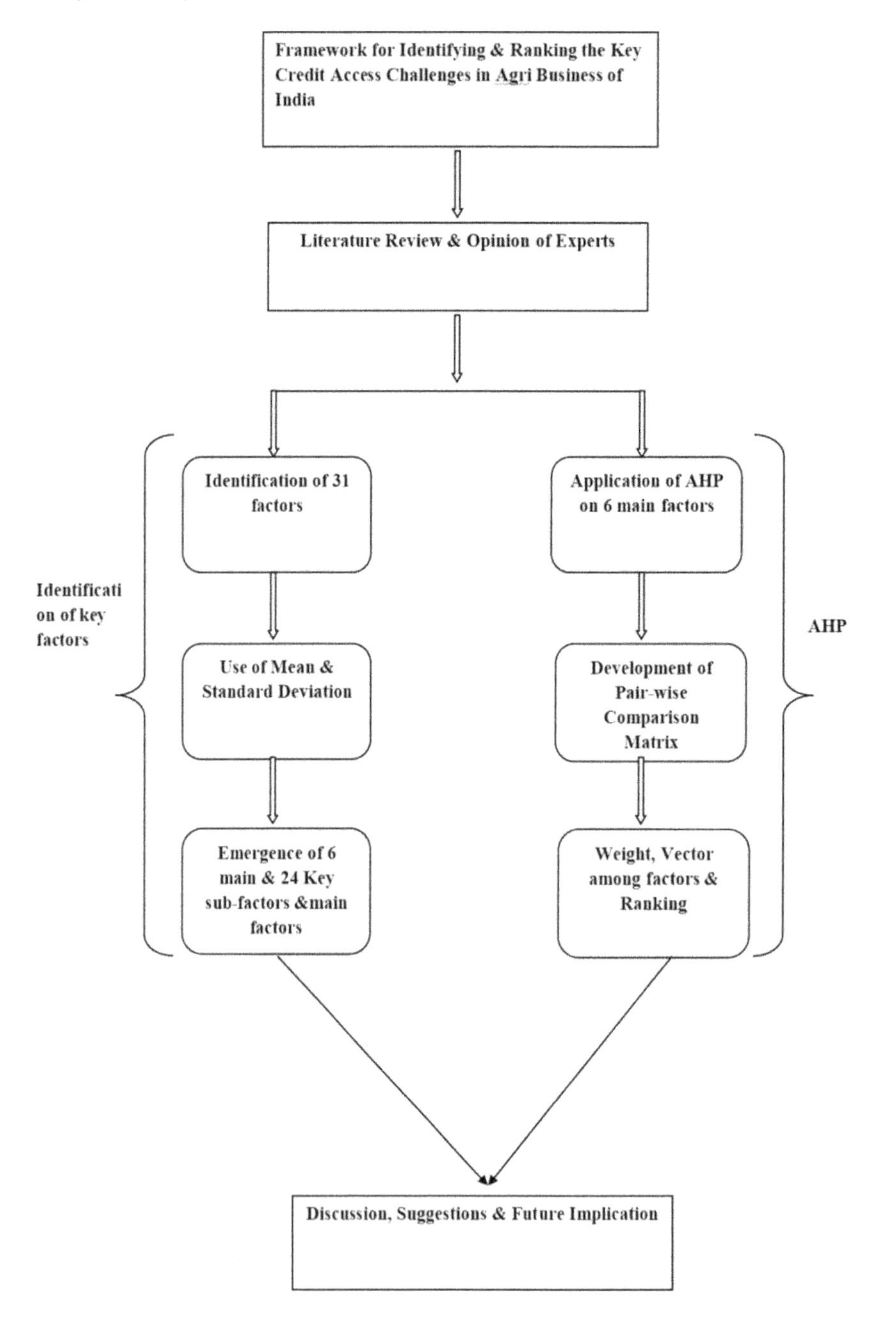

Figure 1 illustrates the comprehensive structure of the study, starts with the methodical progression from a literature review and expert evaluations to the distillation of critical factors and the application of Analytic Hierarchy Process (AHP) for prioritization, which forms the basis for discussion, recommendations for prospective adoption for the development of the agricultural sector of India.

CONCEPTUAL FRAMEWORK

The challenges attached to the provision of rural credit in India are found to be varied and interconnected. In the present discussion from the broad literature review, the six major factors attached to rural credit challenges have been identified. Those include economic challenges, financial challenges, environmental challenges, socio-cultural challenges, technological challenges and political challenges.

Figure 2. Major factors

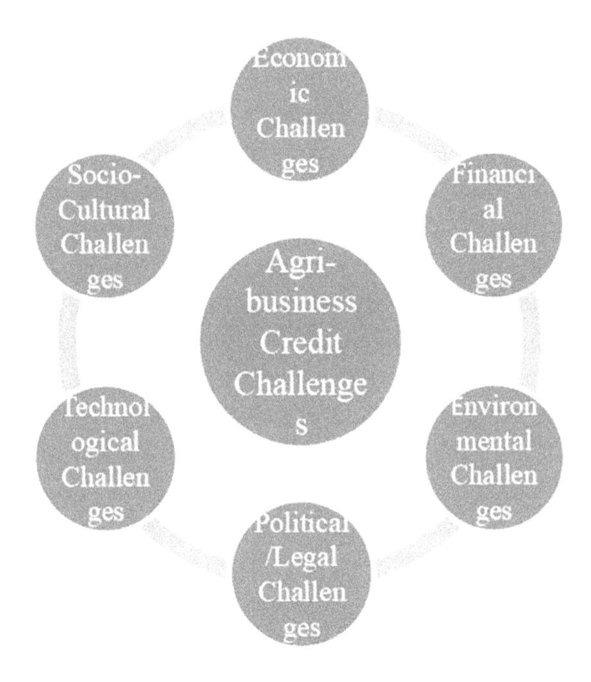

Economic Challenges

Some of the pertinent issues related to economic challenges are delayed and low repayment, high interest rate, mounting over-dues, non-availability of adequate crop insurance, poor economic conditions / less credit worthiness and unscientific agricultural practices. The International Fund for Agricultural Development (IFAD) found that repayment rates among rural borrowers are heavily influenced by agricultural variability. Farmers who face poor harvests or market downturns often struggle to meet their loan obligations (IFAD, 2019). The World Bank reported that high-interest rates are common obstacles in rural credit systems, often stemming from the perceived higher risks of lending in rural areas and the additional costs of servicing remote locations (World Bank, 2017). The Asian Development Bank (ADB) highlighted that accumulating overdue loans can destabilize lenders, thereby reducing the availability of credit in rural areas. Furthermore, the Food and Agriculture Organization (FAO) noted that inadequate crop insurance exacerbates economic vulnerability among rural farmers, impacting their creditworthiness and access to loans (FAO, 2020). According to the International Labour Organization

(ILO), poor economic conditions in rural areas marked by low wages and limited job opportunities also undermine creditworthiness, reducing the effectiveness of rural credit systems (ILO, 2019). Additionally, the Consultative Group on International Agricultural Research (CGIAR) observed that unscientific agricultural practices lower crop yields and increase the risk of loan defaults, further threatening the sustainability of rural credit facilities (CGIAR, 2018).Addressing these economic challenges requires some of the initiatives, including improving rural infrastructure, providing training and education on modern agricultural practices, ensuring access to fair financial products, and implementing effective risk management strategies like crop insurance.

Financial Challenges

The financial factor has been found to be vital for creating agri-business credit challenges. Various sub-factors under this point include, In-proper financial inclusion, Inadequate ROI, Lack of prudent policy, Lack of prudent policy, Poor supply chain management, Inability to offer collateral.The World Bank's “Global Findex Database” indicates that financial inclusion is still a major issue in rural and agricultural areas, where many people struggle to access basic financial services like credit and insurance (World Bank, 2018).The Food and Agriculture Organization (FAO) found that the inherent risks in agriculture lead to unpredictable returns, impacting farmers' profitability and their ability to secure and repay credit (FAO, 2020).The International Monetary Fund (IMF) report emphasizes the need for strong policy frameworks to maintain financial stability, particularly in sectors like agriculture where risk factors are significant. Policies should strike a balance between ensuring financial access and managing risk (IMF, 2017).The United Nations Food Systems Summit (UNFSS) reports that inefficiencies in agricultural supply chains cause post-harvest losses and reduced profitability, undermining farmers' ability to fulfill their financial commitments (UNFSS, 2021).A World Bank study on agricultural finance highlighted that a lack of collateral is a major obstacle to obtaining credit for smallholder farmers and rural entrepreneurs, as lenders often perceive them as higher-risk borrowers (World Bank, 2017).By tackling these issues, agri-business credit facilities can become more accessible, reliable, and supportive of the agricultural sector's growth and sustainability.

Environmental Challenges

The environmental factor plays an important role in identifying and mitigating credit related issues. These factors include, Adverse & sensitive political Environment, Uncertain & Unpredictable Climate, Lack of Crop diversification, Operational & Market Risks and Economic slow downA World Bank report highlights that political instability and governance issues can significantly hinder agricultural productivity and rural development. When political environments are unstable, credit markets become more volatile, making it harder for rural communities to access financial services (World Bank, 2019). The Intergovernmental Panel on Climate Change (IPCC) has detailed how climate change affects agriculture. Unpredictable weather patterns, rising temperatures, and more frequent extreme weather events pose considerable risks to agricultural businesses (IPCC, 2018). The Food and Agriculture Organization (FAO) underscores the critical role crop diversification plays in ensuring agricultural resilience. A lack of crop diversification can lead to economic instability among farmers, increasing their vulnerability to credit risks (FAO, 2020). The Organization for Economic Co-operation and Development (OECD) explores the impact of operational and market risks in agriculture on farmers' financial stability. These risks can lead to reduced incomes, making it harder for farmers to access credit (OECD, 2019). A study by the International Monetary Fund (IMF) indicates that economic slowdowns have ripple effects on rural economies. A decrease in consumer spending can lead to lower market prices for agricultural products, affecting farmers' financial stability and creditworthiness (IMF, 2018). Understanding these factors can help policymakers and stakeholders devise solutions that improve the sustainability and resilience of rural credit systems.

Political/ Legal Challenges

The political or legal challenges are found to be nascent due to the emerging issues such as lack of transparency, presence of redtapism, prevalence of corruption and political and administrative failure. In the present study some more factors like inadequate budget allocation, unstable government, slow legal procedure cumbersome process of credit access are some of the factors have been included under political and legal challenges. The World Bank underscores that sufficient budget allocation is crucial for sustaining rural development, cautioning that underfunding can impede the growth and stability of rural economies (World Bank, 2020). Without adequate financial support, rural credit systems may be unable to meet the demands of farmers and small businesses. An International Monetary Fund (IMF) report suggests that political instability can have a detrimental impact on economic growth, discouraging investment in rural areas and thus affecting credit markets and reducing financial inclusion (IMF, 2019). The Organization for Economic Co-operation and Development (OECD) finds that complex legal frameworks can create significant obstacles for rural businesses, thereby deterring investment in rural credit systems. Simplifying these frameworks is key to ensuring easier access to credit (OECD, 2021).A World Bank report on business environments highlights that excessive bureaucratic red tape can stifle economic activity, posing challenges for rural entrepreneurs seeking credit and other financial services. Minimizing administrative barriers is essential for a more accessible credit system (World Bank, 2018). The International Finance Corporation (IFC) emphasizes that simplifying the process for accessing credit is central to boosting financial inclusion. If the process is overly complex, it can dissuade rural borrowers from applying, leading to a reduction in the reach of rural credit facilities (IFC, 2017). By allocating adequate budgets, ensuring government stability, streamlining legal procedures, reducing red tape, and simplifying credit access, policymakers can enhance the effectiveness of rural credit facilities. This, in turn, can promote rural development and economic growth.

Technological Challenges

In the present, study the factors like in-adequate general and technical infrastructure unskilled or less skilled persons, lack of commercial agricultural practices, existence of technological barriers has been taken under technological challenges. The World Bank identifies poor infrastructure as a major obstacle to rural development. The absence of adequate roads, electricity, and reliable communication networks significantly hinders agricultural productivity and the ability to transport products to market (World Bank, 2020). In addition, a report from the Food and Agriculture Organization (FAO) states that inadequate technical infrastructure, such as irrigation and storage facilities, can lead to reduced crop yields and increased post-harvest losses, impacting farmers' profits (FAO, 2019). The International Labor Organization (ILO) emphasizes that skills development is crucial for rural and agricultural sectors. A shortage of skilled workers can limit the adoption of modern agricultural techniques and decrease the efficiency of rural businesses (ILO, 2018). Similarly, the Organization for Economic Co-operation and Development (OECD) indicates that investing in skills training can substantially boost productivity and economic outcomes in rural areas (OECD, 2019). The Food and Agriculture Organization (FAO) observes that commercial agriculture can stimulate rural economies by increasing productivity and opening market opportunities. A lack of commercial practices can trap rural areas in subsistence farming, restricting economic growth (FAO, 2020). The World Bank also suggests that promoting commercial agriculture can lead to increased incomes and improved financial stability for rural communities (World Bank, 2019). The International Fund for Agricultural Development (IFAD) warns that technological barriers can pose significant impediments to rural development. Without access to modern technologies, rural businesses may find it hard to compete and innovate (IFAD, 2018). Hence, overcoming technological barriers is essential in enhancing productivity and sustainability in agriculture (WEF, 2020).

Socio-Cultural Challenges

The Socio-cultural challenges are the matter of concern, when the challenges related to agribusiness is concerned. Some of the major factors those have been extracted from the literature review are disputed land ownership Improper information about credit access, Inadequate training provision, Lack of faith and motivation. Majority are small and Marginal farmers Rural borrowers often rely on informal sources for credit-related information, which can lead to misunderstandings about terms, interest rates, or repayment schedules. This misinformation can contribute to poor financial decisions and a higher risk of default (Chakrabarty, 2013). Government-sponsored credit programs and subsidies sometimes fail to reach rural areas due to inadequate outreach, pushing rural borrowers towards informal or high-interest credit sources, increasing their financial burden (World Bank, 2015). Many rural borrowers lack basic financial knowledge, including concepts like interest rates, loan terms, or budgeting, which can result in mismanagement of funds and increase the risk of default (FAO, 2019). Without proper training on how to use credit efficiently, farmers and rural entrepreneurs may not get the most from their loans, leading to lower returns and more challenges in loan repayment (USAID, 2017) Additionally, rural areas often face unstable conditions such as market fluctuations, adverse weather, or political uncertainty, which can undermine motivation for investment and, in turn, affect creditworthiness, increasing the risk of default (IFAD, 2016).Small-scale farmers' income is primarily from agricultural production, they're more susceptible to risks from adverse weather, pests, or other uncertainties, which can affect their ability to repay loans during difficult seasons (FAO, 2018). Furthermore, these small and marginal farmers often lack sufficient collateral to secure formal loans, leading them to informal lenders who charge higher interest rates, exacerbating their credit risks (NABARD, 2020).

Table 1. Factors and sub-factors

Economic Challenges (ECC-6)		Financial/Capital Challenges (FCC-6)	
Delayed and low repayment	ECC1	I. In-proper financial inclusion	FCC1
High interest rate	ECC2	II. Inadequate ROI	FCC2
Mounting over-dues	ECC3	III. Lack of prudent policy	FCC3
Non availability of adequate crop insurance	ECC4	IV. Lack of public Investment	FCC4

continued on following page

Table 1. Continued

Poor economic Condition/ Less credit worthiness	ECC5	V. Poor supply Chain Management.	FCC5
Unscientific agricultural practices	ECC6	VI. Inability to offer collateral	FCC6
Environmental Challenges (EC-5)		**Political and Legal Challenges(PLC-5)**	
Adverse & sensitive political Environment	EC1	Inadequate budget allocation	PLC1
Uncertain & Unpredictable Climate	EC2	Unstable Government	PLC2
Lack of Crop diversification	EC3	Slow & Legal procedure	PLC3
Operational & Market Risks	EC4	Redtapism	PLC4
Economic slow down	EC5	Cumbersome process of credit access	PLC5
Technological Challenges (TC-4)		**Socio-Cultural Challenges (SCC-5)**	
Inadequate general & technical infrastructure	TC1	Disputed land ownership	SCC1
Unskilled or less skilled personnel	TC2	In-proper information about credit access	SCC2
Lack of commercial agricultural practices	TC3	Inadequate training provision	SCC3
Technological Barriers.	TC4	Lack of faith & motivation	SCC4
		Majority are small and Marginal farmers	SCC5

Here the study identified 31 factors from the extensive literature review which are very crucial challenges faced in the context of agribusiness credit or loan provisions. Moreover, factors have been found from both demand side and supply side interventions. On the basis of opinions collected from 220 respondents, out of which 200 are farmers, agricultural producers, entrepreneurs and the remaining 20 are experts, who are working as senior level professional in the banking and non-banking financial Companies. Set of Questionnaires were given to them to and they were asked to fill up using the 5 point Likert scale. On the basis of high mean and less SD, 24 factors are selected and grouped under 6 heads on the basis of their similarity nature which list is provided in the study. Then the study proceeds to rank the key challenges to know which one is the topmost challenge and which one is the least important challenge. For prioritizing and ranking the challenging factors AHP technique has used here. The analytic hierarchy process (AHP) is a process that uses hierarchical decomposition to deal with complex information in multi-criterion decision making. The analytic hierarchy process (AHP) helps -to deal with complex information in multi-criterion decision making.

ANALYSIS

In recent years, the AHP model has been broadly used to indicate the factors responsible for wider coverage of rural credit facilities, their overall impact on rural business practices and major barriers found in rural India. For instances, this method was adopted to observe the early warning method of credit risk of agricultural enterprises in Guizhou Province of China to suggest early warning credit risk for the enterprises,(Zhang, M., Li, N., & Zhang, C. 2023).Similarly, this model has also been used to study the credit scoring model for Farmer lending decisions in Rural China, (Mao, J. et al., 2020).The same model has also been used by Gunes, E., & Movassaghi, H. (2017) to investigate the structure of the agricultural credit market in Turkey and identify factors that influence farmers' preference among alternative lenders. Another interesting discussion has been observed to evaluate the level of satisfaction of borrowers with the products and services of microfinance institutions (MFI) at different criterion levels, with the application of the AHP model (Hassan, M., et al. 2023).

Table 2. Descriptive statistics of factors

Factors	Mean	SD	N
1.Delay low payment (ECC1)	4.65	0.835	220
2. High Interest rate (ECC2)	4.1	0.875	220
3.Mounting over-dues (ECC3)	3.55	0.845	220
4. Non availability of adequate Crop Insurance (ECC4)	4.15	0.816	220
5.Poor economic Condition/ Less credit worthiness (ECC5)	4.4	.62	220
6. Unscientific agriculture practice (ECC6)	3.9	.78	220
7.Adverse & Sensitive political Environment (EC1)	4.5	0.826	220
8. Uncertain & Unpredictable Climate (EC2)	4.1	0.731	220
9. Lack of Crop diversification (EC3)	3.15	.905	220
10. Operational & Market Risks (EC4)	3.2	0.665	220
11.Economic slow down (EC5)	4.35	0.902	220
12. In-proper Financial inclusion (FCC1)	4.0	.821	220--
13. Inadequate ROI (FCC2)	4.32	0.713	220
14. Lack of prudent policy(FCC3)	3.1	0.632	220
15 .Lack of public Investment(FCC4)	4.23	0.846	220
16. Poor supply Chain Mgt. (FCC5)	3.7	1.765	220
17 .Inability to offer collateral (FCC6)	4.12	0.187	220
18. Inadequate budget allocation (PLC1)	3.45	0.034	220
19. Unstable Government (PLC2)	4.43	0.912	220
20. Slow & Legal procedure (PLC3)	4.13	0.88	220
21. Redtapism(PLC4)	3.98	0745	220
22. Cumbersome process of credit access (PLC4)	4.73	0.586	220
23. Inadequate general & technical infrastructure (TC1)	4.32	0.713	220
24. Un skilled or less skilled personnel (TC2)	3.225	0.773	220
25. Lack of commercial agricultural practice (TC3)	4.82	0.840	220
26.Existence of Technological Barriers (TC4)	3.53	0.786	220
27. Disputed land ownership (SCC1)	3.35	0.023	220
28. In-proper information about credit access (SCC2)	3.15	1.198	220
29. Inadequate training provision (SCC3)	4.76	0.852	220
30. Lack of faith & motivation (SCC4)	4.77	0.793	220
31. Majority are Small & Marginal Farmers (SCC5)	3.6	.0709	220

To identify the most impactful challenges in accessing credit in agri business, this study has utilized the Analytic Hierarchy Process (AHP) Method. The AHP method is a structured MCDM technique used for analyzing complex situations for taking the decisions. Table 2 presented below offers a comprehensive comparative analysis of various challenges which are crucial to the farmer for accessing the credit for their agri business.

Table 3. Pair-wise comparison matrix of challenging factors

Criteria	FCC	SCC	TC	ECC	PLC	EC
FCC	1	4	3	1/2	6	3
SCC	1/4	1	2	1/3	6	4
TC	1/3	1/2	1	1/5	3	2
ECC	2	3	5	1	5	4
PLC	1/6	1/6	1/3	1/5	1	1/2
EC	1/3	1/4	1/2	1/4	2	1

Each challenging factors shown in the above table have been evaluated against the others, which provides a numerical representation of their relative importance or impact as perceived by the expert who are working and associated with financial institutions providing loans to farmers or owners of agri business. The values in the table are indicative of the comparative influence or priority of one criterion over another, thereby offering an in-depth understanding of how the farmers are facing the challenges when accessing the credit form Banks and other financial institutions.

Table (4) analyzes each criterion for its relative weight and importance in the context of credit access from banks, leading to an overall ranking that reflects their perceived significance.

Table 4. Criteria weightage and ranking of challenging factors

Criteria	FCC	SCC	TC	ECC	PLC	EC	Weights	Ranking
FCC	0.237	0.442	0.249	0.189	0.257	0.202	0.256	2
SCC	0.059	0.110	0.166	0.126	0.257	0.270	0.172	3
TC	0.079	0.055	0.083	0.075	0.129	0.135	0.102	4
ECC	0.473	0.331	0.416	0.377	0.214	0.270	0.324	1
PLC	0.039	0.018	0.028	0.075	0.043	0.034	0.047	6
EC	0.079	0.028	0.042	0.094	0.086	0.067	0.070	5

λmax- 6.535, C.I- 0.089, C.R-0.0676, Random Index (6) - 1.35

Interpretation: The table's comprehensive results provides a detailed and clear understanding of the relative importance of the various factors that act as barriers or challenging to the farmers while procuring loan from Banks or Financial institutions. At the forefront the first noticeable factors is the Economic Challenges (EEC) ranked as 1st having the weightage (0.324), followed by Financial Challenge (FC) with the weightage (0.265) ranked as 2nd . These two challenges are most striking point among all the challenges faced by the farmers and owners of agri-business to avail the loan or credit from the Banks. It signifies that both form supply and demand side strong emphasis should put on these two factors as comparison to others. Challenges like Socio cultural factors (SCC) emphasizes a critical role in credit access ranked as 3rd with the weightage (0.270) and remaining challenging factors are technological Challenge as 4th (0,102), Environmental Challenge (EC) ranked as 5th with weightage (0.070) and finally Political & Legal Challenge (PLC) is the 6th ranked challenges having the weightage (0.047) reflects the least contributing challenging factors which acts as barrier for getting the loan for agricultural purpose. The analysis & findings suggest that Govt. Banks and farmers should focus on these key challenges to overcome the difficulties in accessing the credit for the development of agriculture sector of India.

CONCLUSION AND SUGGESTIONS

The whole study depicts the fact that the provision of rural credit plays a crucial role in augmenting sustainable business practices in agricultural sector. The deep understanding of the study outcome will enhance the credit inflow to the agri-industries and will systematically reduce or mitigate the borrowers' loan default and credit evaluation cost of lenders. Apart from the various factors responsible for the agri-business credit provision as discussed in the study, the following suggestions can be taken to improve agricultural institutional support. The outcome of the present study indicates that the economic factor has been found to be the most important challenge which should be addressed by the government and policy makers. The special focus should be given for the challenges like delayed and low repayment, high interest rate, mounting over-dues, and non availability of an adequate crop insurance, poor economic condition, less credit worthiness and unscientific agricultural practices for the growth and development of agricultural sector of India. It further would increase the overall socio-economic conditions of rural India. At the same time the Government, policy makers, banks and financial institutions should develop a model which can be utilized for easy disbursing the loan to the agricultural sector of India.

FUTURE SCOPE AND IMPLICATION

The outcome of this study will facilitate the catering of easy accessibility of credit which in turn will work for the development of the rural agricultural sector. This document will be helpful to MSMEs, small agri-business, rural agri-entrepreneurs and producers to overcome the hurdles while getting the credit facility from financial institutions. It will work as a driver for paradigm shift of the Indian agricultural sector into most developed, sustainable and profitable sector which in turn will be helpful in pursuing the SDG goal 2030. Finally, the policy makers can refer it for designing the policy for maintaining the coordination among financial markets, agri-entrepreneurs and Government with an objective to influence and motivate the investors to come forward for massive scale of investment in agricultural sectors of India.

REFERENCES

Adam, C., Bevan, D., Gollin, D., & Mkenda, B. (2012). *Transportation costs, food markets and structural transformation: The case of Tanzania (Working Paper)*. International Growth Center.

Aditya, K. S., Jha, G. K., Sonkar, V. K., Saroj, S., Singh, K. M., & Singh, R. K. P. (2019). Determinants of access to and intensity of formal credit: evidence from a survey of rural households in eastern India. *Agricultural economics research review, 32*(conf), 93-102.

Ahmed, J. U., Mpanme, D., Momin, C. C., Shamsan, A. H., & Singh, K. D. (2023). Factors Affecting Access to Agricultural Finance in India: An Empirical Validation from Farmersâ€™ Perspectives. *International Journal of Regional Development*, 10(1), 1–41.

Chakrabarty, K. 2. (2013). Financial Inclusion in India: Journey so far and way forward. *Key note address at Finance Inclusion Conclave Organised by CNBC TV, 18.*

Deininger, K. (2003). *Land Policies for Growth and Poverty Reduction*. World Bank Publications.

Devaraja, T. S. (2011). *Rural credit in India-An overview of history and perspectives.*

FAO. (1999). *Better Practices in Agricultural Lending*. UN Food and Agricultural Organization.

FAO. (2017). *Addressing Land Tenure Issues in Development Cooperation*. Food and Agriculture Organization.

Finagina, O., Prodanova, L., Zinchenko, O., Buriak, I., Gavrylovskyi, O., & Khoroshun, Y. (2021). Improving investment management in agribusiness. *Estudios de Economía Aplicada*, 39(5). 10.25115/eea.v39i5.4981

Gunes, E., & Movassaghi, H. (2017). Agricultural Credit Market and Farmers 'Response: A Case Study of Turkey. *Turkish Journal of Agriculture-Food Science and Technology*, 5(1), 84–92. 10.24925/turjaf.v5i1.84-92.951

Harisha, B. N. (2017). Agricultural credit in India: Issues and challenges, 2(6).

Hassan, M., Iqbal, S., Garg, H., Hassan, S. G., & Yan, Y. (2023). An Integrated FCEM-AHP Approach for Borrower's Satisfaction and Perception Analysis of Microfinance Institution. *CMES-Computer Modeling in Engineering & Sciences, 134*(1).

Linh, T. N., Long, H. T., Chi, L. V., Tam, L. T., & Lebailly, P. (2019). Access to rural credit markets in developing countries, the case of Vietnam: A literature review. *Sustainability (Basel)*, 11(5), 1468. 10.3390/su11051468

Mao, J., Zhu, Q., Wachenheim, C. J., & Hanson, E. D. (2020). A Credit Scoring Model for Farmer Lending Decisions in Rural China. *International Journal of Agricultural Management*, 8(4), 134–141.

NABARD. (2020). *National Bank for Agriculture and Rural Development Report*. National Bank for Agriculture and Rural Development.

Nyoni, J. (2021). Achieving Economic Growth and Economic Development through Agriculture Transformation: a Framework for Enhancing Agriculture Production and Agriculture Producvity. Academia Letters, 2-5.

Puhazhendhi, V., & Jayaraman, B. (1999). Rural credit delivery: Performance and challenges before banks. *Economic and Political Weekly*, 175–182.

Sahni, M. (2020). Challenges of agriculture credit in India. [IERJ]. *International Education and Research Journal*, 6(9), 29–31.

Shobha, K., & Siji, K. (2018). Problems Faced by Farmers in Accessing Agricultural Credit with Special Reference to Malur in Kolar District of Karnataka. *IJCRT,6*(1).

USAID. (2018). *Land and Conflict: A Toolkit for Intervention*. United States Agency for International Development.

World Bank. (2014). *The Importance of Land Tenure*. World Bank Group.

Zhang, M., Li, N., & Zhang, C. (2023, August). Early Warning Method of Credit Risk of Agricultural Enterprises in Guizhou Province Based on AHP. In *Proceedings of the 2nd International Academic Conference on Blockchain, Information Technology and Smart Finance (ICBIS 2023)* (pp. 6). Atlantis. 10.2991/978-94-6463-198-2_66

Chapter 9
Design of Wheels of Agri-Rover for Both Dry and Wet Surfaces (Run-Way)

Ruby Mishra
KIIT University, India

Biswajit Bhattacharjee
KIIT University, India

Siddhant Pani
KIIT University, India

Arkaprava Mukherjee
KIIT University, India

Tamal Dey
KIIT University, India

Anish Pandey
KIIT University, India

Pratyush Chattopadhyay
KIIT University, India

ABSTRACT

The agricultural sector is susceptible to changes in output, methods, and modernization, much like other industry sectors. Farmers' methods of labouring in the fields may be altered by agricultural robots. An agri-rover is most beneficial to farmers since it reduces the time required to plant seeds in the field. It is also possible to automate other farming procedures. The goal of the rover created for this project is to provide farmers with automated planting and ploughing. The wheel and tire system is necessary for easy mobility on uneven, dry, and wet surfaces. Thus, the right wheels are chosen. A problem for this project was the apparatus's weight. The rover's weight was raised by having too many joints and links, which also made manufacturing it more difficult. For the rover to be able to withstand drastic temperature changes throughout its year-round operation, certain materials must be chosen. This is because the rover would function throughout all four seasons, from seed sowing to harvest. Ansys modelling was used to simulate stress, load, and deformation analyses.

DOI: 10.4018/979-8-3693-3583-3.ch009

INTRODUCTION

India has a largely agrarian economy. Farmers are the nation's debtors. According to the Indian Economic Survey of 2020–21, 20.2% of India's GDP came from the agricultural sector, which employed more than 50% of the labour force. Indian farmers are professionals who cultivate crops. In 2020, over 41.49% of the workforce will be employed in agriculture and related industries such as forestry, fishery, and animal husbandry, which together accounted for 17.5% of the GDP in 2016. India has the largest net cropped area in the world, ahead of the US and China (Wikipedia, n.d.). Variable weather and geographic factors pose a big obstacle to enhancing crop productivity in India, making agricultural techniques difficult. Changing weather patterns, various settings, customary methods, and financial losses as a result of insufficient knowledge about crop output are some of these difficulties. One way to get around these obstacles is to use cutting-edge technology. These challenges are overcome by implementing advanced technology in agriculture (Upendra et al., 2020).

In India, the agricultural and farming sectors are rapidly digitizing, much like any other industry. A machine, perhaps a robot, could be programmed to carry out the farmers' duties. The robot can manoeuvre in challenging conditions and bear harsh weather. The expense of inputs, the shortage of skilled personnel, the lack of water resources, and crop inspections are some of the major problems facing Indian agriculture. Advanced mechanization in agriculture sector using robots can be used to tackle this challenge (Bangar et al., 2019).

Agri-rovers are one sort of robot that this project aims to produce a prototype of. The task of sowing will be successfully completed by the rover. To virtually shape the rover, CAD/CAM programs like SolidWorks and Ansys were utilized. The Agri-rover's job of "seedling" is to plant the seeds. A seeder or seed drill is another term for a seedling equipment. A straight furrow in the soil is filled with seeds at precise rates and depths by means of seeders or seeding machines, which are used in agriculture to plant seeds for crops. In order to plant and water plants autonomously, Reddy et al. (2022) created a rover that mimics human behavior. An Arduino Nano, a DHT11 sensor, a WiFi module, and ESP8266 Node MCU controlled DC motors were employed. Their rover runs its entire setup using both solar and DC battery power, depending on availability. The dimensions and mass, material selection, and the rover's multipurpose nature comprised our challenge definition. Heavy weight and numerous joints and linkages made up the original chassis design. The production of a rover like that would be challenging. For the purpose of dispersing the seeds evenly along the field's length, a screw conveyor would be utilized [Figure 1].

Figure 1. Screw conveyor model for seedling mechanism (Dellinger et al., 2016)

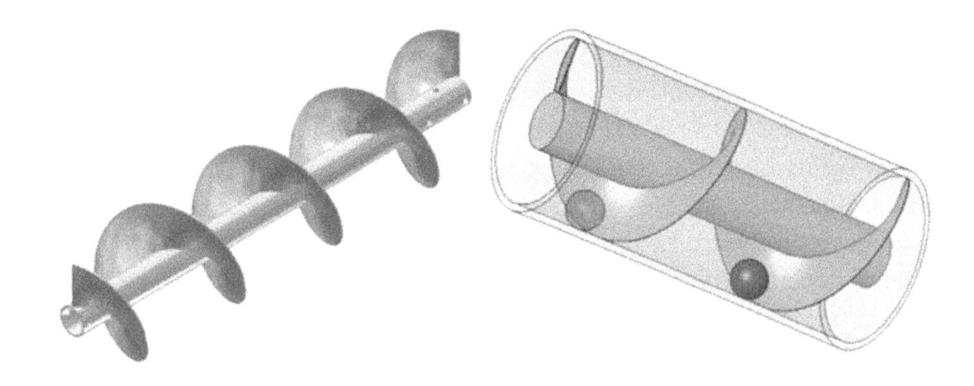

LITERATURE REVIEW

Sieling and Westernberg (1963) found that the Agri-robot, a new automatic plow, once set up properly, does not require an operator. This machine holds the power to address the issue of the shortage of farm workers. Hajjaj and Sahari (2016) investigated the experimental nature of agriculture robots and the reasons associated for agri-robots not being implemented in large scales. Their extensive review and analysis concluded that, practical agriculture robots rely not only on advances in robotics, but also on the presence of support infrastructure. Manderson and Hunt (2013) researched that precision agriculture technology, although a rich approach, having potential to achieve improved production efficiencies, still lacked agriculturally robust sensor delivery systems. Instead, the authors focused on the plausibility of automated ground vehicles as potential alternatives with a wider variety of sensor capabilities. Sowjanya et al (2017) proposed a model for a prototype of multipurpose autonomous agricultural robot with various functions for ploughing, seeding, leveling and water spraying. They also incorporated C programming language for the project. Parameters such as the soil condition, area covered by the robot and weight of the material for leveling were observed for the different conditions. Rogers and Fox (2020) presented a low cost open-source hardware and software platform for automated precision seed planting. It was intended to be used by researchers and practical agri-hackers. The robot platform contains Global Navigation Satellite System (GNSS) to store the geospatial location of each seed to such a point of accuracy so as to enable re-visitation of plants during growth. Bute et al (2018) conducted a project to develop a machine capable of automating operations such as seeding and dig. Additionally, it would keep tabs on the wetness and manual management. They utilized distinct electronics component with micro-controller 8051operated by DC motors. All the parameters were ultimately mechanized using an Arduino. Emmi et al (2014) proposed a system structure for individual agri-robots and robots working in fleets, with the aim to improve reliability, decrease complexity and costs and permit integration of software from different developers. Advantageous results were produced to implement new vehicle controllers and incorporate agricultural implements. Anjikar and Jha (2022) focused on multipurpose agriculture machine having capabilities for seed feeding, spraying of pesticide and fungicide, fertilizers and cutting operation.This paved the way for a more economical and multi-purpose usable equipment for the farmer. They used various past data and compiled them for their research to produce the robot. Bhos et al (2020) proposed a build-up of a robot that would perform tasks such as sowing, ploughing, cutting of grass and water sprinkling. The robot is solar-powered; gets power supply from photovoltaic-panels, so it needn't depend on external power sources. With respect to tractors and electric pumps and other farming equipment, less vitality is acceptable. The robot has efficiently materialized for activities like ploughing, seeding and grass cutting. Trojnacki and Dabek (2019) studied the properties of the mechanical modern wheeled robots notably the kinematics and the dynamics. Robots are organized into groups and for each groups kinematic qualities are analyzed. The chosen properties for the wheeled mobile robots (WMR) were: control DoFs, operational environment, mobility, maneuverability and stability of motion, dead reckoning and complexity. Bruzzone and Quaglia (2012) examined the most advanced ground mobile robot locomotion systems, concentrating on unstructured environment solutions, to assist designers in choosing the best option for certain operating situations. There were three main categories of their analysis: Wheeled, Tracked and Legged. The hybrid categories derived with the combination of the three main categories, along with the three categories were tested for ten parameters, testing their mobility, system complexity and reliablity. Tian et al (2017) contributed to the research of omnidirectional mobile platform with normal or failure mecanum wheels. The motion characteristics of the platform using the six and four mecanum wheels were analyzed. Their findings showed that the six mecanum wheels were more stable than the platform consisting of four mecanum wheels. In their analysis of two modified Mecanum wheel designs intended for challenging terrain, Ramirez-Serrano and Kuzyk (2010) highlighted the benefits and drawbacks of each design over standard Mecanum wheels. The researchers arrived at the conclusion that the proposed two-wheel ideas offer distinct benefits over the conventional Mecanum wheels, particularly in situations where the vehicle is traversing uneven terrain. Due of its omnidirectional and design changes, it is more difficult to develop and produce. K. M. Arjun (2017) explained in India currently ranks among the top two agricultural producers in the world, and agriculture plays a significant role in the country's economy. About 52% of all jobs in India are found in this industry, which also generates about 18.1% of the country's GDP. Elyasi, A., & Teimoury, E. (2022) proposed the seven enforceable policies in the suggested framework help supply chain managers achieve sustainability in this supply chain while simulating the final price of rice in the RSCI. Singh, A.et al (2021) made the Agri-Rover which is a robot that has been designed and studied in an effort to give comprehensive solutions for a variety of agriculturally related tasks. A specific mechanism, which is governed by an Arduino controller, is used by the system to carry out the seed-sowing operation. The Internet of Things (IOT) is also used to regulate the robot's movement.

DESIGN METHODOLOGY

Constraints and Objectives

The Agri-rover's design methodology is separated into two components. Initially, the Agri-rover's chassis, electrical system, sensors, and seedling tool are supposed to be constructed as the body of the vehicle. Second, the proper wheel setup would be installed. The wheel must be able to navigate uneven terrain in order to be considered acceptable. Wet environments and movement on soft terrains are also taken into account. Maintaining the rover's low weight while handling heat and chemical exposure, designing an equidistant seedling mechanism, doing stress analysis on the wheel, and choosing the optimum material for the Agri-rover's wheel and body were the main goals. It was originally intended to have a hefty chassis. Several joints and linkages make up its composition. Manufacturing it required a lot of time and effort. Manufacturing the selected material proved to be challenging and not cost-effective. With so many separate instruments for digging, spraying, and plowing, there were issues with the modular architecture. Regarding the wheel design, both the rover's mobility and the number of degrees of freedom it could achieve were reduced by the inflexible wheel configuration. The rover could not navigate the many hot and dry terrains because it had less traction. The rover was more sensitive and vulnerable to damage in difficult circumstances due to its limited shock resistance. An initial model of the rover body, having isometric, side, top and front views is shown below [Figure 2].

Figure 2. A model of the agri-rover chassis in SolidWorks

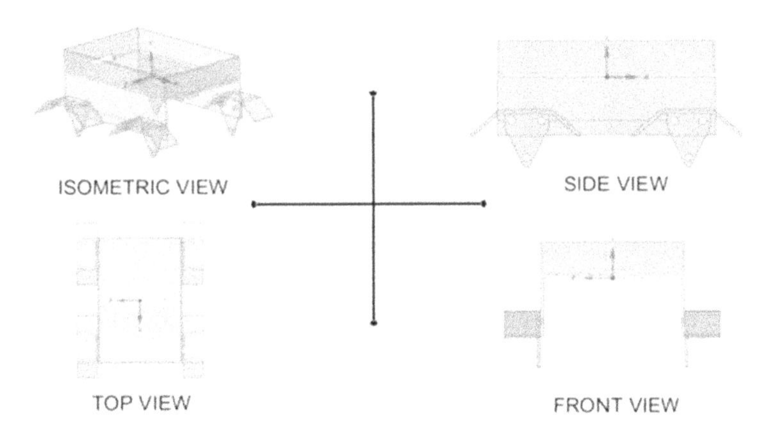

Aluminum scrap was chosen as the material for the rover's chassis. A non-ferrous metal is aluminum. It has a low density (2.7 g/cm3), strong thermal and electrical conductivities, and resistance to corrosion in a few typical conditions, such as the surrounding air (Balasubranian, 2020). It was desired to obtain 5 kg of aluminum scrap. Aluminium has no cost because it was obtained from and machined in a laboratory. The plastic substance acrylonitrile butadiene styrene (ABS), which has the chemical formula (C8H8)x·(C4H6)y·(C3H3N)z, was another substance that was taken into consideration. In thermoplastic polymer form, that is. Impact resistance, toughness, and rigidity are among the advantageous mechanical features of ABS plastic as compared to other common polymers. To enhance the previously listed qualities, a number of changes can be made to the material. The purchase of ABS plastic would have cost Rs. 3200, and the price of 3D printing will make the cost even higher. The problem with aluminum was that it would have made the rover chassis extremely heavy, which went against the goal of creating a lightweight rover. Hence, neither aluminium nor ABS plastic were chosen for the material of the rover body.

Mecanum wheel type were initially selected for the wheels of the rover. Mecanum wheels boast superior omnidirectional capability in comparison to other wheel types for rovers. A conventional six roller mecanum wheel that can traverse laterally is as shown [Figure 3].

Figure 3. A typical mecanum wheel with six rollers (Tian et al., 2017)

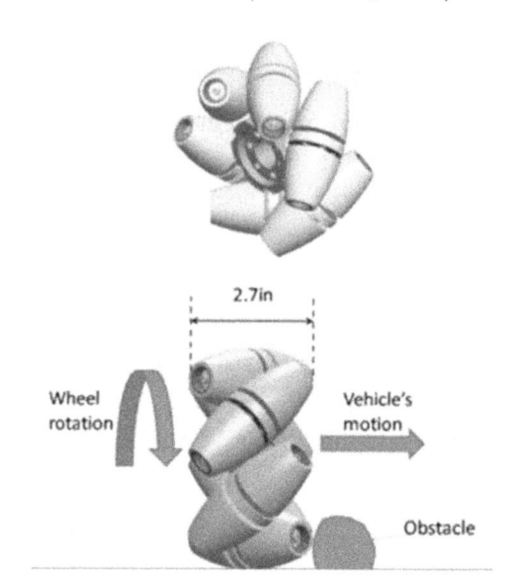

During Mecanum wheel experiments, it was found that the wheels could not effectively go past obstructions when attempting to move laterally (Tian et al., 2017). These tests were conducted with varying loads on irregular surfaces. Mostly the Mecanum wheels were used for rovers moving on smooth surfaces. The major disadvantage that was faced was that the integration of Mecanum wheels would shoot up the cost of the rover, rendering it cost ineffective. That would have been counter to the objective of cheaper and affordable rover to the farmer. Even though, the omnidirectional ability of the Mecanum wheel were superior, it failed to perform under loads of varying degree. Hence, the idea of integrating Mecanum wheel type into the rover system was dropped and a new wheel type was being searched.

Alternative Components and Progress

Alternatively, the material for the chassis body was selected to be polycarbonate. The polycarbonate would be acquired in the form of sheets [Figure 4].

Figure 4. Polycarbonate sheets as material for rover chassis

The polycarbonate material is a kind of thermoplastic polymer group. In their structures, carbonate groups are present. Polycarbonate is 1200 kg/m3 in density, 2.378 GPa in elastic modulus, 0.37 Poisson ratio, 62 MPa tensile strength, and 41 MPa shear strength. Polycaronates are used in engineering for their strength and toughness, and some grades of polycarbonate are also optically transparent. Another advantage, due to which, the polycarbonate was chosen as the material for the rover chassis, is, polycarbonate is available in the form of sheets. They could be procured easily from the market. Since, they are lightweight and also cost less, around Rs. 820, the objective was fulfilled.

The electrical components consist of four motors of 3500 rpm for no load condition, a RF300FA DC Motor, a SG90 servo motor for the seedling purpose, another SG90 servo motor for the moisture sensor, one 9 volt 450 milliampere-hour (mah) battery and another four 1.5 volt, 2000 milliampere-hour and an ARDUINO-UNO. There is also a HC SR04 Ultrasonic Sensor. The requisite code for the functioning of ARDUINO-UNO is:

Computer Code for ARDUINO UNO

```
#include <HCSR04.h>
#include <AFMotor.h>
#include <Servo.h>
AF_DCMotor motor1(1);
AF_DCMotor motor2(2);
AF_DCMotor motor3(3);
AF_DCMotor motor4(4);
Servo myservo;
HCSR04 hc(A0, A1);
int d = 0;
uint8_t i;
int pos = 0;
```

```
void setup() {
  // put your setup code here, to run once:
  myservo.attach(10);
  pinMode(A0, OUTPUT);
  pinMode(A1, INPUT);
  Serial.begin(9600);
motor1.setSpeed(100);motor2.setSpeed(100);motor3.setSpeed(100);mo-
tor4.setSpeed(100); motor1.run(RELEASE);motor2.run(RELEASE);motor3.
run(RELEASE);motor4.run(RELEASE);
}
void loop() {
  // put your main code here, to run repeatedly
  d = hc.dist();
  delay(60);
  Serial.println(d);
  if(d < 10){
    motor1.run(RELEASE);motor2.run(RELEASE);motor3.run(RELEASE);mo-
tor4.run(RELEASE);
    delay(1000);
    motor1.run(FORWARD);motor2.run(FORWARD);motor3.run(BACKWARD);mo-
tor4.run(BACKWARD);
    delay(1000);
    motor1.run(RELEASE);motor2.run(RELEASE);motor3.run(RELEASE);mo-
tor4.run(RELEASE);
    delay(1000);
    if (d < 10) {
    motor1.run(BACKWARD);motor2.run(BACKWARD);motor3.run(FOR-
WARD);motor4.run(FORWARD);
    delay(2000);
    motor1.run(RELEASE);motor2.run(RELEASE);motor3.run(RELEASE);mo-
tor4.run(RELEASE);
    delay(1000);
    }
    if (d < 10) {
    motor1.run(BACKWARD);motor2.run(BACKWARD);motor3.run(FOR-
WARD);motor4.run(FORWARD);
    delay(1000);
    motor1.run(RELEASE);motor2.run(RELEASE);motor3.run(RELEASE);mo-
tor4.run(RELEASE);
    delay(1000);
    }
  }
  else{
    motor1.run(FORWARD);motor2.run(FORWARD);motor3.run(FORWARD);mo-
tor4.run(FORWARD);
    myservo.write(90);
```

```
    delay(1000);
    myservo.write(0);
    delay(1000);
  }
}
```

An assortment of caster wheels was used as the wheel type. Strong, resistant to corrosion, waterproof, and weatherproof are the attributes of the caster wheel type. Furthermore, it is a great facilitator of mobility. Material handling equipment, medical beds, office chairs, and retail carts are just a few of its applications. The Agri-rover prototype can use it. Caster wheels must be applied with precision, taking into account both the weight that they are intended to bear and the terrain on which they will be used. Excellent floor protection and the capacity to roll over a range of obstacles are provided by air-filled rubber wheels. Transferring sensitive material over a range of surfaces is made possible by pneumatic wheels, which cushion and absorb stress. Therefore, pneumatic caster wheels were employed for the function of wheels. In all weather situations, whether wet or dry, rubber material offers superior grip and exceptional friction, enabling control and stopping force. Given enough pressure, it can flex and return to its former shape since it is elastic. Because of their durability, rubber tires can endure a lot of use. Rubber can withstand repeated flexing and rolling since it is a resilient material. Rubber is easier to form into a spherical shape than other materials. AISI 1020 is used for the caster arrangement [Figure 5].

Figure 5. Caster arrangement made of AISI 1020 in SolidWorks

AISI 1020 has the following characteristics: a mass density of 7900 kg/m^3, a yield strength of 351.571 MPa, and a tensile strength of 420.507 MPa. It contains 420 J/kgK of specific heat. The inner rim of the wheel is made of Aluminium 6061 alloy. The alloy has medium to high strength, good corrosion resistance, excellent weldability, machinability and workability. Hence, it was suitable for the inner rim of the wheel.

RESULTS

The simulation results of the rover body or rather the chassis was done in Ansys and the results are shown below [Figure 6].

Figure 6. Analysis of rover chassis under loads: a) Equivalent elastic strain, (b) equivalent stress, (c) total deformation

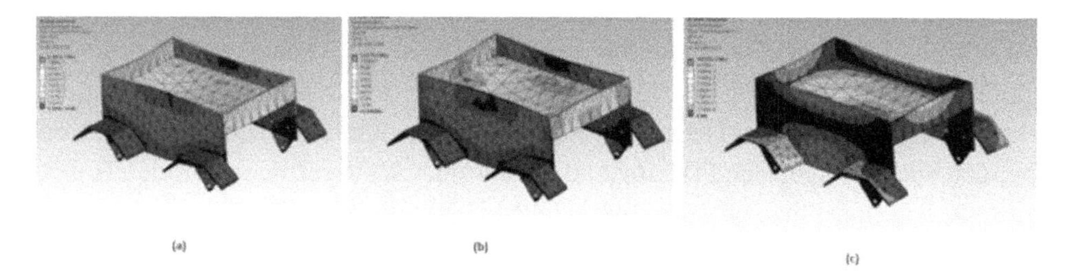

The wheel system of the Agri-rover prototype was tested for the inner rim and the tyre individually. The analysis results for inner rim are shown in figure 7.

Figure 7. Analysis of inner rim: (a) Displacement rate, (b) tensile stress acting perpendicularly to the axis

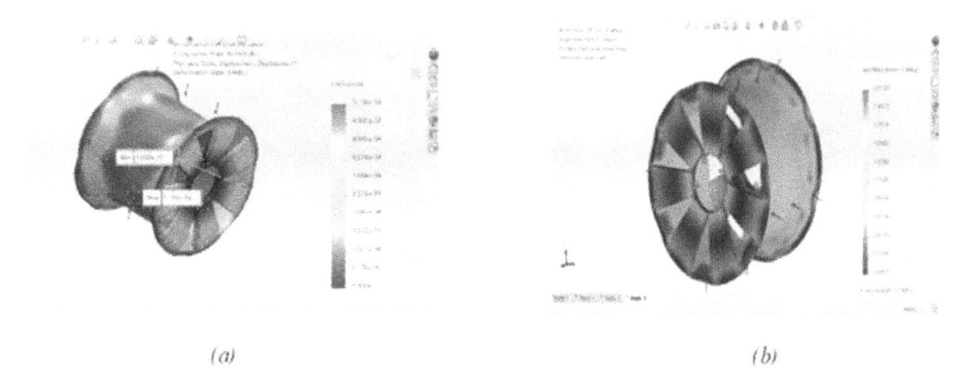

The selected materials for our wheel rim, caster body, and tire were aluminum 6061 alloy, AISI 1020 steel, and natural rubber, respectively. The tensile stress of the rim was accurately determined to be 55 MPa.

The prototype's grippers, which are employed for traction on the wheel surface, are strong enough to sustain a direct impact of a load approaching 50 N.

FUTURE WORK

In future research or engineering work, multiple tools for different agricultural functions can be incorporated. In this project work, only the seedling mechanism is incorporated, i.e, the sowing of seeds work is automated. For later prospects, weeding, harvesting, and ploughing functions can also be automated, all functions being integrated into one single chassis. The physical labour of the farmer is hugely minimized. All the field works could be automated. Further, such automated rovers could be mass produced and with the help of marketing, conventional farming functions could be replaced with automated tools. Concerning the wheel system of the rover prototype, the system could be developed further by incorporating effective braking system, but cheaper, since it is the objective.

CONCLUSION

In this research, the findings of the prototype study indicate that autonomous systems generally offer greater flexibility compared to conventional systems. This flexibility translates into significant advantages such as reduced labour expenses and constraints on daily working hours, leading to substantial enhancements. As a result, many critical operational processes have been successfully automated. The primary goals of achieving a low-cost and efficient Agri-rover have been successfully accomplished. Procured materials are readily available, and the Agri-rover's primary function of seedling is operational. The wheel system is designed. The stress concentration in the rim was found at the inner side of the rim spoke to be 369 Mpa when a direct tensile load of 50N was applied. Stress concentration on the tire arrangement was found to be 31420 Mpa at the junction of the post and the connector. The grippers used for the traction on the wheel surface could withstand a direct impact of load close to 50N and is sufficient enough for our prototype.

The materials finalised for the wheel rim, caster body and tyre where aluminium 6061 alloy, AISI 1020 steel and Natural rubber respectively. This underscores the effectiveness and feasibility of employing autonomous systems in agricultural contexts, promising improved efficiency and productivity in the future.

REFERENCES

Anjikar, A. D. & Jha, V. C. (2022). *Operational Improvement with Advanced Design of Agri Robot in The Era of Agriculture*. Academic Press.

Arun, B. (2017). Indian agriculture-status, importance and role in Indian economy. *Journal for Studies in Management and Planning*, 3(12), 212–213.

Balasubranian, R. (2020). *Callister's Materials Science and Engineering*. Wiley.

Bangar, S., Shelar, P., Alhat, P., & Budgujar, R. (2019). *Multipurpose Agri Robot*. Semantic Scholar.

Bhos, C. D., Deshmukh, S. M., Bhise, P. A., & Avhad, S. B. (2020). Solar Powered Multi-Function Agri-Robot. *International Research Journal of Engineering and Technology (IRJET), 7*(6).

Bruzzone, L., & Quaglia, G. (2012). Review article: Locomotion systems for ground mobile robots in unstructured environments. *Mech. Sci.*, 3(2), 49–62. 10.5194/ms-3-49-2012

Bute, P. V., Deshmukh, S., Rai, G., Patil, C., & Deshmukh, V. (2018). Design and Fabrication of Multipurpose Agro System. *International Journal of Emerging Trends in Engineering Research.*

Dellinger, G., Terfous, A., Garambois, P. A., & Ghenaim, A. (2016). Experimental investigation and performance analysis of Archimedes screw generator. *Journal of Hydraulic Research*, 54(2), 197–209. 10.1080/00221686.2015.1136706

Elyasi, A., & Teimoury, E. (2022). Applying Critical Systems Practice meta-methodology to improve sustainability in the rice supply chain of Iran. *Sustainable Production and Consumption*, 35, 453–468. 10.1016/j.spc.2022.11.024

Emmi, L., Gonzalez-de-Soto, M., Pajares, G., & Gonzalez-de-Santos, P. (2014). New trends in robotics for agriculture: Integration and assessment of a real fleet of robots. *TheScientificWorldJournal*, 2014, 2014. 10.1155/2014/40405925143976

Hajjaj, S. S. H., & Sahari, K. S. M. (2016, December). Review of agriculture robotics: Practicality and feasibility. In *2016 IEEE International Symposium on Robotics and Intelligent Sensors (IRIS)* (pp. 194-198). IEEE. 10.1109/IRIS.2016.8066090

Manderson, A., & Hunt, C. (2013). Introducing the Agri-Rover: An Autonomous on-the-go sensing rover for science and farming. In *Proceedings of the 26th Annual FLRC Workshop Held at Massey University.* Massey University.

Ramirez-Serrano, A., & Kuzyk, R. (2010, March). Modified mecanum wheels for traversing rough terrains. In *2010 Sixth International Conference on Autonomic and Autonomous Systems* (pp. 97-103). IEEE. 10.1109/ICAS.2010.35

Reddy, G. N., Reddy, G. M., Balaji, G., Harish, C. M., & Kethan, C. H. (2022). *IOT Based Seed Planting and Watering Rover*. Academic Press.

Rogers, H., & Fox, C. (2020). An open source seeding agri-robot. *Proceedings of the 3rd UK-RAS Conference.*

Sieling, C., & Westenberg, H. E. (1963). *The Agri-Robot A Revolution in Plowing (No. 630306)*. SAE Technical Paper.

Singh, A., Mishra, R., Sarkar, A., Singh, A. K., Toppo, D., & Kumar, A. (2021). Design and Analysis of Agri-rover for Farming. In *Current Advances in Mechanical Engineering: Select Proceedings of ICRAMERD 2020* (pp. 657-663). Springer Singapore.

Sowjanya, K. D., Sindhu, R., Parijatham, M., Srikanth, K., & Bhargav, P. (2017, April). Multipurpose autonomous agricultural robot. In *2017 International conference of Electronics, Communication and Aerospace Technology (ICECA)* (Vol. 2, pp. 696-699). IEEE. 10.1109/ICECA.2017.8212756

Tian, Y., Zhang, S., Liu, J., Chen, F., Li, L., & Xia, B. (2017). Research on a new omnidirectional mobile platform with heavy loading and flexible motion. *Advances in Mechanical Engineering*, 9(9), 1687814017726683. 10.1177/1687814017726683

Trojnacki, M., & Dąbek, P. (2019). Mechanical properties of modern wheeled mobile robots. *Journal of Automation Mobile Robotics and Intelligent Systems*, 13(3), 3–13. 10.14313/JAMRIS/3-2019/21

Upendra, R. S., Umesh, I. M., Varma, R. R., & Basavaprasad, B. (2020). Technology in Indian agriculture–a review. *Indonesian Journal of Electrical Engineering and Computer Science*, 20(2), 1070–1077. 10.11591/ijeecs.v20.i2.pp1070-1077

Wikipedia. (n.d.). *Agriculture in India*. Wikipedia. https://en.wikipedia.org/wiki/Agriculture_in_India

Chapter 10
Digitalization of SCM in the Agriculture Industry

Debankur Das
KIIT University, India

Anirban Roy
KIIT University, India

Ayan Chaudhuri
KIIT University, India

Sushanta Tripathy
http://orcid.org/0000-0003-2470-4080
KIIT University, India

Deepak Singhal
KIIT University, India

P. Chandrasekhar
KIIT University, India

ABSTRACT

The study examines the impact of digitization on traditional agri-food supply chains, focusing on the utilization of Industry 4.0 and the internet of things (IoT). Precision agriculture, coupled with IoT, addresses challenges in the industry, aiming to enhance productivity. Smart farming leverages drones, AI, big data, and IoT to optimize farm operations, utilizing RFID and barcodes for data collection. IoT sensors enable informed decision-making by monitoring crop conditions. AI and IoT streamline post-harvest processes, enhancing food safety and reducing waste through automation. Industry 4.0 solutions integrate big data, IoT, and mechatronics for real-time monitoring, facilitating agile supply chain management. The study extends its insights to India's agri-food supply chain digitalization, emphasizing hurdles like limited budgets, adverse weather, and farmer reluctance. It underscores cybersecurity, and environmental concerns, and advocates for government support, education, and awareness initiatives.

INTRODUCTION

Understanding Supply Chain: A supply chain is defined as "a group of three or more businesses or people directly involved in the financial, informational, and product flows from a source to a client, both upstream and downstream" (Mentzer et al. 2001). Producers, vendors, transporters, storage facilities, distributors, sellers, and even consumers collectively constitute the supply chain. Each item available in the consumer market undergoes various business-to-business transactions from its raw form to its final state. For instance, when a consumer purchases a bottle of Coca-Cola, they don't acquire it directly

DOI: 10.4018/979-8-3693-3583-3.ch010

from Coca-Cola; instead, the product goes through several intermediary stages involving Coca-Cola, wholesalers, and retailers, until it reaches the end consumer. The intermediary could be a hypermarket or a neighborhood store (Căescu and Dumitru, 2011).

Understanding Digitalized Supply Chain: A fully integrated use of digital technologies like blockchain, AI, and IoT to replace conventional supply chain management techniques is referred to as a "digitized supply chain." Through real-time data sharing and analysis, it connects all stakeholders—from suppliers to consumers—and improves efficiency, visibility, and adaptability. Data analytics, automation, predictive analytics, and demand-driven operations are important components that promote flexibility, cut expenses, and make customer-focused and sustainable practices possible. It's a dynamic approach that helps modern companies stay resilient and competitive in a fast-moving, data-driven global market. These days, discussing digital supply chains would be impossible to do without bringing up the more general subject of Industry 4.0 development (Galati and Bigliardi, 2019). A new ecosystem built on the connections between and among the many functional divisions of an organization is made possible by this new paradigm. Actually, companies are changing their approaches to become more open about all aspects of their operations, including supply chain management (Seyedghorban et. al, 2020). The integration of technologies like Augmented Reality, Robotics, and Cloud Manufacturing is facilitating "smart" information sharing. Strategic initiatives involving the deployment of several technologies to enhance communication and process integration are being driven by the study of this data (Pereira et. al,2017).

Scope of Digitalization: Industry 4.0 and the process of digital transformation hold the potential to evolve a fully digitized supply chain by enhancing transparency through centralized activity, as demonstrated in reference (Preindl et. al, 2020). Additionally, Industry 4.0 can profoundly influence the organization of supply chains and various sustainability-related aspects, as highlighted in reference (Machado et.al, 2020). Reference (Jabbour et. al, 2018) enumerates the ongoing challenges that require careful resolution, including the classification of Industry 4.0 technologies conducive to sustainable operations management decisions, the promotion of collaboration within supply chains, and the establishment of performance facilitators for attainable, incremental objectives. Despite being primarily rooted in production foundations, as suggested by reference (Müller et. al, 2018), Industry 4.0 still lacks comprehensive integration with supply chain management (SCM) from its perspective. Moreover, as (Manavalan and Jayakrishna, 2019) indicate, there is still a dearth of studies on the consequences of the fourth industrial revolution for supply chains. As noted in reference (Ghadge et.al, 2020), traditional supply chains need to quickly adjust in order to successfully integrate the concepts of Industry 4.0 technologies, even as businesses continuously strive to adjust to these new technologies. Only then can they hope to remain competitive in ever-changing and rapidly growing markets. In alignment with this perspective, (Scuotto et. al, 2017) argues that there is inadequate data to substantiate the notion of a digital shift concerning collaboration.

Digitalized Agri-Food Supply Chain: The global agricultural and food sector is highly complex, presenting challenges across various jobs, activities, and procedures. Moreover, it is becoming increasingly inefficient due to rising demands and imposed limitations, highlighting the need for innovative Agro-Food solutions. The Agro-Food sector, like many others, relies heavily on technology for both operational and decision-making processes. Stakeholders in Agro-Food include manufacturers, producers, retailers, and governmental and policy-making bodies, all of which are integral to addressing global challenges and implementing sustainable solutions (Panetto et. al, 2020).

In particular, the food supply chain might be greatly impacted by the Internet of Things (IoT), which depends on sensors to gather data. According to (Li et. al, 2015), today's food supply chain is highly complicated and scattered, with numerous stakeholders, different geographical and temporal dimensions, and extensive operational procedures, all of which pose problems to operational efficiency, public food safety, and quality control. The use of sensors and IoT-enabled devices is key in controlling the food and perishable commodities supply chain since they provide vital data on commodity conditions.

Precision agricultural technologies, frequently discussed in the context of agriculture's digitalization, either reduce overall input costs or enhance productivity and yield (Aiello et. al, 2019). It is expected that rising digital technologies such as real-time monitoring, big data analytics, machine learning, and cloud computing will drive significant changes in agricultural systems (Wolfert et. al, 2017).

This research endeavors to explore various new technological developments that currently assist or have the potential to aid different segments within the entire Agri-Food Supply Chain. The Authors will concentrate on emerging farming methods that promise to optimize the entire supply chain, from processes to customer satisfaction, thereby enhancing productivity.

BACKGROUND

The authors of this chapter examine the digital transformation currently reshaping agricultural supply chain management. While agriculture has historically been slow to adopt new technologies, a confluence of factors is driving a digital revolution in the industry. The Fourth Industrial Revolution (Industry 4.0) is transforming our daily lives, including food, exercise, and personal routines, through developing technology. Furthermore, all sectors of the economy, including industry, agriculture, trade, mining, defense, and politics, face evolving. Opportunities and challenges posed by the advancement of robotic technologies. Digitalization requires a greater emphasis on intellectual capital across several areas, including human life and the economy (Amentae et. al, 2021).

Food system research has long been a priority on the global agenda. Stakeholders in the food sector must identify solutions to transition to more sustainable and resilient food systems.

The United Nations (UN) Food Systems 2021 Summit (The Tokyo Nutrition for Growth (N4G)) emphasized the importance of changing policies and strategies utilized to feed the globe in the 20th century. The Summit concluded that without decisive food system transitions, it is impossible to achieve the goals of universally accessible, affordable, and healthy diets that are environmentally, economically, and socially sustainable (Ringler et. al, 2023).

Recent studies highlight the importance of digitalization for sustainable supply chain innovation.

However, there are limited study findings on SCM sustainability and digital hurdles. Less research is found on sustainable innovation and impediments to SCM digitalization in underdeveloped economies, especially in India. Despite its social importance and the necessity for digitization, SCM has not received adequate attention (Yawar et. al, 2018). There is still more effort to be made to address social barriers to sustainable innovation and digitized supply chain management. By providing a comprehensive background on the driving forces and digital tools, the chapter aims to set the stage for a deeper discussion on the impact of digitalization on agricultural SCM and its far-reaching implications for the future of food production.

Figure 1. Implementation of AI & IoT in major aspects of the agri-food supply chain

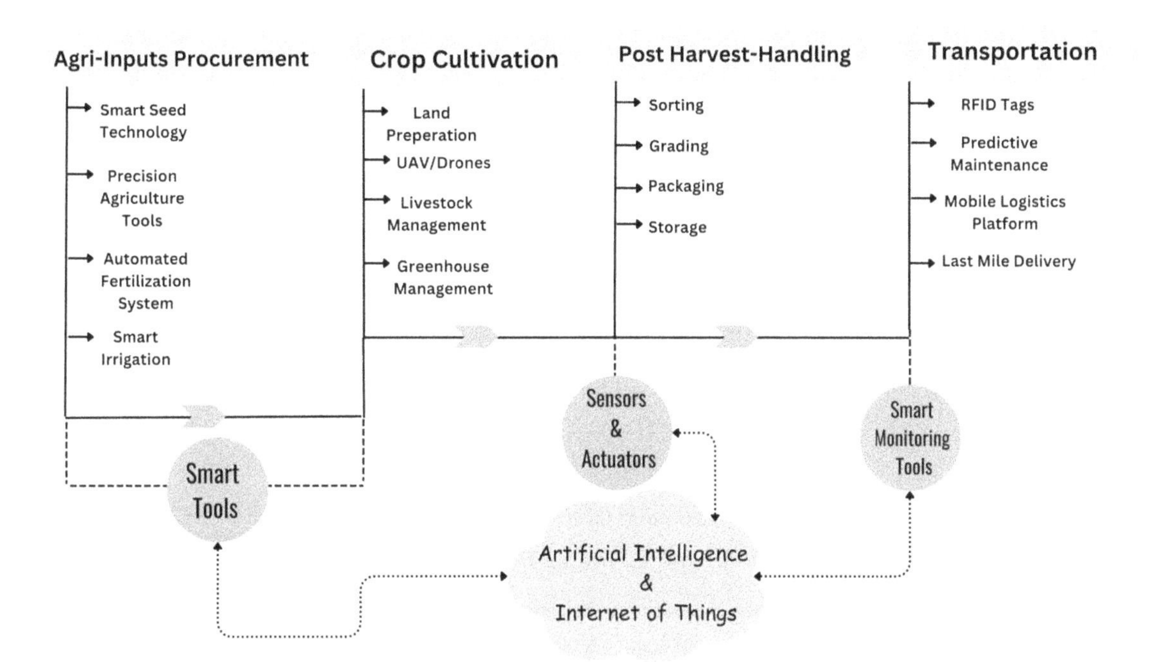

AGRI-INPUTS PROCUREMENT

At the heart of smart farming lies the incorporation of innovative technologies into agricultural methods. Utilizing drones, artificial intelligence, big data, the Internet of Things, satellites, and other advancements transforms traditional farming practices into "smart" agriculture, empowering farmers to optimize productivity and attain superior results. Consequently, agribusiness benefits from enhanced cost-effectiveness through reduced reliance on human labor decreased financial expenditures, and increased production volumes.

Smart Seed Technology: The most common input in agriculture is seed. Merely utilizing high-quality seeds can significantly boost farm yield by 15-20% and achieve food security. The goal of seed enhancement technology is to increase seed performance when used with particular planting tools and under particular conditions.

Numerous cutting-edge international seed companies, like Syngenta, Pioneer, and Monsanto, possess sophisticated breeding information management systems and portable data collection devices. The handheld breeding data collection system swiftly records and transmits field observation data, including pictures and statistics. Thus, following data collection, there is no need to manually restructure the data. Breeders can increase the effectiveness of data collecting by entering images and other details about a plant's progress into the database while they are out in the field (Han et. al, 2017). In these integrated systems, identifying technologies like RFID (Radio Frequency Identification) technology and barcodes are frequently utilized. These systems can provide benefits including fewer data entry errors, automated farming equipment, lower labor expenses, and increased overall output (Samad et. al, 2010).

Field data collecting efficiency can be greatly enhanced with the use of mobile communication technology. Information on farming operations can now be effectively gathered by using portable devices for data gathering and uploading. The present study developed an efficient system for gathering agricultural trait information for precision breeding management using smartphone technology and multi-tag-based identification technologies. This system, known as the Golden Seed Breeding Platform (GSBP), combines an electronic label with a crop breeding data management system. It integrates barcode technology, RFID or Near-Field Communication (NFC), and mobile devices to increase timely data uploading and simplify the data collection procedure (Han et. al, 2017).

Figure 2. Functional framework of IOT-enabled farming and processing systems

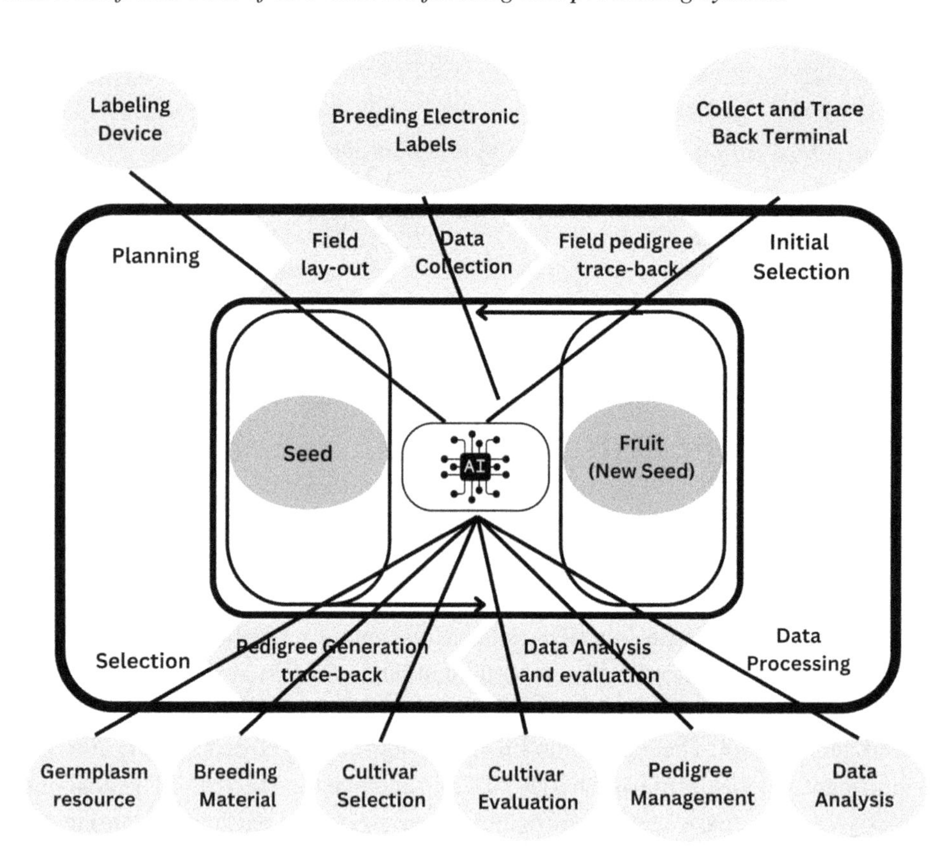

Precision Agriculture Tools: Precision agriculture (PA) uses Internet of Things (IoT) sensors and both near- and far-reaching sensing techniques to monitor crop conditions throughout the growing cycle. PA entails collecting and analyzing a large amount of data about crop health. Plant health is influenced by a variety of elements, including temperature and water availability. Through PA, farmers can identify the specific characteristics necessary for optimal crop health, their precise locations, and the appropriate quantities needed at any given time. This requires gathering extensive data from multiple sources and areas within the field, including soil nutrient levels, insect and weed presence, plant chlorophyll levels, and specific meteorological conditions. Analyzing this data is essential for generating agronomic recommendations. For instance, the chlorophyll concentration, or "greenness," of a plant indicates its nutrient requirements based on its growth stage. The plant's location's soil properties, the weather forecast, and this information are all merged. The total amount of data gathered is then utilized to calculate the appropriate amount of fertilizer to be applied to that plant the following day. Improving yields mostly depends on getting agronomic information to farmers at the appropriate time and making sure they follow the advice (Shafi et. al, 2019).

WSN: A Wireless Sensor Network (WSN), comprising multiple interconnected wireless nodes designed to monitor environmental parameters, serves as the primary source for Precision Agriculture (PA). Each wireless node has a radio transceiver, a microprocessor, one or more sensors, an antenna, and extra circuitry for connecting to a gateway and transferring sensor data (Wang et. al, 2006). Over the last decade, a number of WSN applications have evolved for remote crop health monitoring. An example is a cyber-physical system designed to monitor potato harvests. Intelligent systems are made up of hardware, software, and physical components that work together to detect changes in their surroundings. The system comprises three layers: the physical layer for collecting sensory data, the network layer for transmitting data to the cloud, and the decision layer for processing and analyzing data to derive observations-based decisions (Rad et. al, 2015).

Automated Fertilization System: A fertilizer is a chemically rich synthetic or natural substance that is applied to plants to improve their growth and yield. Fertilization is often accomplished by hand spraying. Sensing abilities are necessary for the ideal method of fertilization, nevertheless, in order to determine the precise location of the fertilizer's demand, the chemical components that are lacking, and the required quantity. To increase productivity, fertilizers must be applied in a very precise quantity. In (Wang et. al, 2011), an automated fertilization system was introduced that measured the fertility of the soil in real-time using sensors. Three components make up the system: decision support, output, and input. Using real-time sensor data, the resolution support element calculated the ideal amount of fertilizer required for plant growth.

Smart Irrigation: Smart irrigation represents a form of artificial irrigation that regulates water usage by identifying areas with the highest water requirements. It stands as a critical element in agriculture, significantly impacting crop productivity, costs, and health. Given that the majority of the world's countries are experiencing a shortage of water, one important component of smart irrigation is preventing water waste. A wireless sensor network was used to provide an energy-efficient irrigation system for growing crops that efficiently controlled water flow based on surrounding variables (Nikolidakiss et. al, 2015). This system used previous data together with sensor data for heat, moistness, and atmospheric pressure to determine the amount of water required for typical irrigation.

Smart Pest Control: The primary reason for the agriculture sector's low productivity is pest infestations. These pests cause a number of dangerous diseases that hinder the growth of the plants. On the other hand, disease prediction gives farmers advance notice, allowing them to take the necessary actions to promptly contain the disease. Electronic gadgets that allow people to recognize traps within a certain range are the foundation of pest control systems (Mahlein et. al, 2012). A plant disease and pest prediction system based on Internet of Things technology was introduced with the aim of reducing the overuse of pesticides and fungicides (Lee et. al, 2017). To establish a correlation between pest proliferation and weather patterns, weather condition monitoring sensors are utilized. These sensors measure various weather parameters including temperature, dew, humidity, and wind speed to monitor weather conditions. The sensors are placed in orchards, and the cloud receives the data that these sensors gather. The concerning state of the pest attack on the crops is communicated to the farmer.

CROP CULTIVATION

Through the introduction of smarter apps, an AI-powered IoT ecosystem offers enormous potential to improve farming practices by controlling and enhancing precision. The extent of these technologies' recent improvements is infinite in agricultural techniques since they can automate difficult jobs with little assistance from hand (Subeesh and Mehta, 2021).

Land Preparation: In contemporary agriculture, digitalized land preparation is becoming more and more important, completely changing how farmers approach the cultivating process. This technology revolution optimizes farming techniques by utilizing data, automation, and precision. Digitalized land preparation is crucial to farming since it can improve production, sustainability, and efficiency. Farmers may make well-informed judgments on soil quality, moisture levels, and terrain by utilizing digital tools and data analytics. This allows them to use resources like water, fertilizers, and pesticides sparingly. It also makes soil health management possible, which lessens the impact on the environment and protects priceless resources. The need for productive and sustainable agriculture is a global concern, but digitalized land preparation is a game-changer that gives farmers the ability to meet these needs while streamlining their operations.

Navigation and Performance of Tractors: To carry out a variety of agricultural chores, modern farmers employ a variety of farm machinery and equipment. The tractor-implement system's performance monitoring is an essential component of the agricultural mechanization system. In industrialized nations, tractors are now equipped with sensors and an inbuilt navigation system that monitors every aspect of the field, both macro and micro. These days, with farms becoming more and more networked and Internet-capable, it is possible to fully harness the capabilities of the Internet of Things and related technologies to monitor tractor performance. A general system comprising hardware and software components keeps an eye on the tractor's navigation and performance. The necessary sensors for determining fuel flow, power consumption, and geolocation are included in the hardware part. The data collected by these sensors is transferred to a printed circuit board (PCB) or processing unit. The data is then sent to the network using an appropriate communication technique, such as a Low Powered Wide Area Network (LPWAN). On the software side, online applications are combined with a scalable, real-time database (Civele, 2019). Major tractor manufacturers, such as John Deere and Case IH, have launched autonomous tractors to worldwide markets. Case IH's prototype tractor uses cameras and LiDAR (Light Imaging, Detection, and Ranging) technologies to accurately identify obstacles. Other object identification models used in agricultural applications include SSDs (Liu et. al, 2016) and Faster RCNN (Ren et. al, 2016).

UAV/Drones: Initially employed for military objectives, drones have progressively transitioned into the agricultural domain. Drone technology has made significant progress in automating various agricultural operations, including pesticide application and land monitoring. One category of Unmanned Aerial Vehicles (UAVs) that aren't piloted by humans is agricultural drones. The ground station is in charge of utilizing protocols like Mavenlink to facilitate communication with the drone. Most ground control stations include an interface for monitoring drones. Hardware components are critical in managing the UAVs' roll, pitch, and yaw. The drone device is normally made up of a central processing unit, a collection of sensors (such as a laser, radar, camera, gyroscope, accelerometer, compass, and GPS receiver for environmental data), and actuators and motors that perform critical operations. Communication is facilitated through remote control and radio frequency range communication. Equipped with thermal and multispectral sensors, UAVs can survey hectares of fields in a single journey (Hoummaidi et. al, 2021).

Figure 3. High-level architecture of drones

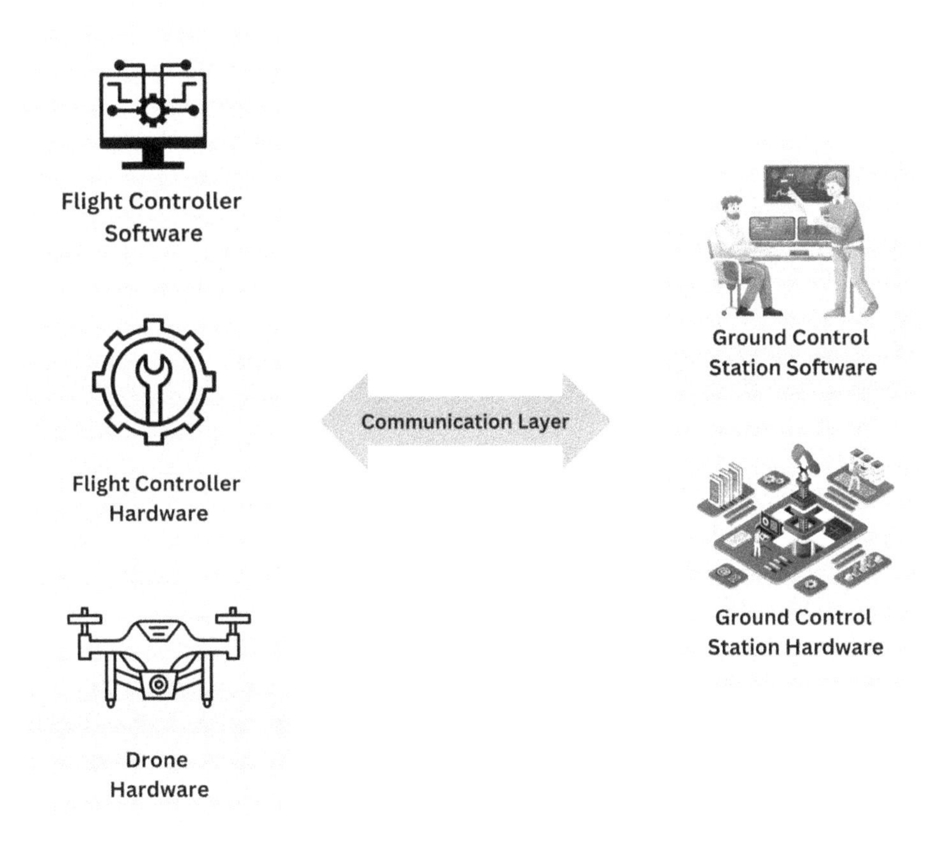

Livestock Management: Livestock is essential to suburban development and means of maintenance. Research indicates that a significant yield gap exists because farmers continue to use antiquated methods. The main issue is how to coordinate and manage this cattle industry's expansion so that the yield gap is as small as possible. As a result, there is a chance for research using cutting-edge technology that could lead to a sustainable solution. By automating every step of the monitoring, analysis, and decision-making process, the Precision animal Farming (PLF) system ensures the health and welfare of the animals. By encouraging bird mobility, autonomous robots can improve the health of birds. The trash on the chicken floor can be aerated by sophisticated robots. This will lessen the likelihood of infections and stop illnesses like Salmonella. The feed intake can be carefully controlled by the precision feeding systems. More viable eggs are produced by precision-fed broiler breeders, according to recent studies (Zuidhof et. al, 2017). Cattle condition data can be gathered by automated systems that use sensors such as temperature, pedometer, accelerometer, and more. Their patterns and duration of activity differ from a normal condition during estrus and calving events. An artificial intelligence model can utilize the collected data to analyze patterns and detect deviations in behavior, indicating potential occurrences such as calving or estrus events in the cow. Hourly rumination time, feeding time, and resting time are the three most significant criteria that can be used to assess the condition of the cattle. These are simply derived from the data collected by the accelerometer. The cattle's accelerometers on their necks and legs are used to extract the rumination and feeding times as well as the standing times. To identify whether a cow is in estrus, binary classification techniques such as logistic regression can be used (Benaissa et. al, 2020). Due to the complexity of decision-making in an IoT-based livestock management system, data analysis is frequently performed on remote servers or in the cloud. This is due to the limited processing capacity and high energy consumption of the microcontroller embedded within the collar or ear, especially during complex calculations. Livestock automation systems rely heavily on data, with most livestock data being unstructured and capable of fitting into various categories such as text, image, or audio. Statistical approaches typically perform poorly on noisy, unstructured data. While multi-layer perceptron feed-forward neural networks are useful for basic classification and pattern recognition (Galan et. al, 2018), Convolutional Neural Networks (CNN) and their variants outperform in image processing tasks.

Greenhouse Management: Maintaining the environmental conditions within the greenhouse is a time-consuming task because of the numerous variables involved. This is one of the areas where technological involvement can make farming easier than it has ever been, as these climate swings can also harm crops. Sensors can be used to measure the parameters of the local climate and current greenhouse environment. Numerous nodes in a wireless sensor network can be used for sensing, acting, and informing stakeholders about information. Typically, the design consists of a base station that is in charge of monitoring and wireless sensor network data handling sub-systems (Akkaş and Sokullu, 2017). A cloud server or remote server is better for storing data. The system's overall design has a striking resemblance to previous Internet of Things-based monitoring systems. These systems can be used for a considerable amount of actuation in addition to monitoring. The systems gather soil moisture data and can be linked using any kind of connectivity protocol, such as Bluetooth or Zigbee. When the moisture content falls below the threshold, the linked micro-irrigation system may be activated. The same exercises can be done with regard to humidity and temperature.

A public dashboard with remote access from any location can be created using the data. Planning for the next days can be facilitated by providing summary statistics on a daily basis (Ulla et. al 2018). It is advised to assess the microclimatic conditions within the greenhouses before beginning the real process of growing crops there. This aids in the best possible design of greenhouses that use less energy and have adaptable climate control systems. The gathering and analysis of data facilitates the monitoring and planning of cost-cutting measures for cooling systems, which are frequently within the reach of growers. It is possible to establish a correlation between crop growth and climatic changes using IoT and decision support systems. This correlation offers growers enhanced insights into crop growth rates before production. A model is constructed that mimics the comfort ratio for various crop growth phases by processing field data, such as temperature, humidity, vapor pressure, sun radiation, etc. With this, improved control recommendations for efficiently managing the greenhouses may be planned (Shamshiri et. al, 2020).

POST-HARVEST HANDLING

A crucial stage in the agricultural supply chain, post-harvest handling is being revolutionized by AI (Artificial Intelligence) and IoT (Internet of Things). Numerous advantages are provided by these technologies, such as increased food safety, decreased waste, and increased efficiency. AI makes it possible to automate processes like grading, sorting, and quality control, guaranteeing that only the best produce makes it to market. On the other hand, IoT devices offer real-time monitoring of storage parameters including humidity, temperature, and airflow, which helps to reduce losses and stop spoiling. In addition to streamlining operations, the combination of AI and IoT provides farmers and agribusinesses with insightful data that helps them make better decisions about how to allocate resources and forecast markets. The integration of AI and IoT in post-harvest processing becomes vital for improving food security and lowering agriculture's environmental footprint as the urge to feed a growing global population sustainably mounts. Companies in the metal-mechanical sector that offer technical solutions to the industry are involved in post-harvest activities. These businesses provide ways to preserve the quality of grains or cereals following harvesting so that retailers, the agroindustry, and end users can use them (Hossen et. al, 2020)

Sorting & Packaging: AI finds utility across food processing, storage, and delivery, where robotics and intelligent drones play pivotal roles in streamlining packing processes and ensuring efficient food delivery (Castillo and Meliif, 1970) (Bera, 2021) (Tyagi et. al, 2021). Within food processing, one of the most time-consuming tasks is accurately sorting and packaging food items. AI-based systems can effectively handle such labor-intensive tasks, minimizing errors and enhancing production rates in the industry. However, the irregular shapes, colors, and sizes of fruits and vegetables pose challenges for designing AI-based systems, necessitating a substantial amount of data to train these systems effectively (Tripathi et. al, 2020) (Dewi et. al, 2020).

AI-based decision-making systems use a variety of instruments and techniques to examine all aspects of food products, including fruits and vegetables. These include high-resolution cameras, laser technology, X-ray systems, and infrared spectroscopy. TOMRA, for instance, has been demonstrated to improve sorting and grading issues for potatoes by 5–10% (Pnishchuk, 2020) (Bünger, 2021). Sorting machine systems can be controlled using different methods, such as Arduino controllers, Raspberry Pi, Peripheral Interface Controllers (PIC) microcontrollers, or Programmable Logic Controllers (PLC) (Sheth et. al, 2010) (Kunhimohammed et. al, 2015) (Kumar and Kayalvizhi, 2015) (Sobota et. al, 2013). Arduino controllers, known for their versatility, are used for various tasks, including robotic device control, sensor management, and vision modules (Silva et. al, 2017) (Salman and Abdelaziz, 2020) (Nosirov et. al, 2020). In sorting systems, Arduino controllers automate conveyor systems driven by DC motors, stepper motors, and servo motors (Krishnan et. al, 2016). Further, it was demonstrated that a distributed computer vision system for fruit grading and inspection, which was used as a distributed network architecture to communicate with a camera unit that captured photos of fruits on conveyors and quickly transmitted them to the computer system (Yogitha and Sakthivel, 2014). The suggested system uses image processing. It primarily functions by acquiring the fruit's color and geometric details. Many processes are employed to do this, including pixel averaging, filling, filtering, edge detection, erosion, dilating, filling, and histogram color values.

(Aziz et. al, 2021) focuses on the Creation of an Artificial Intelligence-Powered Smart Sorting Device for Chili Fertigation Sectors. This project's implementation of picture pre-processing entails a number of crucial processes. Using a VGA camera module, photos of chili samples are first taken, and they are subsequently cropped into the region of interest (ROI). Image scaling is done to increase processing speed and standardize image dimensions. To improve data detail, the photos are transformed from RGB to the CIE lab color space. The Color Thresholder Application in MATLAB is used to achieve color segmentation, which isolates the chile from the background. High-pass filters are used in image sharpening to improve image edges and details. Finding the RGB values' mean, and standard deviation is the process of color extraction, which helps with machine learning—specifically, training the ANN algorithm for quicker and more accurate categorization. These pre-processing procedures are essential to raising the caliber and effectiveness of this project's feature extraction procedure.

Storage: Wireless sensor nodes can be used to monitor the quality of farm goods in storage facilities. Temperature and humidity detectors ensure that storage conditions remain optimal. Variations in sensor readings over time can be slight, prompting the calculation of cumulative values to assess humidity and temperature fluctuations in the storage environment. Each factor is assigned a threshold determined by the type of agricultural product stored. Sensor nodes link to the internet via a gateway, and data is aggregated by a distant database server. Farmers can locally monitor storage conditions by integrating data insights with a graphical user interface. In developed countries, farmers extensively employ this method, particularly in seed potato storage systems (Tervonen, 2018). Ensuring food safety involves identifying and removing pesticide residues. Convolutional Neural Network (CNN) models are effective in detecting pesticide residues in hyperspectral images of fruits and vegetables (Jiang et. al, 2019). Cold storage management systems are designed to operate within controlled environments. Long-term storage of agricultural produce is another advantage of automated cold storage systems. With the exception of automation using IoT, the system architecture is unchanged from a typical storage system. connecting a suitable alarm system or smartphone app that can support system control and local monitoring (Kumar et. al, 2018).

TRANSPORTATION

In today's industrial revolution, when firms attempt to meet consumer demands by manufacturing a diverse range of items quickly, traditional supply chain management systems are prone to unsafe delivery during the supply chain management (SCM) process (Abdirad and Krishnan, 2020). Epidemics, local or regional geographical conditions, political policies, societal norms, and stakeholder attitudes within SCM can all have a negative impact on the delivery of valuable goods to their intended destinations, including challenges such as floods, earthquakes, rain, and extreme weather conditions (Yoon et. al, 2020) (Ivanov, 2020) (Sawik, 2019) (Alora and Barua, 2019) (Ahlqvist et. al, 2020) (Luo et. al, 2019). As a result, SCM stakeholders such as farmers, enterprises, consumers, and third-party intermediaries must reduce risks in order to create resilient, cost-effective, and flexible solutions that meet the needs of modern SCM (Bottani et. al, 2019).

Given computational methods' capacity to track shipments efficiently, they prove helpful (Hatefi et. al, 2019). Recent trends show the increasing popularity of the Internet of Things (IoT), cloud computing, and blockchain technologies in handling the transportation of perishable commodities (Layaq et. al, 2019). These technologies have greatly contributed to the digitization of the entire supply chain management process, allowing stakeholders along the producer-consumer supply chain to monitor the delivery process at each stage, thanks to developments in connectivity and communication (Irfan et. al, 2019). Addressing current corporate demands and objectives, supply chain nodes offer transaction visibility, transparency, dynamic mobility, and adaptability in SCM (Zagurskiy and Titova, 2019). Moreover, the reliability and security provided by a robust SCM contributes to swift commercial prospects for both producers and consumers (Saberi et. al, 2019).

Considering the challenges outlined in modern SCM mechanisms (Wong et. al, 2020) (Wamba et. al, 2020) several questions arise: How should SCM be structured to ensure prompt shipment of agri-food goods? How can SCM stakeholders' expectations be agilely met while ensuring real-time tracking and the preservation of goods' safety to prevent spoilage? How can manufacturers and customers maintain trust while maintaining secured access to consignment log files free of tampering by arbiters?

Mechatronics: The initial component of Industry 4.0 comprises two aspects: automation and robotics (Bader and Rahimifard, 2020). Automation involves employing diverse automatic mechanized technologies, while robotics encompasses the development and deployment of robots to execute specific tasks. Every firm that wants to save expenses while increasing production volume and speed also wants to embrace automation and robotics. Automation and robotics-powered processes also result in increased productivity and a better working ambiance (Duong et. al, 2020). Robotics, a type of automation, gives firms production flexibility, which is critical when quick changes are necessary to satisfy client needs (Echelmeyer et. al, 2008). Simple automation can make a big difference in warehouse logistics by helping move things more efficiently. One example of this is the usage of automated guided vehicles (AGV). AGV utilization enables lower labor and operating expenses in addition to higher output. The ability to manage the path of the products using AGVs is perhaps its greatest advantage; this allows for easy facility rearranging as needed (Karabegovi´c et. al, 2015).

Swarm robots that are mounted on railings and move around merchandise are another application for robotics (Dekhne et. al, 2019). Additionally, robotics is used in product distribution, packing, sorting, and transportation (Karabegovi´c et. al, 2015). These mobile robots are equipped with proximity sensors and warning systems to guarantee the safety of their operating environment, which may include areas where human labor is present. Drones can also be employed for inventory management and surveillance within production facility warehouses.

Big Data: The food logistics industry might gain a lot from the advances in big data. Because so many food products are handled, a lot of data is produced in the field of food logistics. This data can be used to obtain information on their origin, distribution chain, dimensional qualities (size, weight), quality (appearance), freshness, and environmental conditions (temperature, humidity, microbiological activity, etc.) (Jagtap and Duong, 2019). Big data analysis, for example, can optimize truck delivery routes by identifying or altering routes based on traffic congestion or road accidents, minimizing food deterioration as a result of delayed transit caused by traffic. Big data can also assist in real-time planning and estimating capacity availability for deliveries. Supermarket chains with numerous locations around the nation might leverage big data to implement different pricing strategies for comparable products based on the incomes of their customers, the employment rate, and the crime rate in the area in which they operate.

Internet of Things (IoT): The Internet of Things (IoT) is a network of physical items, or "things," that are integrated with sensors, actuators, and software to enable data sharing with other devices via the network. To counteract the creation of counterfeit alcohol, MovingNet, a vehicular ad hoc sensor network with several sensors, was installed on a public transportation system (Ramesh and Das, 2011). A study done on IoT integration into a pig supply chain's enterprise resource planning (ERP) system was shown to build an early warning system for pork quality, resulting in lower logistical costs and increased circulation efficiency (Ma et. al, 2012). IoT with a greater connection can optimize operations by reducing operating costs and performing predictive maintenance, hence improving food quality, safety, and operational efficiency (Bader and Jagtap, 2020) (Jagtap et. al, 2019). For example, as shown in Figure 4, IoT is used in food delivery, where a food truck outfitted with numerous sensors monitors its location, temperature, stock levels, and maintenance requirements, sending out notifications in the event of traffic congestion.

Figure 4. IoT-powered food delivery vehicle

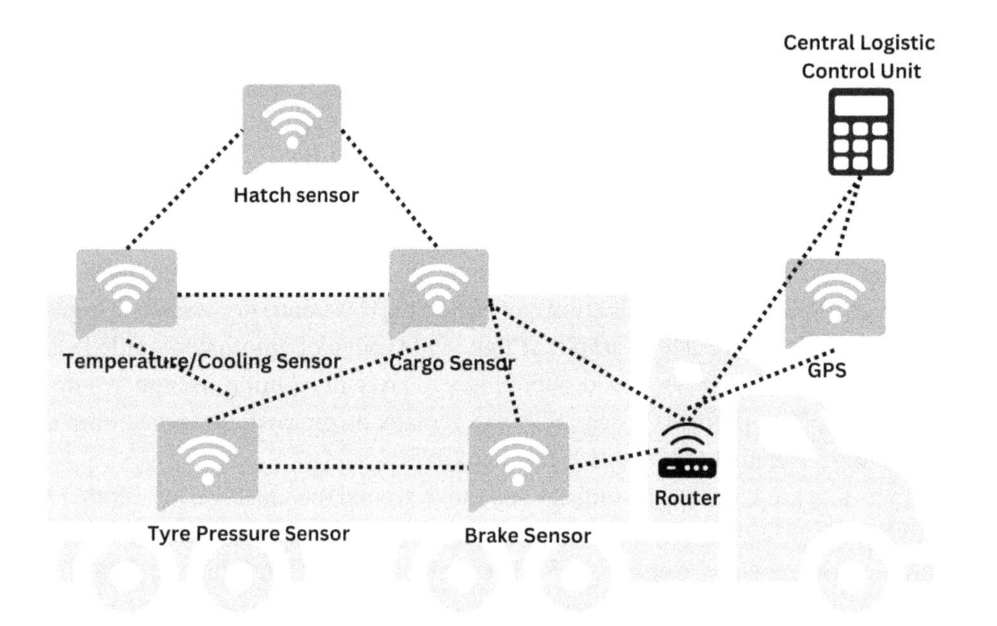

DISCUSSION

This chapter is a review of the technology available and the potential of future technology that can ideally digitalize the Agri-Food Supply Chain. Special emphasis on the word 'ideal' is important. Even the best agri-food supply chain actors have yet to effectively implement more than half of the solutions listed above. A robust financial foundation is essential for fully integrating cutting-edge technologies into all aspects of the supply chain. In addition, external factors such as India's tropical weather, and many undeveloped roads lead to the failure of implementation of route optimization techniques using Artificial Intelligence. The data that needs to be collected in order to process India's vast and complex road network would take profound research over many years to perfectly implement a groundbreaking solution. Another key point at the very start of the supply chain is the fear of acceptance of new technology. Indian farmers and laborers take pride in managing the work manually, however long it takes. On top of that, the skeptical fear of losing jobs is also prevalent. Although digitization and automation will create job opportunities in advanced sectors, it will take a profoundly long time to educate the farmers and workers on programming, robotics, quality control, and other innovative systems. Campaigns to raise awareness and educate people about the possible benefits and appropriate ways to use these technologies are necessary. A large percentage of agriculture is concentrated in rural areas, which frequently have unreliable internet access. The deployment of Internet of Things devices becomes difficult in the absence of a strong network. The government's support for the local farmers can give them the initial financial support to headstart with the implementation of innovative technologies by handing low-interest rate loans and lines of credit that will impact the agricultural economy and affect the country's GDP in the long term. Also with the implementation of new technologies comes a cost of vulnerability to hackers, malware, and other malicious elements, so it is equally important to educate the farmers about the importance of data security and how to protect privacy and data theft. Another important issue that might arise giving emphasis on the problems of stubble burning happening in the northern states of India especially Haryana and Punjab that have even affected the air quality of the national capital region of New Delhi, is the problem of E-Waste. Effective waste management is currently a pressing issue, especially considering the limited and rudimentary technologies in use today. Hence, it is imperative to focus on developing sustainable practices for the disposal and recycling of electronic components to minimize environmental impact. These difficulties cover socioeconomic, technological, infrastructure, and regulatory issues that may push back the digitalization of the Agri-Food Industry in India.

FUTURE SCOPE AND LIMITATION

The authors lay a strong foundation for understanding the digitalization of agricultural SCM by focusing on crucial areas like input procurement, cultivation, and post-harvest handling. However, the future of this field offers exciting possibilities for further exploration. Expanding the scope to encompass subsequent stages of the food supply chain – processing, distribution, retail, consumer, and waste management – would provide a more comprehensive picture. This could involve delving into how blockchain ensures farm-to-fork transparency, how e-commerce platforms reshape distribution, and how AI and data analytics optimize demand forecasting and inventory management based on consumer behavior.

Furthermore, incorporating waste management and sustainability considerations would strengthen the analysis. The chapter could investigate how digital tools promote sustainable practices like precision agriculture and explore how digital platforms connect farmers with waste management solutions. Additionally, examining how digital tools like microfinance platforms facilitate access to credit for smallholder farmers and the role of big data and sensor technology in precision agriculture could provide valuable insights for financial inclusion and optimizing farm practices.

It is crucial to acknowledge the limitations of the current scope. The digital divide and infrastructure issues faced by many agricultural workforces, particularly in developing economies, necessitate exploring solutions like low-cost technologies or government initiatives. Data security and privacy concerns around data collection in agriculture must also be addressed. Discussing secure data storage and user privacy protocols would strengthen the work. Finally, acknowledging the potential cost barriers associated with implementing advanced digital solutions, alongside exploring cost-effective alternatives and challenges of scaling these technologies across diverse farm sizes and locations, would provide a more balanced perspective. By outlining these future research directions and limitations, the chapter demonstrates a comprehensive understanding of the digitalization of agricultural SCM and paves the way for further exploration in this dynamic field.

CONCLUSION

In conclusion, a revolutionary change from conventional to digitalized approaches that make use of cutting-edge technologies characterizes the evolution of the agri-food supply chain. Integrating Industry 4.0 solutions such as big data, mechatronics, and the IoT has emerged as a critical facilitator for tackling supply chain management difficulties. This paradigm change could lead to increased productivity, openness, and cooperation in the agri-food industry.

The chapter emphasizes how important technology is, especially in precision agriculture, where AI-powered IoT is revolutionizing everything from livestock management to soil preparation. Applications for smart farming, like drones and cutting-edge seed technology, support sustainable farming and higher productivity. Examining post-harvest management highlights the benefits of AI and IoT, which promote food safety, waste minimization, and operational effectiveness.

The chapter goes on to discuss the particular difficulties the Indian agri-food supply chain faces, highlighting the necessity of funding, public awareness initiatives, and cybersecurity training. The obstacles presented by topography and customary agricultural methods demand a planned and incremental approach, with government support in the form of low-interest loans suggested as a spur for the adoption of new technologies. The effects on the environment, especially from burning stubble and electronic garbage, highlight how crucial sustainable behaviors are.

All things considered, the texts under evaluation present a dynamic picture of the agri-food industry in which technology is a beneficial catalyst for development. Digitalization has many potential advantages, but there are drawbacks as well. To solve these, a comprehensive strategy combining stakeholders, education, and sustainable practices is needed. The path to an entirely digitalized agri-food supply chain necessitates teamwork, flexible thinking, and a dedication to striking a balance between technological advancement and socioeconomic and environmental factors.

REFERENCES

Abdirad, M., & Krishnan, K. (2021). Industry 4.0 in logistics and supply chain management: A systematic literature review. *Engineering Management Journal*, 33(3), 187–201. 10.1080/10429247.2020.1783935

Abdul Aziz, M. F., Bukhari, W. M., Sukhaimie, M. N., Izzuddin, T. A., Norasikin, M. A., Rasid, A. F. A., & Bazilah, N. F. (2021). Development of smart sorting machine using artificial intelligence for chili fertigation industries. *Journal of Automation Mobile Robotics and Intelligent Systems*, 15(4), 44–52.

Ahlqvist, V., Norrman, A., & Jahre, M. (2020). *Supply chain risk governance: towards a conceptual multi-level framework.*

Aiello, G., Giovino, I., Vallone, M., Catania, P., & Argento, A. (2018). A decision support system based on multisensor data fusion for sustainable greenhouse management. *Journal of Cleaner Production*, 172, 4057–4065. 10.1016/j.jclepro.2017.02.197

Akkaş, M. A., & Sokullu, R. (2017). An IoT-based greenhouse monitoring system with Micaz motes. *Procedia Computer Science*, 113, 603–608. 10.1016/j.procs.2017.08.300

Alora, A., & Barua, M. K. (2019). An integrated structural modelling and MICMAC analysis for supply chain disruption risk classification and prioritisation in India. *International Journal of Value Chain Management*, 10(1), 1–25. 10.1504/IJVCM.2019.096538

Amentae, T. K., & Gebresenbet, G. (2021). Digitalization and future agro-food supply chain management: A literature-based implications. *Sustainability (Basel)*, 13(21), 12181. 10.3390/su132112181

Bader, F., & Jagtap, S. (2020). Internet of things-linked wearable devices for managing food safety in the healthcare sector. In *Wearable and Implantable Medical Devices* (pp. 229–253). Academic Press. 10.1016/B978-0-12-815369-7.00010-0

Bader, F., & Rahimifard, S. (2020). A methodology for the selection of industrial robots in food handling. *Innovative Food Science & Emerging Technologies*, 64, 102379. 10.1016/j.ifset.2020.102379

Benaissa, S., Tuyttens, F. A. M., Plets, D., Trogh, J., Martens, L., Vandaele, L., Joseph, W., & Sonck, B. (2020). Calving and estrus detection in dairy cattle using a combination of indoor localization and accelerometer sensors. *Computers and Electronics in Agriculture*, 168, 105153. 10.1016/j.compag.2019.105153

Bera, S. (2021). An application of operational analytics: for predicting sales revenue of restaurant. *Machine learning algorithms for industrial applications*, 209-235.

Bottani, E., Murino, T., Schiavo, M., & Akkerman, R. (2019). Resilient food supply chain design: Modelling framework and metaheuristic solution approach. *Computers & Industrial Engineering*, 135, 177–198. 10.1016/j.cie.2019.05.011

Bünger, L. (2021). *Robotic waste sorting*. Worcester Polytechnic Institute.

Căescu, Ş. C., & Dumitru, I. (2011). Particularities Of The Competitive Environment In The Business To Business Field. *Management & Marketing, 6*(2).

Castillo, O., & Meliif, P. (1970). Automated quality control in the food industry combining artificial intelligence techniques with fractal theory. *WIT Transactions on Information and Communication Technologies, 10.*

Civele, C. (2019). Development of an IOT based tractor tracking device to be used as a precision agriculture tool for Turkey's agricultural tractors. *Sch. J. Agric. Vet. Sci*, 6, 199–203.

Dekhne, A., Hastings, G., Murnane, J., & Neuhaus, F. (2019). Automation in logistics: Big opportunity, bigger uncertainty. *The McKinsey Quarterly*, 24.

Dewi, T., Risma, P., & Oktarina, Y. (2020). Fruit sorting robot based on color and size for an agricultural product packaging system. *Bulletin of Electrical Engineering and Informatics*, 9(4), 1438–1445. 10.11591/eei.v9i4.2353

Duong, L. N., Al-Fadhli, M., Jagtap, S., Bader, F., Martindale, W., Swainson, M., & Paoli, A. (2020). A review of robotics and autonomous systems in the food industry: From the supply chains perspective. *Trends in Food Science & Technology*, 106, 355–364. 10.1016/j.tifs.2020.10.028

Echelmeyer, W., Kirchheim, A., & Wellbrock, E. (2008, September). Robotics-logistics: Challenges for automation of logistic processes. In *2008 IEEE International Conference on Automation and Logistics* (pp. 2099-2103). IEEE. 10.1109/ICAL.2008.4636510

El Hoummaidi, L., Larabi, A., & Alam, K. (2021). Using unmanned aerial systems and deep learning for agriculture mapping in Dubai. *Heliyon*, 7(10), e08154. 10.1016/j.heliyon.2021.e0815434703924

Galati, F., & Bigliardi, B. (2019). Industry 4.0: Emerging themes and future research avenues using a text mining approach. *Computers in Industry*, 109, 100–113. 10.1016/j.compind.2019.04.018

Ghadge, A., Er Kara, M., Moradlou, H., & Goswami, M. (2020). The impact of Industry 4.0 implementation on supply chains. *Journal of Manufacturing Technology Management*, 31(4), 669–686. 10.1108/JMTM-10-2019-0368

Gutierrez-Galan, D., Dominguez-Morales, J. P., Cerezuela-Escudero, E., Rios-Navarro, A., Tapiador-Morales, R., Rivas-Perez, M., Dominguez-Morales, M., Jimenez-Fernandez, A., & Linares-Barranco, A. (2018). Embedded neural network for real-time animal behavior classification. *Neurocomputing*, 272, 17–26. 10.1016/j.neucom.2017.03.090

Han, Y. Y., Wang, K. Y., Liu, Z. Q., Zhang, Q., Pan, S. H., Zhao, X. Y., & Wang, S. F. (2017). A crop trait information acquisition system with multitag-based identification technologies for breeding precision management. *Computers and Electronics in Agriculture*, 135, 71–80. 10.1016/j.compag.2017.01.004

Hatefi, S. M., Moshashaee, S. M., & Mahdavi, I. (2019). A bi-objective programming model for reliable supply chain network design under facility disruption. *International Journal of Integrated Engineering*, 11(6), 80–92. 10.30880/ijie.2019.11.06.009

He, J., Wang, J., He, D., Dong, J., & Wang, Y. (2011). The design and implementation of an integrated optimal fertilization decision support system. *Mathematical and Computer Modelling*, 54(3-4), 1167–1174. 10.1016/j.mcm.2010.11.050

Hossen, M. A., Talukder, M. R. A., Al Mamun, M. R., Rahaman, H., Paul, S., Rahman, M. M., Miaruddin, M., Ali, M. A., & Islam, M. N. (2020). Mechanization status, promotional activities and government strategies of Thailand and Vietnam in comparison to Bangladesh. *AgriEngineering*, 2(4), 489–510. 10.3390/agriengineering2040033

Irfan, M., Wang, M., & Akhtar, N. (2019). Impact of IT capabilities on supply chain capabilities and organizational agility: A dynamic capability view. *Operations Management Research : Advancing Practice Through Research*, 12(3-4), 113–128. 10.1007/s12063-019-00142-y

Ivanov, D. (2020). Predicting the impacts of epidemic outbreaks on global supply chains: A simulation-based analysis on the coronavirus outbreak (COVID-19/SARS-CoV-2) case. *Transportation Research Part E, Logistics and Transportation Review*, 136, 101922. 10.1016/j.tre.2020.10192232288597

Jagtap, S., & Duong, L. N. K. (2019). Improving the new product development using big data: A case study of a food company. *British Food Journal*, 121(11), 2835–2848. 10.1108/BFJ-02-2019-0097

Jagtap, S., Rahimifard, S., & Duong, L. N. (2022). Real-time data collection to improve energy efficiency: A case study of food manufacturer. *Journal of Food Processing and Preservation*, 46(8), e14338. 10.1111/jfpp.14338

Jiang, B., He, J., Yang, S., Fu, H., Li, T., Song, H., & He, D. (2019). Fusion of machine vision technology and AlexNet-CNNs deep learning network for the detection of postharvest apple pesticide residues. *Artificial Intelligence in Agriculture*, 1, 1–8. 10.1016/j.aiia.2019.02.001

Karabegović, I., Karabegović, E., Mahmić, M., & Husak, E. J. A. I. P. E. (2015). The application of service robots for logistics in manufacturing processes. *Advances in Production Engineering & Management*, 10(4), 185–194. 10.14743/apem2015.4.201

Kottalil, A. M., Krishnan, B. B., Anto, A., & Alex, B. (2016). Automatic sorting machine. *Journal for Research*, 2(04).

Kumar, K., & Kayalvizhi, S. (2015). Real Time Industrial Colour Shape and Size Detection System Using Single Board. *International Journal of Science, Engineering and Technology Research (IJSETR), 4*(3).

Kumar, T. A., Lalswamy, B., Raghavendra, Y., Usharani, S. G., & Usharani, S. (2018). Intelligent food and grain storage management system for the warehouse and cold storage. *Int. J. Res. Eng. Sci. Manag*, 1(4), 130–132.

Kunhimohammed, C. K., Saifudeen, K. M., Sahna, S., Gokul, M. S., & Abdulla, S. U. (2015). *Automatic color sorting machine using TCS230 color sensor and PIC microcontroller*. Research Gate.

Layaq, M. W., Goudz, A., Noche, B., & Atif, M. (2019). Blockchain technology as a risk mitigation tool in supply chain. *Int. J. Transp. Eng. Technol*, 5(3), 50–59. 10.11648/j.ijtet.20190503.12

Lee, H., Moon, A., Moon, K., & Lee, Y. (2017, July). Disease and pest prediction IoT system in orchard: A preliminary study. In *2017 Ninth International Conference on Ubiquitous and Future Networks (ICUFN)* (pp. 525-527). IEEE. 10.1109/ICUFN.2017.7993840

Li, S., Xu, L. D., & Zhao, S. (2015). The internet of things: A survey. *Information Systems Frontiers*, 17(2), 243–259. 10.1007/s10796-014-9492-7

Liu, D., Mishra, A. K., & Yu, Z. (2016). Evaluating uncertainties in multi-layer soil moisture estimation with support vector machines and ensemble Kalman filtering. *Journal of Hydrology (Amsterdam)*, 538, 243–255. 10.1016/j.jhydrol.2016.04.021

Lopes de Sousa Jabbour, A. B., Jabbour, C. J. C., Godinho Filho, M., & Roubaud, D. (2018). Industry 4.0 and the circular economy: A proposed research agenda and original roadmap for sustainable operations. *Annals of Operations Research*, 270(1-2), 273–286. 10.1007/s10479-018-2772-8

Luo, L., Shen, G. Q., Xu, G., Liu, Y., & Wang, Y. (2019, March). Stakeholder-associated supply chain risks and their interactions in a prefabricated building project in Hong Kong. *Journal of Management Engineering*, 35(2), 05018015. 10.1061/(ASCE)ME.1943-5479.0000675

Ma, C., Li, Y., Yin, G., & Ji, J. (2012, July). The monitoring and information management system of pig breeding process based on internet of things. In *2012 Fifth International conference on information and computing science* (pp. 103-106). IEEE. 10.1109/ICIC.2012.61

Machado, C. G., Winroth, M. P., & Ribeiro da Silva, E. H. D. (2020). Sustainable manufacturing in Industry 4.0: An emerging research agenda. *International Journal of Production Research*, 58(5), 1462–1484. 10.1080/00207543.2019.1652777

Mahlein, A. K., Oerke, E. C., Steiner, U., & Dehne, H. W. (2012). Recent advances in sensing plant diseases for precision crop protection. *European Journal of Plant Pathology*, 133(1), 197–209. 10.1007/s10658-011-9878-z

Manavalan, E., & Jayakrishna, K. (2019). A review of Internet of Things (IoT) embedded sustainable supply chain for industry 4.0 requirements. *Computers & Industrial Engineering*, 127, 925–953. 10.1016/j.cie.2018.11.030

Mentzer, J. T., DeWitt, W., Keebler, J. S., Min, S., Nix, N. W., Smith, C. D., & Zacharia, Z. G. (2001). Defining supply chain management. *Journal of Business Logistics*, 22(2), 1–25. 10.1002/j.2158-1592.2001.tb00001.x

Müller, J. M., & Voigt, K. I. (2018). The impact of industry 4.0 on supply chains in engineer-to-order industries-an exploratory case study. *IFAC-PapersOnLine*, 51(11), 122–127. 10.1016/j.ifacol.2018.08.245

Nikolidakis, S. A., Kandris, D., Vergados, D. D., & Douligeris, C. (2015). Energy efficient automated control of irrigation in agriculture by using wireless sensor networks. *Computers and Electronics in Agriculture*, 113, 154–163. 10.1016/j.compag.2015.02.004

Nosirov, K., Begmatov, S., Arabboev, M., Kuchkorov, T., Chedjou, J. C., Kyamakya, K., & Abhiram, K. (2020). The greenhouse control based-vision and sensors. In *Developments of Artificial Intelligence Technologies in Computation and Robotics: Proceedings of the 14th International FLINS Conference (FLINS 2020)* (pp. 1514-1523). World Scientific. 10.1142/9789811223334_0181

Onishchuk, M. O. (2020). *Opto-mechanical sorting of municipal solid waste* [Doctoral dissertation, BHTY].

Panetto, H., Lezoche, M., Hormazabal, J. E. H., Diaz, M. D. M. E. A., & Kacprzyk, J. (2020). Special issue on Agri-Food 4.0 and digitalization in agriculture supply chains-New directions, challenges and applications. *Computers in Industry*, 116, 103188. 10.1016/j.compind.2020.103188

Pereira, A. C., & Romero, F. (2017). A review of the meanings and the implications of the Industry 4.0 concept. *Procedia Manufacturing*, 13, 1206–1214. 10.1016/j.promfg.2017.09.032

Preindl, R., Nikolopoulos, K., & Litsiou, K. (2020, January). Transformation strategies for the supply chain: The impact of industry 4.0 and digital transformation. In Supply Chain Forum [). Taylor & Francis.]. *International Journal (Toronto, Ont.)*, 21(1), 26–34.

Rad, C. R., Hancu, O., Takacs, I. A., & Olteanu, G. (2015). Smart monitoring of potato crop: A cyber-physical system architecture model in the field of precision agriculture. *Agriculture and Agricultural Science Procedia*, 6, 73–79. 10.1016/j.aaspro.2015.08.041

Ramesh, M. V., & Das, R. N. (2012). *A public transport system based sensor network for fake alcohol detection.* In Wireless Communications and Applications: First International Conference, ICWCA 2011, Sanya, China. 10.1007/978-3-642-29157-9_13

Ren, S., He, K., Girshick, R., & Sun, J. (2015). Faster r-cnn: Towards real-time object detection with region proposal networks. *Advances in Neural Information Processing Systems*, 28.

Ringler, C., Agbonlahor, M., Baye, K., Barron, J., Hafeez, M., Lundqvist, J., & Uhlenbrook, S. (2023). Water for food systems and nutrition. *Science and Innovations for Food Systems Transformation, 497.*

Saberi, S., Kouhizadeh, M., Sarkis, J., & Shen, L. (2019). Blockchain technology and its relationships to sustainable supply chain management. *International Journal of Production Research*, 57(7), 2117–2135. 10.1080/00207543.2018.1533261

Salman, A. D., & Abdelaziz, M. A. (2020). Mobile robot monitoring system based on IoT. *Journal of Xi'An University of Architecture & Technology*, 12(3), 5438–5447.

Samad, A., Murdeshwar, P., & Hameed, Z. (2010). High-credibility RFID-based animal data recording system suitable for small-holding rural dairy farmers. *Computers and Electronics in Agriculture*, 73(2), 213–218. 10.1016/j.compag.2010.05.001

Sawik, T. (2019, July). Two-period vs. multi-period model for supply chain disruption management. *International Journal of Production Research*, 57(14), 4502–4518. 10.1080/00207543.2018.1504246

Scuotto, V., Caputo, F., Villasalero, M., & Del Giudice, M. (2017). A multiple buyer–supplier relationship in the context of SMEs' digital supply chain management. *Production Planning and Control*, 28(16), 1378–1388. 10.1080/09537287.2017.1375149

Seyedghorban, Z., Tahernejad, H., Meriton, R., & Graham, G. (2020). Supply chain digitalization: Past, present and future. *Production Planning and Control*, 31(2-3), 96–114. 10.1080/09537287.2019.1631461

Shafi, U., Mumtaz, R., García-Nieto, J., Hassan, S. A., Zaidi, S. A. R., & Iqbal, N. (2019). Precision agriculture techniques and practices: From considerations to applications. *Sensors (Basel)*, 19(17), 3796. 10.3390/s1917379631480709

Shamshiri, R. R., Bojic, I., van Henten, E., Balasundram, S. K., Dworak, V., Sultan, M., & Weltzien, C. (2020). Model-based evaluation of greenhouse microclimate using IoT-Sensor data fusion for energy efficient crop production. *Journal of Cleaner Production*, 263, 121303. 10.1016/j.jclepro.2020.121303

Sheth, S., Kher, R., Shah, R., Dudhat, P., & Jani, P. (2010). Automatic sorting system using machine vision. In *Multi-Disciplinary International Symposium on Control, Automation & Robotics*.

Silva, S., Duarte, D., Barradas, R., Soares, S., Valente, A., & Reis, M. J. C. S. (2017). Arduino recursive backtracking implementation, for a robotic contest. In: *Human-Centric Robotics*, (pp. 169–178). World Scientific. 10.1142/9789813231047_0023

Sobota, J. PiŜl, R., Balda, P., & Schlegel, M. (2013). Raspberry Pi and Arduino boards in control education. *IFAC Proceedings Volumes, 46*(17), 7-12.

Subeesh, A., & Mehta, C. R. (2021). Automation and digitization of agriculture using artificial intelligence and internet of things. *Artificial Intelligence in Agriculture*, 5, 278–291. 10.1016/j.aiia.2021.11.004

Tervonen, J. (2018). Experiment of the quality control of vegetable storage based on the Internet-of-Things. *Procedia Computer Science*, 130, 440–447. 10.1016/j.procs.2018.04.065

Tripathi, S., Shukla, S., Attrey, S., Agrawal, A., & Bhadoria, V. S. (2020). Smart industrial packaging and sorting system. *Strategic system assurance and business analytics*, 245-254.

Tyagi, N., Khan, R., Chauhan, N., Singhal, A., & Ojha, J. (2021). E-rickshaws management for small scale farmers using big data-Apache spark. []. IOP Publishing.]. *IOP Conference Series. Materials Science and Engineering*, 1022(1), 012023. 10.1088/1757-899X/1022/1/012023

Ullah, M. W., Mortuza, M. G., Kabir, M. H., Ahmed, Z. U., Supta, S. K. D., Das, P., & Hossain, S. M. D. (2018). *Internet of things based smart greenhouse: remote monitoring and automatic control. DEStech Trans*. Environ. Energy Earth Sci.

Wamba, S. F., Queiroz, M. M., & Trinchera, L. (2020). Dynamics between blockchain adoption determinants and supply chain performance: An empirical investigation. *International Journal of Production Economics*, 229, 107791. 10.1016/j.ijpe.2020.107791

Wang, N., Zhang, N., & Wang, M. (2006). Wireless sensors in agriculture and food industry—Recent development and future perspective. *Computers and Electronics in Agriculture*, 50(1), 1–14. 10.1016/j.compag.2005.09.003

Wolfert, S., Ge, L., Verdouw, C., & Bogaardt, M. J. (2017). Big data in smart farming–a review. *Agricultural Systems*, 153, 69–80. 10.1016/j.agsy.2017.01.023

Wong, L. W., Tan, G. W. H., Lee, V. H., Ooi, K. B., & Sohal, A. (2020). Unearthing the determinants of Blockchain adoption in supply chain management. *International Journal of Production Research*, 58(7), 2100–2123. 10.1080/00207543.2020.1730463

Yawar, S. A., & Seuring, S. (2018). The role of supplier development in managing social and societal issues in supply chains. *Journal of Cleaner Production*, 182, 227–237. 10.1016/j.jclepro.2018.01.234

Yogitha, S., & Sakthivel, P. (2014, March). A distributed computer machine vision system for automated inspection and grading of fruits. In *2014 International Conference on Green Computing Communication and Electrical Engineering (ICGCCEE)* (pp. 1-4). IEEE. 10.1109/ICGCCEE.2014.6922281

Yoon, J., Talluri, S., & Rosales, C. (2020). Procurement decisions and information sharing under multi-tier disruption risk in a supply chain. *International Journal of Production Research*, 58(5), 1362–1383. 10.1080/00207543.2019.1634296

Zagurskiy, O. N., & Titova, L. L. (2019). Problems and prospects of blockchain technology usage in supply chains. *Journal of Automation and Information Sciences*, 51(11), 63–74. 10.1615/JAutomatInf-Scien.v51.i11.60

Zuidhof, M. J., Fedorak, M. V., Ouellette, C. A., & Wenger, I. I. (2017). Precision feeding: Innovative management of broiler breeder feed intake and flock uniformity. *Poultry Science*, 96(7), 2254–2263. 10.3382/ps/pex01328159999

Chapter 11
Harnessing Agricultural Data:
Advancing Sustainability Through the Application of Find S Algorithm

C. V. Suresh Babu
http://orcid.org/0000-0002-8474-2882
Hindustan Institute of Technology and Science, India

Yadavamuthiah K.
Hindustan Institute of Technology and Science, India

Sathiyanarayana S.
Hindustan Institute of Technology and Science, India

Sheldon Mathew
Hindustan Institute of Technology and Science, India

ABSTRACT

This chapter emphasizes the Find S algorithm to explore the use of cutting-edge computational techniques to improve agri-business sustainability. Precision farming, data analytics, and machine learning are all combined in "agriculture 4.0" to maximize productivity and promote sustainability. Case examples from real-world situations are provided to illustrate the usefulness of using cutting-edge computational techniques in agriculture. The chapter also covers the value of cooperation and government assistance, tackles issues related to technological adoption, and provides solutions for broader acceptance in the farming community. This chapter intends to contribute to the continuing conversation on data-driven decision-making and productivity improvement in the agricultural sector by offering insightful information to researchers, practitioners, policymakers, and stakeholders who are interested in using computational methods to improve sustainability in the industry.

DOI: 10.4018/979-8-3693-3583-3.ch011

INTRODUCTION

Particularly in nations like India, the agriculture industry faces a variety of difficulties, from maintaining environmental sustainability to ensuring economic viability. The idea of "Agriculture 4.0," which integrates cutting-edge technology like machine learning, data analytics, and precision agriculture to transform farming methods, has come to light as a potential answer in recent years (Araújo et al., 2021). The application of sophisticated computer techniques is at the core of this technological growth, and the Find S algorithm has emerged as a potent instrument for the analysis of agricultural data (Majumdar, Naraseeyappa, & Ankalaki, 2017). With an emphasis on dairy and animal farming specifically, this chapter aims to investigate the Find S algorithm's transformational potential in tackling the sustainability issues that are common in India's agriculture industry.

The agriculture industry in India has several obstacles, such as the depletion of natural resources, the effects of climate change, and the requirement to satisfy the expanding needs of an expanding populace (Nedumaran & Manida, 2019). Dairy and cattle farming, which provide vital sources of income for millions of farmers nationwide, are crucial in this regard. But conventional agricultural methods frequently find it difficult to adjust to changing economic and environmental conditions, calling for creative alternatives that might maximise output while reducing negative environmental effects. A paradigm change towards a more data-driven and technologically advanced agricultural method is symbolised by agriculture 4.0, which provides opportunity to solve these issues in a sustainable way.

A key component of Agriculture 4.0 is the application of cutting-edge computational techniques to extract knowledge from massive volumes of agricultural data (Bujang & Bakar, 2019). The Find S method was developed in the field of machine learning and has gained popularity due to its effectiveness in finding patterns in data through quick searching. By evaluating data on soil health, crop output, weather patterns, and animal management, this algorithm has the potential to optimise farming operations in the agricultural sector (Pretty, 2008). Farmers may increase productivity, save resource consumption, and eventually advance sustainability in their operations by utilising the power of data analytics to influence their decision-making.

The Find S algorithm and other computational techniques have enormous promise, but achieving it will entail coordinated efforts from many different stakeholders, including researchers, legislators, software developers, and farmers themselves. Working together is essential to creating accessible, farmer-specific tools and technology that are easy to use. In order to provide the infrastructure, financial incentives, and legal framework required to promote the adoption of cutting-edge agricultural technology, government assistance is also essential. We can quicken the shift to a more resilient and sustainable agriculture industry by tackling the obstacles related to technology adoption and creating a positive environment.

We will explore the useful uses of the Find S algorithm in dairy and cattle production in this chapter, based on actual case studies and research findings (Eastwood, 2008). We will also talk about how crucial government assistance and teamwork are to promoting technology adoption and removing obstacles to its use. We hope to add to the continuing conversation about sustainable agricultural methods and how technology will affect agriculture in India and beyond by illuminating the transformational potential of sophisticated computational tools in the field.

REVIEW OF LITERATURE

Abubacker and Raheem (2024) deliver a pioneering research on smart agriculture in Malaysia, highlighting creative solutions that use ubiquitous computing to drive long-term economic growth. Despite restrictions such as data accessibility issues, technical infrastructure discrepancies, and socioeconomic considerations, the article proposes ground-breaking solutions. The paper suggests real-time monitoring and decision support systems for agricultural techniques by integrating IoT sensors, drones, and mobile apps. Furthermore, it emphasises long-term economic growth by merging technical advances with environmentally responsible farming methods. The study advocates for multidisciplinary collaboration to promote information sharing and stakeholder involvement, hence promoting the co-creation of specialised smart agriculture solutions.

Ingram and Maye (2020) provide light on the constraints and advances that digitization has brought to agriculture. While noting issues such as the digital divide, information quality, and infrastructural constraints, the report highlights a number of unique characteristics. For starters, digital platforms improve knowledge sharing by facilitating real-time information exchange and collaboration among agricultural stakeholders. Second, bespoke decision support tools use farm-specific data to make personalised suggestions that improve production and sustainability. Finally, digitization makes data-driven research easier, allowing academics to analyse enormous datasets and get practical insights for tackling complicated agricultural concerns. These advancements demonstrate the revolutionary power of digital technology in altering the future of agriculture.

Yang et al. (2024) present a comprehensive review that underscores both limitations and innovations surrounding the application of question answering systems in agriculture. While recognizing challenges like data quality, technical hurdles, and user acceptance, the study highlights several innovative aspects. Firstly, these systems serve as intelligent decision support tools, leveraging natural language processing and machine learning to provide tailored insights for optimizing farming practices and resource management. Secondly, they enhance knowledge accessibility by offering user-friendly interfaces for seamless access to agricultural information, empowering farmers with timely guidance. Lastly, integration with IoT and big data analytics enables context-aware responses, facilitating data-driven decision-making and precision agriculture practices for sustainability. These innovations signify the transformative potential of question answering systems in revolutionizing agricultural management and production.

Alazzai et al. (2024) discuss the constraints and developments around smart agricultural solutions, with a special emphasis on AI and IoT technology for crop management. The report recognises resource restrictions, data privacy problems, and technical dependency as potential limitations, but it also identifies some new methods. To begin, AI-powered decision support systems make personalised suggestions for optimising planting, irrigation, fertilisation, and pest management tactics, resulting in higher agricultural yields and resource efficiency. Second, IoT-enabled precision agricultural systems use sensor networks to collect real-time data on soil moisture, temperature, and crop health, allowing for proactive interventions to improve productivity and sustainability. Finally, the study emphasises the creation of scalable and cost-effective solutions customised to smallholder farmers and resource-constrained communities, democratising access to AI and IoT technologies for enhancing agricultural practices and strengthening climate resilience.

Coelho et al. (2024) discuss the limits and advancements in green manuring as a sustainable soil management method in tropical areas. While noting issues such as geographical distinctiveness, implementation problems, and concerns about long-term impacts, the study emphasises numerous novel characteristics. First, green manuring is a sustainable soil management approach that improves soil fertility, reduces dependency on synthetic fertilisers, and promotes general soil health in tropical agricultural systems (Suresh Babu, C. V., Mahalashmi, et. al 2023). Second, it contributes to climate change mitigation by sequestering carbon, lowering greenhouse gas emissions, and increasing resistance to harsh weather events. Finally, the study emphasises information exchange and capacity development measures to encourage farmers to use green manure, hence encouraging innovation and collaboration in tropical agriculture.

Balasubramanian (2024) investigates both the limitations and breakthroughs of robotic farming systems for agricultural sustainability evaluations. While acknowledging obstacles such as technical reliance, data interpretation issues, and environmental concerns, the paper emphasises some novel elements. To begin, robotic farming enables precision agriculture approaches by utilising sensors, drones, and autonomous machinery for focused interventions and data-driven decision-making to maximise harvests while minimising environmental effect. Second, robotic systems' real-time monitoring and feedback mechanisms allow for proactive management methods and quick reaction to developing difficulties, hence improving agricultural sustainability and resilience. Finally, the use of AI and machine learning algorithms improves predictive analytics and decision support, allowing for proactive risk management and adaptive management techniques in the face of climate change and uncertainty. These ideas highlight the transformational power of

Vemuri et al. (2023) investigates the constraints and advances around the incorporation of IoT technology into smart agricultural methods. Recognising obstacles such as technological accessibility, data security concerns, and integration issues, the research emphasises numerous unique components (Suresh Babu, C. V. & Rahul. A., 2024). For starters, IoT enables real-time monitoring and management by giving farmers quick access to data on soil conditions, crop health, and environmental variables. This enables proactive decision-making and precision management approaches, which optimise resource utilisation while increasing agricultural output and sustainability. Second, IoT-powered predictive analytics and decision support systems use data trends to estimate crop yields, weather patterns, and offer best farming practices, providing farmers with actionable insights for increasing yields while reducing environmental impact. Finally, IoT devices' remote monitoring and automation capabilities simplify agricultural operations.

Neethirajan (2023) investigates the constraints and advances associated with the integration of metaverse-based solutions for cattle welfare. While noting problems such as technological hurdles, data privacy issues, and ethical considerations, the study emphasises numerous novel characteristics. To begin, metaverse technologies provide virtual monitoring and simulation tools for real-time assessment of livestock conditions, providing chances to optimise animal care procedures and increase overall welfare. Second, metaverse platforms enable remote training and education programmes by delivering immersive learning experiences that help livestock carers improve their knowledge and abilities in animal welfare procedures. Finally, the metaverse enables collaborative decision-making processes by bringing stakeholders together in virtual spaces to exchange ideas, discuss best practices, and co-create solutions to difficult issues in cattle management. These innovations represent the transformational

Sundararajan et al. (2024) investigate the constraints and advances associated with the use of green-synthesised nanomaterials in sustainable agricultural production systems. While noting problems such as environmental issues, regulatory hurdles, and technical limits, the paper emphasises numerous novel characteristics. To begin, green-synthesized nanomaterials provide environmentally friendly solutions for sustainable agriculture by utilising eco-friendly synthesis technologies to reduce environmental effect and encourage ecologically sustainable activities. Second, these nanomaterials improve nutrient efficiency and absorption in crops, increasing yield while minimising nutrient losses and supporting resource conservation. Finally, using green-synthesized nanomaterials into integrated pest management techniques allows for more focused pest control, lowering the need for chemical pesticides while also reducing environmental pollution and human health hazards. These developments highlight the promise of green nanotechnology in reducing global concerns and improving sustainable crops.

In order to promote sustainable agricultural development, this research study integrates findings from a wide range of studies on smart agriculture, emphasizing creative approaches that make use of digital technologies. The study showcases innovative approaches like AI-powered decision support tools, real-time monitoring systems, and IoT-enabled precision agriculture, while highlighting the revolutionary potential of interdisciplinary collaboration. Important research gaps are noted despite noteworthy advancements: these include the need for customized solutions for smallholder farmers, more in-depth studies of adoption hurdles, and thorough evaluations of the effects on the environment and society. Researchers can help optimize the advantages of digital agriculture while guaranteeing its moral and just use by filling in these gaps.

THEORETICAL BACKGROUND

Foundations of Data Analysis in Agriculture: A vast amount of data is produced by agriculture from a variety of sources, such as sensors, satellites, and farm management software. This data includes a lot of different information, such crop health, weather, soil composition, and yield measures. In agriculture, data analysis tools are essential for deriving significant insights from this large and heterogeneous dataset. These methods cover a variety of procedures such as transformation, statistical analysis, visualisation, and data cleansing. By using these methods, scientists and industry professionals may find patterns, relationships, and trends in agricultural data, which helps with well-informed decision-making and enhanced farm management strategies (Delgado, Short Jr, Roberts, & Vandenberg, 2019). Statistical analysis, for instance, may highlight connections between crop yields and weather patterns, assisting farmers in making the most of irrigation and planting schedule decisions. Additionally, data visualisation methods can offer understandable depictions of intricate agricultural data.

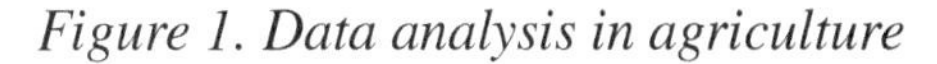

Figure 1. Data analysis in agriculture

(created in https://designer.microsoft.com/image-creator)

Machine Learning Principles and Applications: Agricultural data analysis and decision-making may benefit greatly from machine learning's robust tools. For applications including crop categorization, disease detection, and yield prediction, supervised learning algorithms—models trained on labelled data—are widely employed in agriculture (Suresh Babu, C. V, Swapna et. al. 2023). In contrast, unsupervised learning methods help with tasks like soil categorization and anomaly detection by allowing for pattern identification and grouping in unlabelled data. By gaining knowledge from feedback over time, reinforcement learning may optimise agricultural techniques and enable independent decision-making in dynamic contexts.

Concept Learning and the Find S Algorithm: The process of generalising rules or patterns from particular cases is known as concept learning, and it is essential to machine learning. In particular, agriculture can benefit greatly from the application of the classic concept learning algorithm, Find S. Finding the most precise hypothesis that minimises mistakes and fits all positive examples of a notion is its area of expertise (Iannone, Palmisano, & Fanizzi, 2007). When used in agriculture, the Find S algorithm helps with tasks like crop classification, determining ideal growth conditions, and detecting problems with the crops based on patterns in the data that are noticed.

Optimization Techniques for Agriculture: The process of generalising rules or patterns from particular cases is known as concept learning, and it is essential to machine learning. In particular, agriculture can benefit greatly from the application of the classic concept learning algorithm, Find S. Finding the most precise hypothesis that minimises mistakes and fits all positive examples of a notion is its area of expertise (Jain et al., 2018). When used in agriculture, the Find S algorithm helps with tasks like crop classification, determining ideal growth conditions, and detecting problems with the crops based on patterns in the data that are noticed.

Integration of Computational Methods in Agriculture: A comprehensive strategy for tackling agricultural problems is provided by the combination of data analysis, machine learning, and optimisation approaches. Through the integration of these techniques, scholars and professionals may evaluate intricate farming systems, forecast future results, and enhance the process of making decisions (Kumar et al., 2015). By using an integrated strategy, farmers may better use data-driven insights to increase crop yields, lower expenses, and manage risks, all of which contribute to the long-term sustainability and resilience of agricultural systems.

Future Directions and Challenges: Advanced computational techniques have promise for agriculture, but there are still a number of obstacles to overcome. Adoption is severely hampered by problems with interoperability, scalability, and data quality. Furthermore, special attention must be paid to ethical issues pertaining to algorithmic bias, data privacy, and socioeconomic consequences. It will need an interdisciplinary effort involving data scientists, engineers, legislators, and agricultural experts to address these issues. In order to secure the appropriate and fair application of computational methods in agriculture for the sake of environmental sustainability and global food security, research and development activities must be sustained (Suresh Babu, C.V. 2023).

RESEARCH METHODOLOGY

Figure 2. Architecture

Figure 3. Sample code

```
1  import files
2  import pandas as pd
3  import numpy as np
4  uploaded = files.upload()
5  data = pd.read_csv('exno1.csv')
6  concepts = np.array(data)[:, :-1]
7  target = np.array(data)[:, -1]
8  def train(con, tar):
9      # Initialize the specific hypothesis to the first positive instance
10     for i, val in enumerate(tar):
11         if val == 'yes':
12             specific_h = con[i].copy()
13             break
14     # Compare with other positive instances to generalize the hypothesis
15     for i, val in enumerate(con):
16         if tar[i] == 'yes':
17             for x in range(len(specific_h)):
18                 # If attributes do not match, make them more general
19                 if val[x] != specific_h[x]:
20                     specific_h[x] = '?'
21     return specific_h
22 print(train(concepts, target))
```

The research methodology for developing a monitoring platform using the Find S algorithm is a structured approach designed to ensure the effectiveness and reliability of the platform. It begins with a clear definition of objectives, aiming to create a comprehensive agricultural monitoring system that leverages advanced computational methods. This involves identifying specific aspects of agriculture to be monitored and optimized, such as crop health, soil conditions, irrigation efficiency, and pest management.

The strategies used for data collecting are essential for obtaining the information required to meet the goals of the study. Numerous sources are used in these techniques, such as sensor data, satellite imaging, meteorological data, and old agricultural records. In addition to guaranteeing compliance with the monitoring platform and enabling smooth integration of data sources, the methodology describes techniques for effectively gathering, storing, and managing data.

The platform depends on preprocessing and data cleaning techniques to guarantee the accuracy and consistency of the data it uses. This includes addressing missing values, standardising variables, detecting outliers, and eliminating noise. These stages set the stage for accurate and trustworthy insights to be derived using the Find S algorithm by readying the raw data for analysis.

A fundamental part of the technique is the implementation of the Find S algorithm, which allows the platform to evaluate preprocessed data and identify certain patterns or hypotheses related to agricultural surveillance. The method may be used to categorise crops, determine ideal growth conditions, and find abnormalities in agricultural data; examples are given to demonstrate this.

Another important component of the technique is integration with monitoring devices, which guarantees smooth communication and interoperability between the platform and drones, satellite systems, or IoT sensors. Through this connection, real-time data collecting from agricultural fields is made possible, improving the platform's capacity to give users accurate and timely information.

The creation of dashboards and visualisation tools is emphasised as a way to provide user-friendly presentations of analysed data and insights. Users are empowered to make educated decisions by using interactive maps, charts, and graphs that make it easier for them to explore and interpret agricultural data.

The addition of a decision support system expands the usefulness of the platform by offering recommendations that may be put into practice based on data analysis. To assist users in making well-informed decisions, recommendations are provided for crop management techniques, irrigation timing, pest control strategies, and nutrient supplementation.

Implementing testing and validation protocols is essential to guaranteeing the accuracy and operation of the platform in actual agricultural environments. These methods assess the platform's functionality and confirm that the Find S algorithm supports decision-making processes with accuracy and efficacy.

An iterative approach is used throughout the development process to collect input from stakeholders and users and to continuously enhance the functionality and usability of the platform. By using an iterative improvement method, the platform is guaranteed to stay adaptable to customer demands and equipped to tackle changing difficulties in agriculture. The ultimate goal of the study technique is to implement an intuitive monitoring platform that facilitates informed agricultural decision-making, hence improving agricultural systems' resilience, sustainability, and production.

FINDINGS AND RESULTS

Technological Advancements in Agriculture Management

- Focus on the significant strides made possible by the Find S algorithm-powered monitoring platform.
- Discussion of how advanced technology contributes to sustainable agriculture practices and improved management strategies.
- Exploration of the platform's role in driving innovation within the agricultural sector.

Data driven Insights for Enhanced Farming Practices

- Analysis of the platform's capability to uncover patterns, correlations, and trends in agricultural datasets.
- Examination of how these insights inform crucial factors such as crop health, soil conditions, irrigation efficiency, and pest control.
- Discussion on the implications of data-driven decision-making for optimizing farming practices and promoting sustainability.

Efficacy and Impact of the Monitoring Platform

- Evaluation of performance tests highlighting the platform's accuracy, efficiency, and efficacy.
- Overview of concrete benefits observed in agricultural operations, including improvements in crop management, soil health, and resource allocation.

- Examination of realworld case studies and user feedback demonstrating the platform's practical value and its influence on simplifying farming operations, increasing production, and supporting sustainable agriculture practices.

DISCUSSION

Utilizing Data-Driven Insights for Enhanced Agricultural Decision-Making

- Discussion on how research findings offer significant implications for agricultural management and sustainability.
- Examination of how agricultural stakeholders can leverage data from the monitoring platform to make informed decisions about crop management, resource allocation, and environmental stewardship.
- Exploration of specific examples where farmers can optimize irrigation schedules, detect crop illnesses early, and enhance soil health using the platform's advice.
- Highlighting the potential for increased productivity and resilience in agriculture through the adoption of data-driven decision-making.

Policy Implications and Promoting Sustainable Practices

- Analysis of how policymakers can utilize data-driven insights to create more effective agriculture policies and programs.
- Exploration of the role of advanced computational tools in promoting sustainable practices and ensuring food security.
- Discussion on how the project helps address critical agricultural concerns such as resource optimization, environmental consequences, and adaptation to climate change.
- Emphasizing the importance of evidence-based policies in driving sustainable agricultural practices and rural development.

Challenges and Opportunities in Agricultural Data Analysis

- Identification of significant limitations and challenges faced during the study, including data availability, processing resources, and algorithmic complexity.
- Discussion on the impact of these limitations on the breadth and accuracy of the monitoring platform's analysis.
- Exploration of potential solutions, including collaboration among researchers, policymakers, and industry stakeholders to improve data gathering infrastructure, invest in computing resources, and develop more efficient algorithms for agricultural analysis.

- Highlighting the need for ongoing research and innovation to overcome challenges and unlock the full potential of agricultural data analysis for sustainable development.

FUTURE SCOPE FOR FURTHER STUDY

The findings and discussion presented herein underscore the profound implications of leveraging agricultural data to advance sustainability through the application of the Find S algorithm. Through technological advancements, data-driven insights, and real-world case studies, the monitoring platform has demonstrated its efficacy in enhancing farming practices, optimizing resource allocation, and promoting environmental stewardship. However, challenges such as data availability and algorithmic complexity highlight the need for ongoing research and innovation to unlock the platform's full potential. Looking ahead, future studies can focus on enhancing functionality, integrating advanced technologies, and addressing ethical and social implications to ensure the broader adoption and impact of data-driven agricultural practices. By embracing these opportunities for further study, we can pave the way for a more sustainable and resilient agricultural future.

CONCLUSION

In conclusion, the research culminating in the development and implementation of the monitoring platform driven by the Find S algorithm underscores its pivotal role in advancing agricultural management practices and fostering sustainability in the agricultural sector. By meticulously analyzing agricultural data, the platform has provided invaluable insights into crucial aspects of agricultural systems, ranging from crop health to soil conditions and irrigation efficiency. These insights hold profound implications for decisionmaking processes among stakeholders, empowering them to optimize resource allocation, enhance productivity, and mitigate environmental impacts. Despite encountering challenges such as data availability and computational complexity, the study's significance lies in its contribution to the field of agricultural monitoring and precision agriculture, highlighting the transformative potential of integrating advanced computational methods with traditional farming practices. Looking forward, future research endeavors will focus on enhancing the platform's functionality, scalability, and applicability, thereby ensuring its broader adoption and impact on agricultural sustainability and resilience.

REFERENCES

Alazzai, W. K., Obaid, M. K., Abood, B. S. Z., & Jasim, L. (2024). Smart Agriculture Solutions: Harnessing AI and IoT for Crop Management. In *E3S Web of Conferences* (*Vol. 477*, p. 00057). EDP Sciences.

Araújo, S. O., Peres, R. S., Barata, J., Lidon, F., & Ramalho, J. C. (2021). Characterising the agriculture 4.0 landscape—Emerging trends, challenges and opportunities. *Agronomy (Basel)*, 11(4), 667. 10.3390/agronomy11040667

Balasubramanian, S. (2025). A comprehensive assessment of agricultural sustainability through robotic farming. *Journal ID*, 4625, 6352.

Bujang, A. S., & Bakar, B. H. A. (2019). *Agriculture 4.0: Data-driven approach to galvanize Malaysia's agro-food sector development. FFTC Agriculture Policy Platform*. FFTCAP.

. Coelho, F. C., Prins, C. L., da Rocha, J. G. D. G., de Jesus, V. P., Teixeira, N. S., Eiras, P. P., & Vaz, A. S. (2024). *Advancing agricultural sustainability in tropical climates: Harnessing the potential of green manuring*.

Delgado, J. A., Short, N. M.Jr, Roberts, D. P., & Vandenberg, B. (2019). Big data analysis for sustainable agriculture on a geospatial cloud framework. *Frontiers in Sustainable Food Systems*, 3, 54. 10.3389/fsufs.2019.00054

Eastwood, C. R. (2008). *Innovative precision dairy systems: A case study of farmer learning and technology co-development*. University of Melbourne, Melbourne School of Land and Environment.

Iannone, L., Palmisano, I., & Fanizzi, N. (2007). An algorithm based on counterfactuals for concept learning in the semantic web. *Applied Intelligence*, 26(2), 139–159. 10.1007/s10489-006-0011-5

Ingram, J., & Maye, D. (2020). What are the implications of digitalisation for agricultural knowledge? *Frontiers in Sustainable Food Systems*, 4, 66. 10.3389/fsufs.2020.00066

Jain, R., Malangmeih, L., Raju, S. S., Srivastava, S. K., Immaneulraj, K., & Kaur, A. P. (2018). Optimization techniques for crop planning: A review. *Indian Journal of Agricultural Sciences*, 88(12), 1826–1835. 10.56093/ijas.v88i12.85423

. Kumar, S., Shamim, M., Bansal, M., Aggarwal, R. P., & Gangwar, B. (2015). *Emerging trends and statistical analysis in computational modeling in agriculture*.

Majumdar, J., Naraseeyappa, S., & Ankalaki, S. (2017). Analysis of agriculture data using data mining techniques: Application of big data. *Journal of Big Data*, 4(1), 20. 10.1186/s40537-017-0077-4

Nedumaran, G., & Manida, M. (2019). Impact of FDI in agriculture sector in India: Opportunities and challenges. *International Journal of Recent Technology and Engineering*, 8(3), 380–383.

Neethirajan, S. (2023). Harnessing the Metaverse for Livestock Welfare: Unleashing Sensor Data and Navigating Ethical Frontiers. *Preprints,* 2023040409.

Pretty, J. (2008). Agricultural sustainability: Concepts, principles and evidence. *Philosophical Transactions of the Royal Society of London. Series B, Biological Sciences*, 363(1491), 447–465. 10.1098/rstb.2007.216317652074

Sundararajan, N., Habeebsheriff, H. S., Dhanabalan, K., Cong, V. H., Wong, L. S., Rajamani, R., & Dhar, B. K. (2024). Mitigating global challenges: Harnessing green synthesized nanomaterials for sustainable crop production systems. *Global Challenges (Hoboken, NJ)*, 8(1), 2300187. 10.1002/gch2.20230018738223890

Suresh Babu, C. V. (2023). *Artificial Intelligence and Expert Systems*. Anniyappa Publications.

Suresh Babu, C. V., Mahalashmi, J., Vidhya, A., Nila Devagi, S., & Bowshith, G. (2023). Save soil through machine learning. In Habib, M. (Ed.), *Global Perspectives on Robotics and Autonomous Systems: Development and Applications* (pp. 345–362). IGI Global. 10.4018/978-1-6684-7791-5.ch016

Suresh Babu, C. V., & Rahul, A. (2024). Securing the Future: Unveiling Risks and Safeguarding Strategies in Machine Learning-Powered Cybersecurity. In Almaiah, M., Maleh, Y., & Alkhassawneh, A. (Eds.), *Risk Assessment and Countermeasures for Cybersecurity* (pp. 80–95). IGI Global. 10.4018/979-8-3693-2691-6.ch005

Suresh Babu, C. V., Swapna, A., Chowdary, D. S., Vardhan, B. S., & Imran, M. (2023). Leaf disease detection using machine learning (ML). In Khang, A. (Ed.), *Handbook of Research on AI-Equipped IoT Applications in High-Tech Agriculture* (pp. 188–199). IGI Global. 10.4018/978-1-6684-9231-4.ch010

Vemuri, N., Thaneeru, N., & Tatikonda, V. M. (2023). Smart farming revolution: Harnessing IoT for enhanced agricultural yield and sustainability. *Journal of Knowledge Learning and Science Technology, 2*(2), 143-148..

Yang, T., Mei, Y., Xu, L., Yu, H., & Chen, Y. (2024). Application of question answering systems for intelligent agriculture production and sustainable management: A review. *Resources, Conservation and Recycling*, 204, 107497. 10.1016/j.resconrec.2024.107497

Chapter 12
Harvesting Insights Unveiling the Interplay of Climate, Pesticides, and Rainfall in Agricultural Yield Optimization

Dwijendra Nath Dwivedi
http://orcid.org/0000-0001-7662-415X
Krakow University of Economics, Poland

Ghanashyama Mahanty
http://orcid.org/0000-0002-6560-2825
Utkal University, India

Shafik Khashouf
University of Liverpool, UK

ABSTRACT

In this study, a wide range of geoFigureical locations are investigated to investigate the complex relationships that exist between agricultural productivity and important environmental parameters. These elements include fluctuations in temperature, patterns of rainfall, and the application of pesticides. Through the utilization of a vast dataset that encompasses yield measures, meteorological conditions, and agricultural practices over a period of several years, we employ sophisticated statistical and machine learning techniques in order to uncover the subtle linkages that regulate crop output. The findings of our study indicate that there are substantial correlations between the outcomes of yields and particular environmental parameters. These findings show the major impact that sustainable farming practices and climate adaptation methods have on the efficiency of agricultural production. The findings highlight the significance of integrated resource management and the requirement for precision agriculture

DOI: 10.4018/979-8-3693-3583-3.ch012

INTRODUCTION

Agriculture, which is at the center of global food security, is inextricably connected to both the problems that the 21st century presents and the solutions that this century offers. In light of the fact that the globe is struggling to meet the urgent requirements of an ever-increasing population, the search for environmentally responsible agriculture practices has never been more vital. The focus of this study work is to investigate the intricate relationship that exists between environmental elements, including climate variability, rainfall patterns, and the application of pesticides, and the effects that these factors have on agricultural productivity. In order to develop policies that not only increase agricultural yield but also protect the environment and maintain environmental s sustainability over the long term, it is essential to have a solid understanding of these relationships. Temperature increases, changing precipitation patterns, and an increase in the frequency of extreme weather events all contribute to the disruption of crop production because of climate change, which poses a tremendous challenge to agricultural stability all over the world. Because of the complex relationship that exists between climate conditions and agricultural output, it is necessary to conduct an in-depth investigation into the aspects of temperature changes and rainfall variability that have an impact on yield.

Although pesticides play an important role in the management of diseases and pests, they can have a negative impact on agricultural production. On the one hand, they make a considerable contribution to the conservation and increase of yields; on the other hand, their excessive use and misuse pose serious dangers to the health of the environment, including the deterioration of soil, the contamination of water, and the loss of biodiversity. This article examines the equilibrium that exists between the positive and negative impacts that pesticide application has on crop yields, with a particular focus on the necessity of using pesticides in a prudent manner within the context of integrated pest management (IPM) frameworks. The amount of rainfall is a significant factor in determining agricultural productivity, particularly in rain-fed farming systems, which are predominantly used on the majority of the cropland across the world. An in-depth investigation of the influence that climate change has on crop yields is required because of the unpredictability of rainfall patterns, which is frequently made worse by climate change. The purpose of this project is to investigate the ways in which the variability of rainfall affects agricultural outcomes and the ways in which adaptive water management systems can offset any negative consequences.

Objectives of the Research

In this paper, we hope to:

- Determine the extent to which temperature shifts, patterns of precipitation, and the application of pesticides have an impact on agricultural yields in a variety of geoFigureical areas and crop kinds using quantitative analysis.
- Determine the most important patterns and trends that can be used to inform the development of sustainable agriculture methods and procedures.
- It is important to provide insights into adaptive techniques that can improve crop tolerance to climate shocks and lessen reliance on chemical pesticides.

The Organization of the Paper

The structure of the paper is as follows, beginning with this introduction: In the second section, a literature study is conducted on the effects that climate change, the usage of pesticides, and the fluctuation of rainfall have had on agriculture. In the third section, the methodology is discussed, which includes the data sources, analytical techniques, and the framework for evaluating the correlations between environmental parameters and yield. This section presents the findings, focusing on the most important findings and the consequences of those findings. In the fifth section, the findings are discussed in relation to the current body of literature concerning sustainable agriculture and its significance. Last but not least, the conclusion of the report is included in Section 6, which includes a summary of the findings, recommendations for policy, and proposals for further research.

LITERATURE STUDY

Researchers have tried to use classical analytics in various ways to build models for pest forecasting in the previous several decades. Despite the fact that several of such programs saw only limited field application. One of the main reasons why these types of research have their limits is because pest cycles are extremely complicated natural phenomena. When it came to predicting pest life cycles, traditional analytics, which mostly dealt with linear or less sophisticated models, fell short. Even after accounting for a variety of agri-hetroginites, these analytical models were only applicable to a small region. Farmland size, agricultural methods, manure use, and pesticide chemical usage are examples of these agri-heterogeneities. Hundreds of models will be required to cover a large portion of the nation.

As discussed in the Introduction to Machine Learning, we can now develop several models rapidly thanks to new age technologies. This is all in the name of making a good difference in the lives of people all over the world, especially the many small-scale farmers who are responsible for cultivating tea and other commodities. Numerous studies utilizing machine learning models to study pest life cycles have been conducted in recent years (Zhang et al., 2017). Zhang et al. (2017) used SVM, GRNN, and multilayer feed-forward neural networks (MLFN) to forecast where D. superans would be most prevalent. They found that SVM was the most effective and accurate model for predicting where insect pests would be most prevalent. Xiao et al. (2018) conducted an additional study that modeled cotton pests using extensive data of eight diverse parameters: maximum and minimum temperatures, relative humidity in the morning and evening, rainfall, wind speed, daylight hours, and evapotranspiration. With an area under the curve (AUC) of 0.97, they discovered that LSTM neural nets deliver great results. The aforementioned research shows that various modelling approaches work well for certain pests. Which modeling approach would provide good results is also heavily influenced by the data employed. Despite this, most tea is grown by small-scale farmers who lack the resources to collect comprehensive data on a variety of factors (Parasar et al., 2019). We employ various Neural Network models for tea pest Looper predictions, taking the above into account. Among the most damaging pests is the tea looper (Reference). In addition, we make use of the sparse data set consisting just of weather conditions, which are accessible to even the most modest tea plantations. The market offers a variety of instruments, each with its own set of advantages and disadvantages. The stacking model and SAS Deep Learning Python (DLPy) were our tools of choice for this task. For this study's time series analysis, we opted to employ SAS Visual Forecasting. The following methods are employed to generate automated forecasts: Stacked Model,

Panel Series Neural Network, Seasonal Model, Temporal Aggregation Model, RNN Forecasting, and Auto-Forecasting. No human intervention is required. From less need for expensive expert resources to more consistent analyses and results, automation suggests various advantages. It also takes significantly less time to create and monitor the full forecast model. Past efforts have been made to utilize pest forecasting in order to comprehend the characteristics of infestations and so achieve more effective pest control in tea production. This approach aims to reduce expenses and minimize the amount of pesticides used (Kallor et al., 2020; Azrag et al., 2018). Several of the initial predictions were basic correlation studies or regression models (Ahmed et al., 2012). In recent years, as new technologies have advanced, people have come to recognize that approaches such as Neural Networks are more suitable for modeling the nonlinear phenomena of pest infestations (Zhang et al., 2017). This is because the majority of phenomena in nature exhibit strong nonlinearity and involve intricate interactions between multiple components. Zhang et al. (2017) assert that the Support Vector Machine Model is a dependable predictive technique for accurately and effectively forecasting the occurrence areas of insect pests.

However, in a recent study by Xiao et al. (2018), LSTM models were employed to model insect infestation in Cotton. The researchers utilized data from eight distinct criteria to forecast the likelihood of pests and diseases. The weather features are Maximum Temperature (MaxT in °C), Minimum Temperature (MinT in °C), Relative Humidity in the morning (RH1 in %), Relative Humidity in the evening (RH2 in %), Rainfall (RF in mm), Wind Speed (WS in comp), Sunshine Hour (SSH in hours), and Evaporation (EVP in mm). The testing results revealed that LSTM outperformed in predicting the occurrence of pests and diseases in cotton fields, achieving an impressive Area Under the Curve (AUC) of 0.97. Approximately 70% of the world's tea production is contributed by tiny tea planters, as mentioned above. These smaller firms lack the technological capabilities to collect data for several of the factors stated before. There is a significant disparity in the implementation of scientific methods for tea growing among small tea growers (Parasar et al., 2020).

Based on the aforementioned factors, we have chosen to utilize various Neural Network models for predicting pests in the tea business. Our main objective is to assess several neural network models specifically for the purpose of detecting and addressing Helopeltis infestation. However, the tea sector faces a hurdle in utilizing neural nets to construct complicated nonlinear models due to the lack of detailed data on various weather aspects, particularly for small tea growers. For accurate assessment of pest infestation, it is necessary to calculate the precise intensity of pests caught in traps placed at different locations within the tea garden. Subsequently, the collected samples are extrapolated to represent the entire tea garden. Similarly, small-scale cultivators sometimes lack the necessary resources to quantify characteristics like as evaporation or wind speed. We have chosen to utilize only two variables, namely the daily maximum and lowest temperatures, as well as the amount of rainfall, in order to simulate the extent of infestation by Helopeltis. There are multiple instruments accessible in the market, each with its own advantages and disadvantages. We opted to utilize SAS Deep Learning Python (DLPy) and a stacked model for this purpose.

DATA AND UNIVARIATE ANALYSIS

This comprehensive dataset is an amalgamation of several key agricultural and environmental variables, meticulously compiled to explore the multifaceted influences on agricultural yield. The dataset encompasses data from various geoFigureic locations, spanning multiple years, providing a rich temporal and spatial perspective on agricultural trends and patterns. The primary components of this dataset include:

Yield Data: Central to the dataset, the yield data captures the productivity of various crops, measured in hectograms per hectare (hg/ha). This metric serves as the primary outcome variable, reflecting the efficiency of agricultural production across different regions and timeframes.

Pesticide Usage: This variable quantifies the application of pesticides in tonnes of active ingredients. Pesticide data offers insights into the intensity of chemical use in agriculture, shedding light on practices that may affect both crop yields and environmental health.

Rainfall Patterns: The dataset incorporates rainfall measurements, expressed in millimeters per year, to assess the impact of water availability on agricultural productivity. Rainfall is a critical environmental factor, especially in rain-fed agricultural systems, where it directly influences crop growth and yield outcomes.

Temperature Data: Including average temperature readings, this variable allows for the examination of climate variability's effect on agriculture. Temperature fluctuations can significantly impact crop physiology, growth phases, and ultimately, yields.

GeoFigureic and Temporal Scope: The dataset spans multiple countries and years, offering a diverse and comprehensive view of global agricultural practices and outcomes. This wide-ranging coverage enables an in-depth analysis of how different environmental and management factors interact across various agricultural contexts.

Univariate Analysis

Univariate analysis involves examining each variable independently to summarize and find patterns in the data. The following visualizations and summaries provide insights into the distributions of key variables within the dataset.

Distribution of hg

Figure 1. Distribution of hg

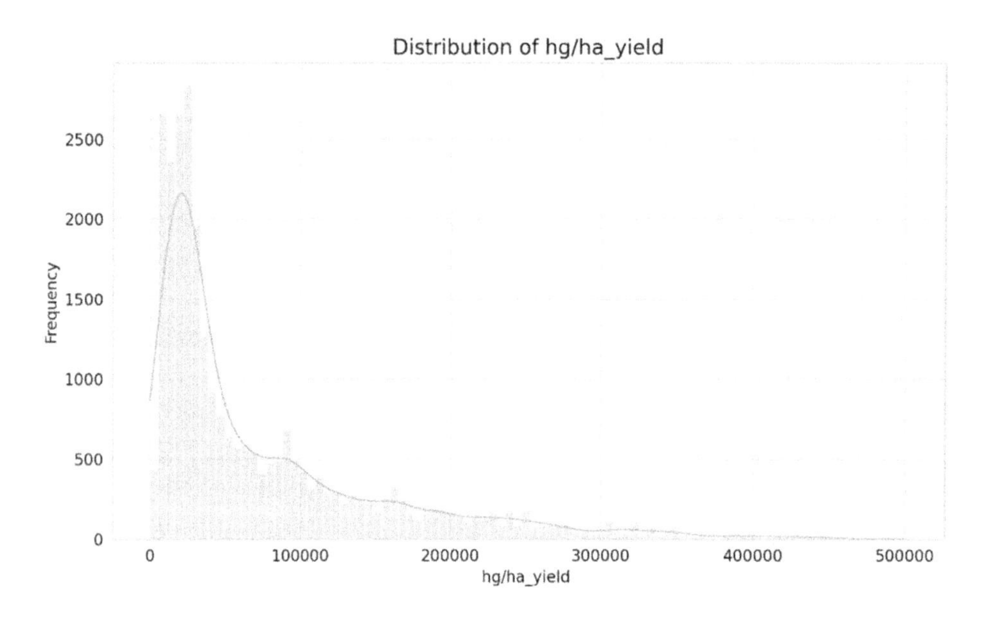

This histogram shows the distribution of values, helping us to understand the spread, central tendency, and outliers within the data. The presence of skewness or kurtosis in the distribution can also provide valuable insights into the underlying data generation process.

Distribution of Average Rainfall

Figure 2. Distribution of rainfall

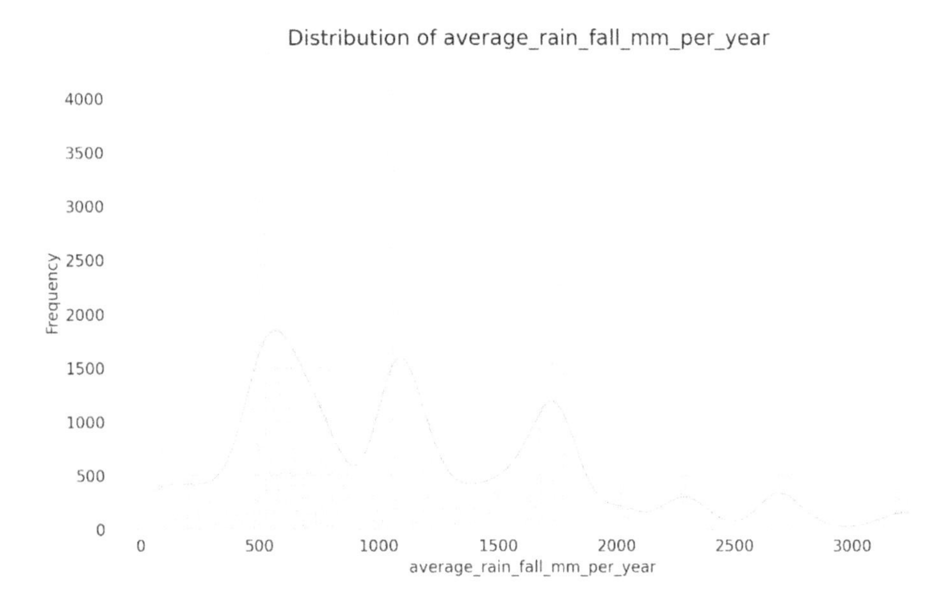

This histogram shows the distribution of values, helping us to understand the spread, central tendency, and outliers within the data. The presence of skewness or kurtosis in the distribution can also provide valuable insights into the underlying data generation process.

Distribution of Pesticides

Figure 3. Distribution of pesticides

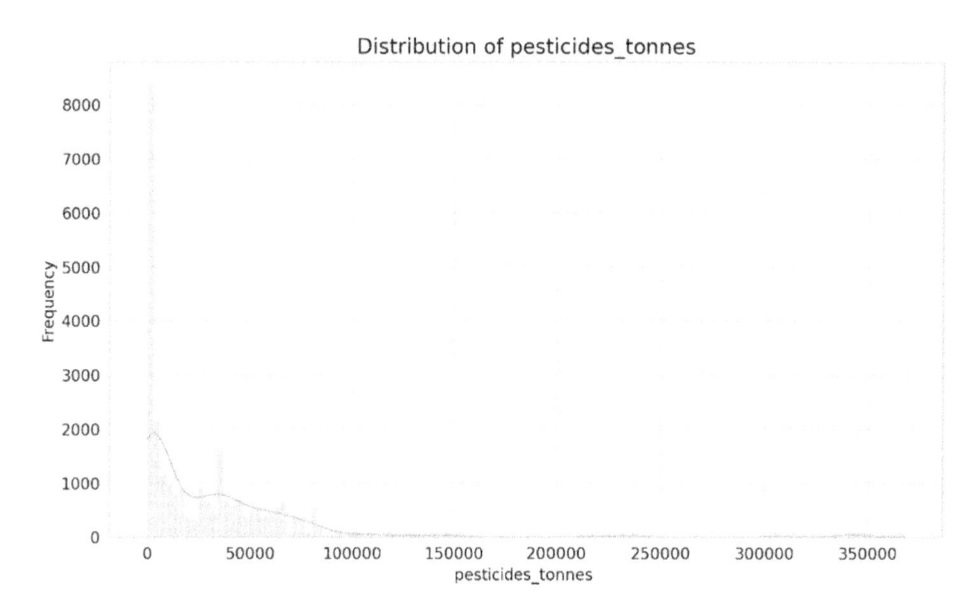

This histogram shows the distribution of values, helping us to understand the spread, central tendency, and outliers within the data. The presence of skewness or kurtosis in the distribution can also provide valuable insights into the underlying data generation process.

Distribution of Average Temperature

Figure 4. Distribution of average temperature

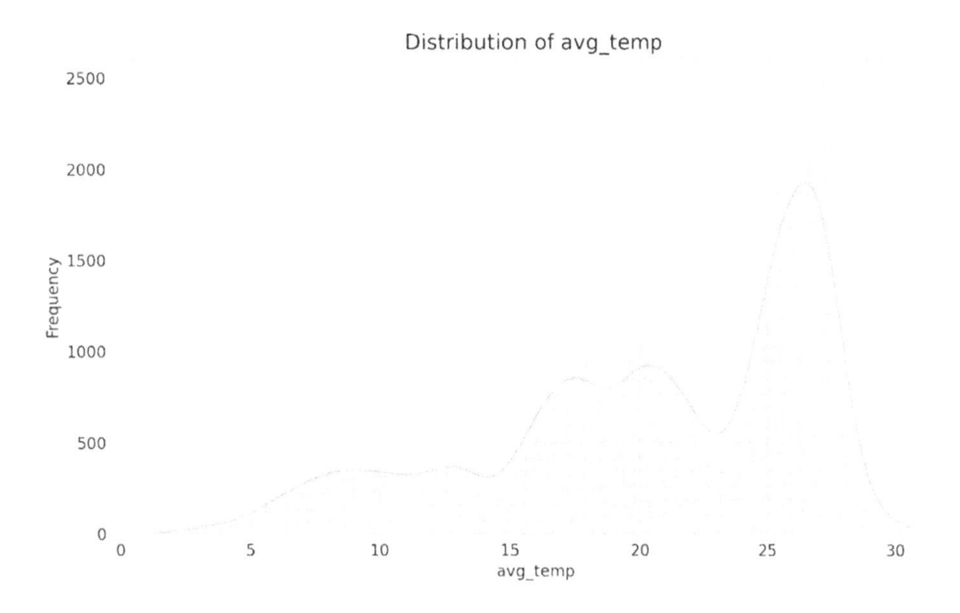

This histogram shows the distribution of values, helping us to understand the spread, central tendency, and outliers within the data. The presence of skewness or kurtosis in the distribution can also provide valuable insights into the underlying data generation process.

Bivariate Analysis

Bivariate analysis explores the relationship between two variables. This section presents scatter plots to visualize how pairs of variables relate to each other, highlighting any potential correlations or trends.

average_rain_fall_mm/year vs. hg/ha_yield

Figure 5. Distribution of erage_rain_fall_mm/year vs. hg/ha_yield

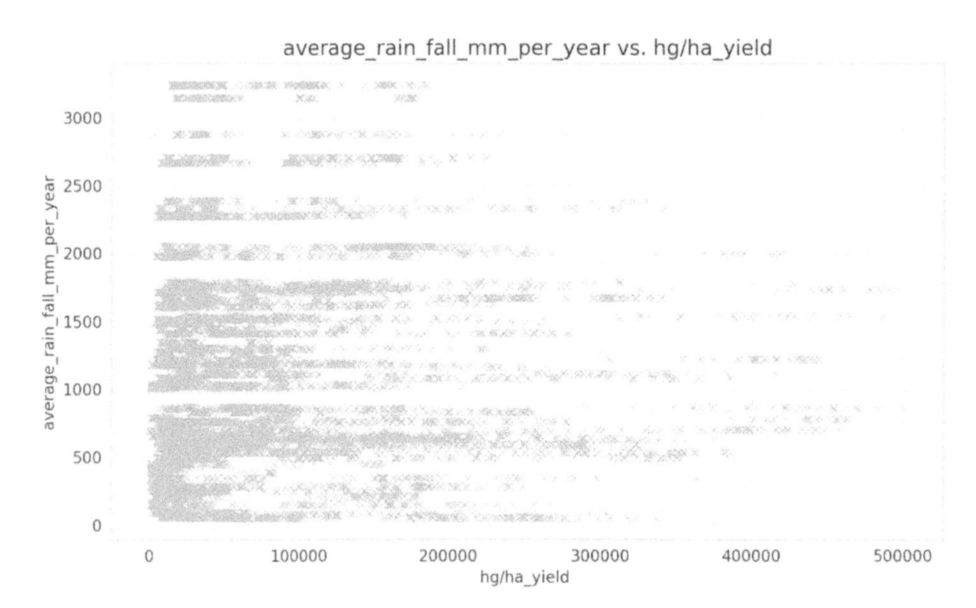

This scatter plot compares average_rain_fall_mm/year against hg/ha_yield, potentially revealing any correlations or patterns between the two variables. A clear trend or clustering within the plot can indicate a relationship worth investigating further.

pesticides_tonnes vs. hg/ha_yield

Figure 6. Distribution of pesticides_tonnes vs. hg/ha_yield

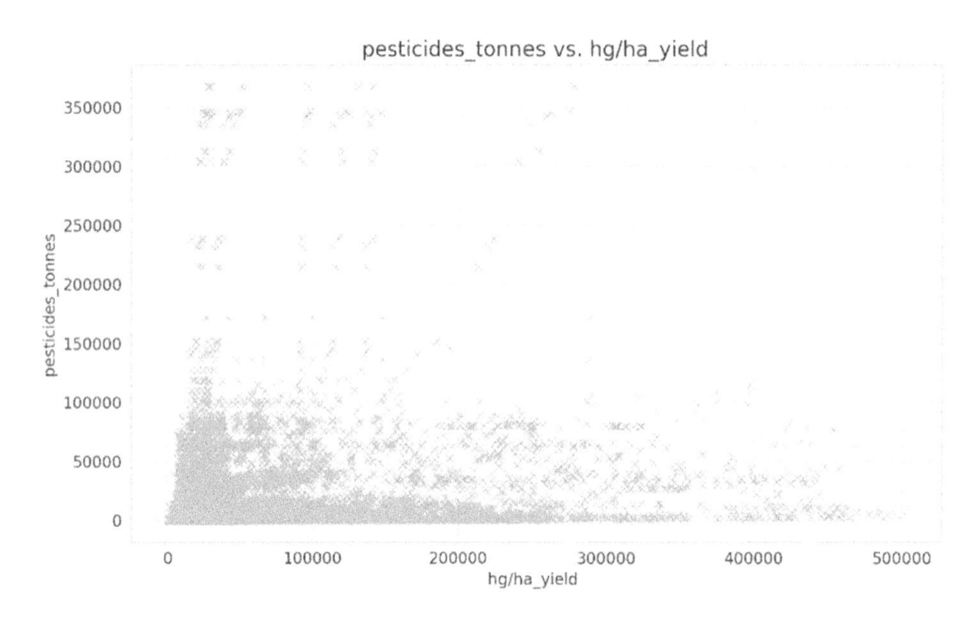

This scatter plot compares pesticides_tonnes against hg/ha_yield, potentially revealing any correlations or patterns between the two variables. A clear trend or clustering within the plot can indicate a relationship worth investigating further.

avg_temp vs. hg/ha_yield

Figure 7. Distribution of avg_temp vs. hg/ha_yield

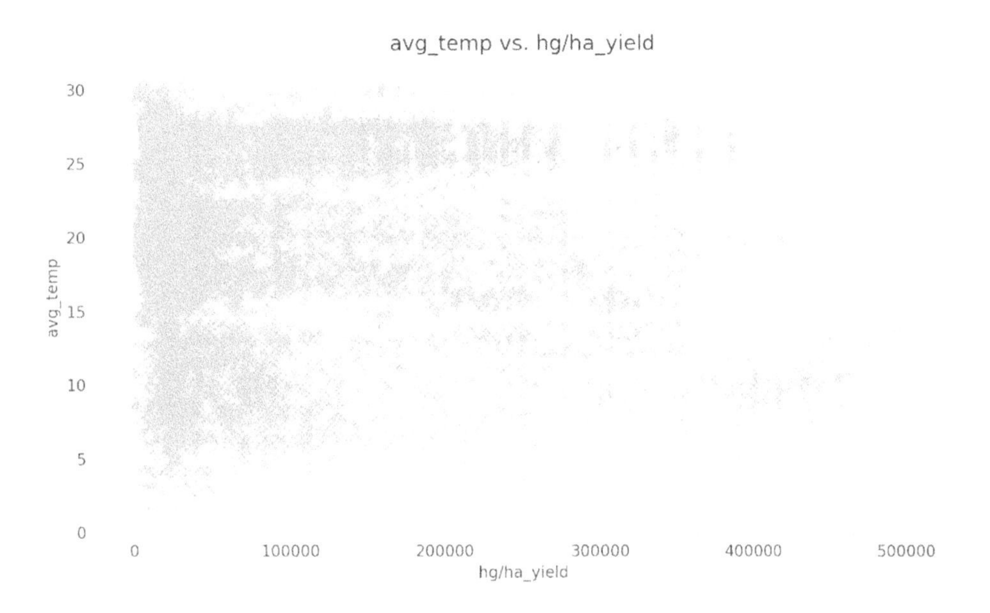

This scatter plot compares avg_temp against hg/ha_yield, potentially revealing any correlations or patterns between the two variables. A clear trend or clustering within the plot can indicate a relationship worth investigating further.

PREDICTIVE MODEL COMPARISON RESULTS

Feature Importance: This plot reinforces the findings that average temperature plays the most critical role in influencing yield, followed by pesticide usage and average rainfall.

Figure 8. Feature importance

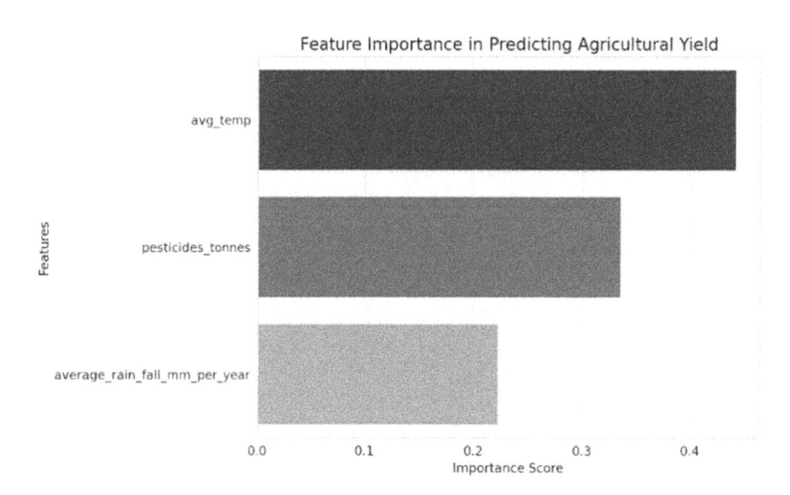

Predictive Model

A variety of regression models were initially selected to predict the agricultural yield, including Linear Regression, Random Forest Regressor, Gradient Boosting Regressor, and Support Vector Regressor. Each model was trained on a split of the dataset, with 80% used for training and 20% for testing, ensuring a robust evaluation of their predictive capabilities.

Model Comparison

The models were evaluated based on their Root Mean Square Error (RMSE) and Coefficient of Determination (R^2). RMSE provides a measure of the error magnitude between the predicted and actual values, while R^2 indicates the proportion of variance in the dependent variable that is predictable from the independent variables. To enhance the analysis, additional models were introduced, including Lasso Regression, Ridge Regression, and ElasticNet, along with an ensemble approach that averaged the predictions from all individual models. This extended analysis aimed to explore a broader range of modeling techniques and potentially uncover more effective prediction strategies. Models with negative R^2 scores, indicating performance worse than a simple mean-based model, were identified and excluded from further consideration. This refinement step focused the analysis on models that provided meaningful predictive insights. An updated comparison of the refined set of models was conducted, with the Gradient Boosting Regressor emerging as the most effective model based on its lower RMSE and higher R^2 score. The ensemble approach showed a slight improvement over some individual models, demonstrating the potential value of combining predictions.

Table 1. Model comparison based on R^2 score

Model	RMSE	R^2	MAE
Linear Regression	84254.64	0.021	-
Gradient Boosting Regressor	79665.09	0.125	-
Lasso Regression	84254.64	0.020	64100.00
Ridge Regression	84754.64	0.021	64239.00
ElasticNet	84983.88	0.020	64099.60
Ensemble Average	83833.74	0.031	62446.00

Residual Plot for the Gradient Boosting Regressor

This plot displays the residuals (differences between actual and predicted values) against the predicted values. A well-performing model would have residuals randomly scattered around the horizontal line (y=0), indicating no pattern in the errors. The presence of any systematic pattern could suggest issues like heteroscedasticity or non-linearity not captured by the model.

Figure 9. Residual plot

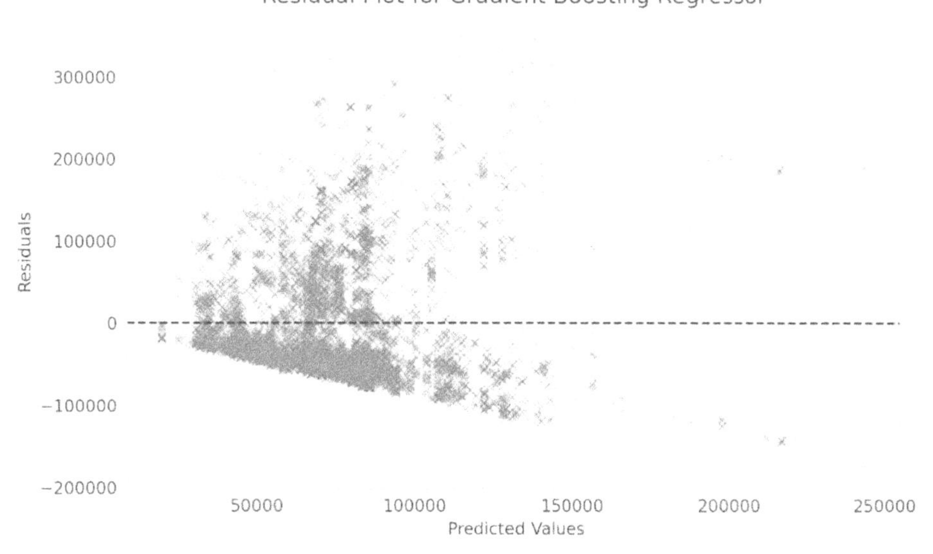

Learning Curve for the Gradient Boosting Regressor: This curve shows the training and cross-validation scores (negative mean squared error) for different sizes of the training data. It helps in understanding the model's behavior with increasing data and identifying if the model is suffering from high variance (overfitting) or high bias (underfitting)

Figure 10. Learning curve

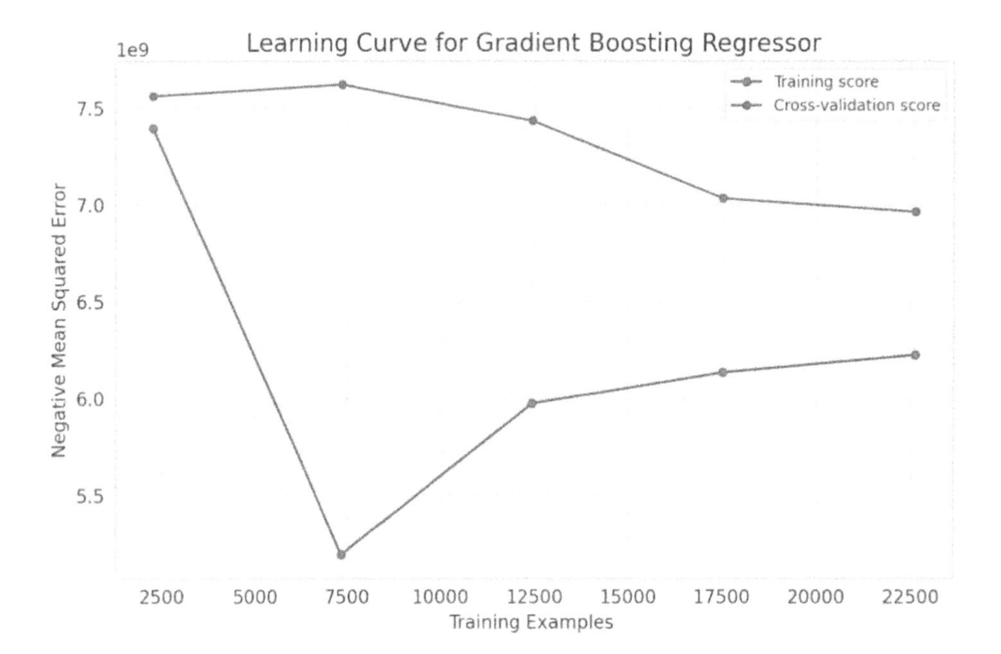

SUMMARY

Climate, pesticide use, and rainfall patterns were the primary foci of this research's multi-pronged investigation into agricultural yield drivers. The study painstakingly assessed the effect of each component on agricultural yield by making use of a large dataset and a variety of cutting-edge statistical and machine learning methods. The results showed that pesticide application rates and temperature fluctuations had a significant effect on production outcomes, highlighting the importance of adaptive farming techniques when dealing with climate change.

Key Findings

- The Centrality of Temperature: The results of the study highlighted the importance of temperature as a key factor in determining yield, with changes in temperature explaining large variations in agricultural output. This exemplifies how the threats to food security from climate change are becoming increasingly severe.
- The complex analysis of pesticide use revealed its duality: although necessary for pest management and crop improvement, the results call for their prudent use to reduce negative effects on the environment.

- The investigation highlighted the complex connection between rainfall patterns and crop yields, highlighting the need for effective water management measures to make agriculture more resilient to unpredictable weather.
- The predictive dynamics controlling agricultural yields were better understood through the comparative analysis of different models. Gradient Boosting Regressors stood out as the most effective model in capturing the intricate interdependencies of the elements that were analyzed.

CONCLUSION

This research sheds light on the complex web of factors that affect agricultural productivity, with climate change being one of the most pressing problems that contemporary agriculture is facing. This study's findings highlight the critical importance of a research- and policy-supported paradigm change towards more resilient and environmentally friendly farming methods. In light of the world's increasing population and the challenges posed by climate change, this study provides valuable insights on how to achieve long-term food security in this unpredictable period.

The research emphasizes the need of sustainable agricultural approaches that combine ecological management with production maximization. The pressing need for agricultural practices that are climate-smart and increase resilience to environmental stresses is highlighted by the significant influence of climate-related variables on yield. In addition, the study's results call for a shift in thinking about pesticide use, with the authors arguing for more eco-friendly integrated pest management strategies. Encourage the cultivation of crop varieties bred for higher resilience to temperature extremes and erratic rainfall. Leverage technology-driven approaches, such as precision agriculture, optimize resource utilization, including water and pesticides, enhancing yield efficiency and environmental sustainability. The study underscores the need for policy frameworks that incentivize sustainable agricultural practices and further research into adaptive strategies that mitigate the adverse impacts of climate change on agriculture.

Further Research

Future investigations could prioritize doing an in-depth examination of the enduring consequences of climate change on microclimatic circumstances in particular places that cultivate crops. The objective is to comprehend the influence of these alterations on the variability of agricultural production. Moreover, the implementation of localized models that integrate more detailed environmental data, such as soil moisture levels and micro-topographical variables, has the potential to greatly improve the precision of yield forecasts. Exploring the efficacy of alternative, eco-friendly pest control technologies, such as biological control agents, could enhance the incorporation of sustainable practices in pest management systems. Conducting comparative research on the resilience of genetically modified crops and traditional crops in different climatic stressors might offer useful insights into adaptive agriculture techniques.

Furthermore, it is crucial to examine the socio-economic effects of precision agriculture technology, particularly in small-scale farming communities in developing nations, to guarantee that technical progress contributes favorably to all agricultural sectors. Machine learning algorithms have the capability to forecast insect outbreaks by analyzing both environmental and biological data. This has the potential to decrease the need for chemical pesticides, resulting in a reduction in environmental degradation. Assessing the carbon footprint linked to various farming methods, such as pesticide usage and emerging technology, will be essential for advancing sustainability. Finally, doing lifecycle evaluations of crops from farm to table, in conjunction with research on governmental regulations impacting the adoption of sustainable farming practices, can yield a thorough comprehension of the systemic modifications necessary to bolster sustainable agriculture in light of global climate change.

REFERENCES

Ariyani, M., Pitoi, M. M., Koesmawati, T. A., Maulana, H., & Endah, E. S. (2020). *Pyrethroid residues on tropical soil of an Indonesian tea plantation: analytical method development, monitoring, and risk assessment* (Vol. 7).

Ayouba, K., & Vigeant, S. (2020). Can We Really Use Prices to Control Pesticide Use? Results from a Nonparametric Model. Springer. 10.1007/s10666-020-09714-w

Bengal, S. W., Dutta, T., & Nayak, C. (2019). S-Transferase Enzyme Activities and Their Correlation with Genotypic Variations Based on GST M1 and GST T1 Loci in Long Term-Pesticide-Exposed Tea Garden. *Toxicology and Environmental Health Sciences*, 11(1), 63–72. 10.1007/s13530-019-0389-1

Bondori, A. (2021). *Modeling farmers ' intention for safe pesticide use: the role of risk perception and use of information sources.* Research gate.

Bonvoisin, T., Utyasheva, L., Knipe, D., Gunnell, D., & Eddleston, M. (2020). *Suicide by pesticide poisoning in India: a review of pesticide regulations and their impact on suicide trends.*

Damalas, C. A. (2021). Farmers ' intention to reduce pesticide use: the role of perceived risk of loss in the model of the planned behavior theory. Research Gate.

Kramer, K. (2020). Seasonal abundance of tea mosquito bug, Looper Antoni Signoret infesting name. *Journal of Entomology and Zoology Studies*, 8(6), 2006–2009. christianaid.org.uk. 10.22271/j.ento.2020.v8.i6aa.8118

Dwijendra, N. (2022). Machine Learning Time Series Models For Tea Pest Helopeltis Infestation In India. *Webology, 19*(2). https://www.webology.org/abstract.php?id=1625

. Dwijendra N. (2022). Benchmarking of traditional and advanced machine Learning modelling techniques for forecasting in book. *Visualization Techniques for Climate Change with Machine Learning and Artificial Intelligence*. Elsevier. 10.1016/B978-0-323-99714-0.00017-0

Dwivedi, D., Kapur, P. N., & Kapur, N. N. (2023). Machine Learning Time Series Models for Tea Pest Looper Infestation in Assam, India. In Sharma, A., Chanderwal, N., & Khan, R. (Eds.), *Convergence of Cloud Computing, AI, and Agricultural Science* (pp. 280–289). IGI Global. 10.4018/979-8-3693-0200-2.ch014

Dwivedi, D., & Vemareddy, A. (2023). Sentiment Analytics for Crypto Pre and Post Covid: Topic Modeling. In Molla, A. R., Sharma, G., Kumar, P., & Rawat, S. (Eds.), Lecture Notes in Computer Science: Vol. 13776. *Distributed Computing and Intelligent Technology. ICDCIT 2023*. Springer. 10.1007/978-3-031-24848-1_21

Dwivedi, D. N., & Anand, A. (2021). Trade Heterogeneity in the EU: Insights from the Emergence of COVID-19 Using Time Series Clustering. *Zeszyty Naukowe Uniwersytetu Ekonomicznego w Krakowie*, 3(993), 9–26. 10.15678/ZNUEK.2021.0993.0301

Dwivedi, D. N., & Anand, A. (2021). The Text Mining of Public Policy Documents in Response to COVID-19: A Comparison of the United Arab Emirates and the Kingdom of Saudi Arabia. *Public Governance / Zarządzanie Publiczne, 55*(1), 8-22. 10.15678/ZP.2021.55.1.02

Dwivedi, D. N., & Anand, A. (2022). A Comparative Study of Key Themes of Scientific Research Post COVID-19 in the United Arab Emirates and WHO Using Text Mining Approach. In Tiwari, S., Trivedi, M. C., Kolhe, M. L., Mishra, K., & Singh, B. K. (Eds.), *Advances in Data and Information Sciences. Lecture Notes in Networks and Systems* (Vol. 318). Springer. 10.1007/978-981-16-5689-7_30

Dwivedi, D. N., & Batra, S. (2024). Case Studies in Big Data Analysis: A Novel Computer Vision Application to Detect Insurance Fraud. In Darwish, D. (Ed.), *Big Data Analytics Techniques for Market Intelligence* (pp. 441–450). IGI Global. 10.4018/979-8-3693-0413-6.ch018

Dwivedi, D. N., Batra, S., & Pathak, Y. K. (2024). Enhancing Customer Experience: Exploring Deep Learning Models for Banking Customer Journey Analysis. In Sharma, H., Chakravorty, A., Hussain, S., & Kumari, R. (Eds.), *Artificial Intelligence: Theory and Applications. AITA 2023. Lecture Notes in Networks and Systems* (Vol. 843). Springer. 10.1007/978-981-99-8476-3_39

Dwivedi, D. N., & Gupta, A. (2022). Artificial intelligence-driven power demand estimation and short-, medium-, and long-term forecasting. In *Artificial Intelligence for Renewable Energy Systems* (pp. 231–242). Woodhead Publishing. 10.1016/B978-0-323-90396-7.00013-4

Dwivedi, D. N., & Mahanty, G. (2024). Mental Health in Messages: Unravelling Emotional Patterns Through Advanced Text Analysis. In Rai, M., & Pandey, J. (Eds.), *Using Machine Learning to Detect Emotions and Predict Human Psychology* (pp. 187–208). IGI Global. 10.4018/979-8-3693-1910-9.ch009

Dwivedi, D. N., Mahanty, G., & Vemareddy, A. (2022). How Responsible Is AI?: Identification of Key Public Concerns Using Sentiment Analysis and Topic Modeling. [IJIRR]. *International Journal of Information Retrieval Research*, 12(1), 1–14. 10.4018/IJIRR.298646

Dwivedi, D. N., Mahanty, G., & Vemareddy, A. (2023). Sentiment Analysis and Topic Modeling for Identifying Key Public Concerns of Water Quality/Issues. In: Harun, S., Othman, I.K., Jamal, M.H. (eds) *Proceedings of the 5th International Conference on Water Resources (ICWR). Lecture Notes in Civil Engineering*. Springer, Singapore. 10.1007/978-981-19-5947-9_28

Dwivedi, D. N., Pandey, A. K., & Dwivedi, A. D. (2023). Examining the emotional tone in politically polarized Speeches in India: An In-Depth analysis of two contrasting perspectives. *SOUTH INDIA JOURNAL OF SOCIAL SCIENCES, 21*(2), 125-136. https://journal.sijss.com/index.php/home/article/view/65

Dwivedi, D. N., & Pathak, S. (2022). Sentiment Analysis for COVID Vaccinations Using Twitter: Text Clustering of Positive and Negative Sentiments. In Hassan, S. A., Mohamed, A. W., & Alnowibet, K. A. (Eds.), *Decision Sciences for COVID-19. International Series in Operations Research & Management Science* (Vol. 320). Springer. 10.1007/978-3-030-87019-5_12

Dwivedi, D. N., & Patil, G. (2023). Climate change: Prediction of solar radiation using advanced machine learning techniques. In Srivastav, A., Dubey, A., Kumar, A., Narang, S. K., & Khan, M. A. (Eds.), *Visualization Techniques for Climate Change with Machine Learning and Artificial Intelligence* (pp. 335–358). Elsevier. 10.1016/B978-0-323-99714-0.00017-0

Dwivedi, D. N., Wójcik, K., & Vemareddyb, A. (2022). Identification of Key Concerns and Sentiments Towards Data Quality and Data Strategy Challenges Using Sentiment Analysis and Topic Modeling. In Jajuga, K., Dehnel, G., & Walesiak, M. (Eds.), *Modern Classification and Data Analysis. SKAD 2021. Studies in Classification, Data Analysis, and Knowledge Organization*. Springer. 10.1007/978-3-031-10190-8_2

Gallia, G. K. N., & Kephaliacos, C. (2021). Ecological - economic modeling of pollination complexity and pesticide use in agricultural crops. *Journal of Bioeconomics*, 23(3), 297–323. 10.1007/s10818-021-09317-9

González-Nucamendi, A., Noguez, J., Neri, L., Robledo-Rella, V., & García-Castelán, R. M. (2023). Predictive analytics study to determine undergraduate students at risk of dropout. *Frontiers in Education*, 8, 1244686. 10.3389/feduc.2023.1244686

Gupta, A. (2021). Understanding Consumer Product Sentiments through Supervised Models on Cloud: Pre and Post COVID. *Webology, 18*(1). .10.14704/WEB/V18I1/WEB18097

Gupta, A., Dwivedi, D.N. & Jain, A. (2021). Threshold fine-tuning of money laundering scenarios through multi-dimensional optimization techniques. *Journal of Money Laundering Control*. 10.1108/JMLC-12-2020-0138

Gupta, A., Dwivedi, D. N., & Shah, J. (2023). Overview of Money Laundering. In: Artificial Intelligence Applications in Banking and Financial Services. *Future of Business and Finance*. Springer, Singapore. 10.1007/978-981-99-2571-1_1

Gupta, A., Dwivedi, D. N., & Shah, J. (2023). Financial Crimes Management and Control in Financial Institutions. In: *Artificial Intelligence Applications in Banking and Financial Services*. pringer, Singapore. 10.1007/978-981-99-2571-1_2

Gupta, A., Dwivedi, D. N., & Shah, J. (2023). Overview of Technology Solutions. In: *Future of Business and Finance*. Springer, Singapore. 10.1007/978-981-99-2571-1_3

Gupta, A., Dwivedi, D. N., & Shah, J. (2023). Data Organization for an FCC Unit. In *Future of Business and Finance*. Springer, Singapore. 10.1007/978-981-99-2571-1_4

Gupta, A., Dwivedi, D. N., & Shah, J. (2023). Planning for AI in Financial Crimes. In: *Future of Business and Finance*. Springer, Singapore. 10.1007/978-981-99-2571-1_5

Gupta, A., Dwivedi, D. N., & Shah, J. (2023). Applying Machine Learning for Effective Customer Risk Assessment. In: *Future of Business and Finance*. Springer, Singapore. 10.1007/978-981-99-2571-1_6

Gupta, A., Dwivedi, D. N., & Shah, J. (2023). Artificial Intelligence-Driven Effective Financial Transaction Monitoring. In: *Future of Business and Finance*. Springer, Singapore. 10.1007/978-981-99-2571-1_7

Gupta, A., Dwivedi, D. N., & Shah, J. (2023). Machine Learning-Driven Alert Optimization. In: *Future of Business and Finance*. Springer, Singapore. 10.1007/978-981-99-2571-1_8

Gupta, A., Dwivedi, D. N., & Shah, J. (2023). Applying Artificial Intelligence on Investigation. In: *Future of Business and Finance*. Springer, Singapore. 10.1007/978-981-99-2571-1_9

Gupta, A., Dwivedi, D. N., & Shah, J. (2023). Ethical Challenges for AI-Based Applications. In: *Future of Business and Finance*. Springer, Singapore. 10.1007/978-981-99-2571-1_10

Gupta, A., Dwivedi, D. N., & Shah, J. (2023). Setting up a Best-In-Class AI-Driven Financial Crime Control Unit (FCCU). In: *Future of Business and Finance*. Springer, Singapore. 10.1007/978-981-99-2571-1_11

Hellas, A., Ihantola, P., Petersen, A., Ajanovski, V., Gutica, M., Hynninen, T., Knutas, A., Leinonen, J., Messom, C., & Liao, S. Predicting academic performance: a systematic literature review. *Proceedings Companion of the 23rd Annual ACM Conference on Innovation and Technology in Computer Science Education*. (175-199). ACM. 10.1145/3293881.3295783

Heshmati, F. (2021). *Simultaneous multi- determination of pesticide residues in black tea leaves and infusion: a risk assessment study*.

Jia, C., Hew, K., Bai, S. & Huang, W. (2021). Adaptation of a conventional flipped course to an online flipped format during the Covid-19 pandemic: Student learning performance and engagement. *Journal of Research on Technology in Education, 54*(2). 10.1080/15391523.2020.1847220

Kang, K., & Wang, S. (2018). Analyze and predict student dropout from online programs. In *Proceedings of the 2nd International Conference on Compute and Data Analysis* (pp. 6-12). ACM. 10.1145/3193077.3193090

Kim, S., Yoo, E., & Kim, S. (2023). *A Study on the Prediction of University Dropout Using Machine Learning*. arXiv preprint arXiv:2310.10987. DOI:/arXiv.2310.1098710.48550

Kwakye, M. O., Mengistie, B., & Ofosu, J. (2019). Pesticide registration, distribution and use practices. *Environment, Development and Sustainability*, 21(6), 2647–2671. 10.1007/s10668-018-0154-7

Lainjo, B. (2023). Mitigating Academic Institution Dropout Rates with Predictive Analytics Algorithms. *International Journal of Education, Teaching, and Social Sciences*.

Lainjo, B., & Tsmouche, H. (2023). Impact of Artificial Intelligence On Higher Learning Institutions. *International Journal of Education, Teaching, and Social Sciences*.

Le, V. S., Lesueur, D., Herrmann, L., Hudek, L., Ngoc, L., & Lambert, Q. (2021). Sus- tainable tea production through agroecological management practices in Vietnam: A review. *Environmental Sustainability*, 4(0123456789), 589–604. 10.1007/s42398-021-00182-w

Lourens, A., & Bleazard, D. (2016). Applying predictive analytics in identifying students at risk: A case study. *South African Journal of Higher Education*, 30(2), 129–142. 10.20853/30-2-583

Mahanty, G., Dwivedi, D. N., & Gopalakrishnan, B. N. (2021). The Efficacy of Fiscal Vs Monetary Policies in the Asia-Pacific Region: The St. Louis Equation Revisited. *Vision (Basel)*, (November). 10.1177/09722629211054148

Mallik, P., & Ghosh, T. (2008). Impact of climate on tea production: A study of the Dooars region in India. *Theor. Apple. Climatol*. 10.1007/s00704-021-03848-x

Moon, M.-H., & Kim, G. (2023). Predicting University Dropout Rates Using Machine Learning Algorithms. *Journal of Economics and Finance Education*, 32(2), 57–68. 10.46967/jefe.2023.32.2.57

Nikolaidis, P., Ismail, M., Shuib, L., Khan, S., & Dhiman, G. (2022). *Predicting Student Attrition in Higher Education through the Determinants of Learning Progress: A Structural Equation Modelling Approach. Sustainability.* MDPI. https://www.mdpi.com/2071-1050/14/20/1358410.3390/su142013584

Niwa, H. (2021). Detection of organic tea farms based on the density of spider webs using aerial photoFigurey with an unmanned aerial vehicle (UAV). *Landscape and Ecological Engineering*, 17(4), 541–546. 10.1007/s11355-021-00454-x

Nurmalitasari, A. (2023). *The Predictive Learning Analytics for Student Dropout Using Data Mining Technique: A Systematic Literature Review. Advances in Technology Transfer Through IoT and IT Solutions.* Springer. . (9-17). 10.1007/978-3-031-25178-8_2

Oqaidi, K., Aouhassi, S., & Mansouri, K. (2022). A Comparison between Using Fuzzy Cognitive Mapping and Machine Learning to Predict Students' Performance in Higher Education *2022 IEEE 3rd International Conference on Electronics, Control, Optimization and Computer Science (ICECOCS).* IEEE. https://ieeexplore.ieee.org/document/9983470/10.1109/ICECOCS55148.2022.9983470

Pal, K. (2019). Bio-control of Pests in Tea: Effect of Environmental. *International Journal of Applied and Computational Mathematics*, 5(3), 1–9. 10.1007/s40819-019-0666-3

Pardo, L. A. (2020). Pesticide exposure and risk of aggressive prostate cancer among private pesticide applicators. *Reading the tea leaves Climate change and the British cuppa.* 10.1186/s12940-020-00583-0

Prasad, K., Roy, S., Sen, S., Neave, S., Nagpal, A., & Pandit, V. (2020). Impact of different pest management practices on natural enemy population in tea plantations of Assam special emphasis on spider fauna. Springer. 10.1007/s42690-020-00111-0

Prasanth, A., & Alqahtani, H. (2023). Predictive Models for Early Dropout Indicators in University Settings Using Machine Learning Techniques. In *2023 IEEE International Conference on Emerging Technologies and Applications in Sensors (ICETAS).* IEEE. 10.1109/ICETAS59148.2023.10346531

Seidel, E., & Kutieleh, S. (2017). Using predictive analytics to target and improve first year student attrition. *Australian Journal of Education*, 61(2), 200–218. 10.1177/0004944117712310

Shafiq, D. A., Marjani, M., Habeeb, R. A. A., & Asirvatham, D. (2022). Predictive Analytics in Education: A Machine Learning Approach. In *2022 3rd International Multidisciplinary Conference on Computer and Energy Science (SpliTech)* (pp. 1-6). IEEE. [DOI:10.1109/MACS56771.2022

Shuqfa, Z., & Harous, S. (2019). Data Mining Techniques Used in Predicting Student Retention in Higher Education: A Survey. *2019 International Conference on Electrical and Computing Technologies and Applications (ICECTA).* IEEE. . https://ieeexplore.ieee.org/document/8959789/10.1109/ICECTA48151.2019.8959789

Soydan, D. K., Turgut, N., Yalç, M., & Turgut, C. (2021). *Evaluation of pesticide residues in fruits and vegetables from the Aegean region of Turkey and assessment of risk to consumers.*

Suganthi, M., Event, S., & Senthilkumar, P. (2020). Comparative bioefficacy of Bacil- lus and Pseudomonas continues against Looper severe in tea (Camellia sinensis (L.) O. Kuntze. *Physiology and Molecular Biology of Plants*, 26(10), 2053–2060. 10.1007/s12298-020-00875-233088049

Sun, R., Yang, W., Li, Y., & Sun, C. (2021). Multi - residue analytical methods for pesticides in teas: A review. *European Food Research and Technology*, 247(8), 1839–1858. 10.1007/s00217-021-03765-3

Tambe, A. B., Mbanga, B. M. R., Nzefa, D. L., & Name, M. G. (2019). Pesticide usage and occupational hazards among farmers working in small-scale tomato farms in Cameroon. Research Gate. 10.1186/s42506-019-0021-x

Tinto, V. (1975). Dropout from Higher Education: A Theoretical Synthesis of Recent Research. *Review of Educational Research*, 45(1), 89–125. 10.3102/00346543045001089

Xie, S. (2019). *Does a dual reduction in chemical fertilizer and pesticides, improve nutrient loss and tea yield and quality? A pilot study in a green tea garden in Shaoxing*. Zhejiang Province. 10.1007/s11356-018-3732-1

Zhang, D. (2019). Detection of systemic pesticide residues in tea products at trace level based on SERS and verified by GC – MS. Springer. 10.1007/s00216-019-02103-7

Zhang, W. Y., Jing, T. Z., & Yan, S. C. (2017). Studies on prediction models of Den- drolimus superans occurrence area based on machine learning. *Journal of Beijing Forestry University*, 39(1), 85–93.

Chapter 13
Identification, Classification, and Grading of Crops Grain Using Computer Intelligence Techniques: A Review

Nabin Kumar Naik
Sambalpur University, India

Prabira Kumar Sethy
http://orcid.org/0000-0003-3477-6715
Guru Ghasidas Vishwavidyalaya, Bilaspur, India

Santi Kumari Behera
Veer Surendra Sai University of Technology, India

ABSTRACT

India is the second-largest food producer globally, trailing only in China. However, significant agricultural losses occur because of the lack of skilled laborers. Harvested commodities often go into waste. Additionally, the imprecise nature of crop identification, classification, and quality inspection, which is influenced by human subjectivity, poses challenges. To address these issues and reduce labor costs, the agricultural sector must embrace automation. Developing an automated system capable of distinguishing between various crops based on their texture, shape, and color is feasible by employing appropriate image-processing techniques and machine-learning methods. This study focuses on advancing the state-of-the-art research in this field. It briefly explores recent research publications' methodologies, comparing them using diverse techniques, such as k-nearest neighbors (KNN), artificial neural networks (ANN), random forest (RF), naive bayes (NB), backpropagation neural networks (BPNN), support vector machines (SVM), and convolutional neural networks (CNN).

DOI: 10.4018/979-8-3693-3583-3.ch013

INTRODUCTION

It is impossible to think about life without agriculture as it is to breathe without oxygen because sustaining food is essential. Agriculture is not only a primary source of food, but also has a huge impact on employment and income across the world. Agriculture is an integral part of the world economy, mainly for developing countries, including India, China, the United States, Brazil, Mexico, Russia, Japan, Germany, and France. Agriculture is the main sector in the world economy as well as the essence of human life. It plays an important role in the economic growth of a country. Owing to India's varied environment, various types of crop grains are always available. After China, it produces different types of crops, such as rice, lentils, and wheat worldwide. Grains are important because they provide the best nutrients, such as carbohydrates. Approximately 58% of India's total land is dependent on agriculture. India is the 6th largest country in terms of food marketing and land production worldwide. The agricultural industry in India is increasing at a high pace and is contributing to global trade in a continuous manner. According to FAO (2021), agriculture employs 67 percent of the total population and accounts for 39.4 percent of GDP and 43 percent of all exports. In India 158.25 million number of people will be employed in the agriculture sector in the financial year 2022. Figure 1 shows the GDP share of agriculture in different countries in 2022 (Anwer et al., 2015).

Figure 1. Top 20 countries on the basis of GDP share of agriculture

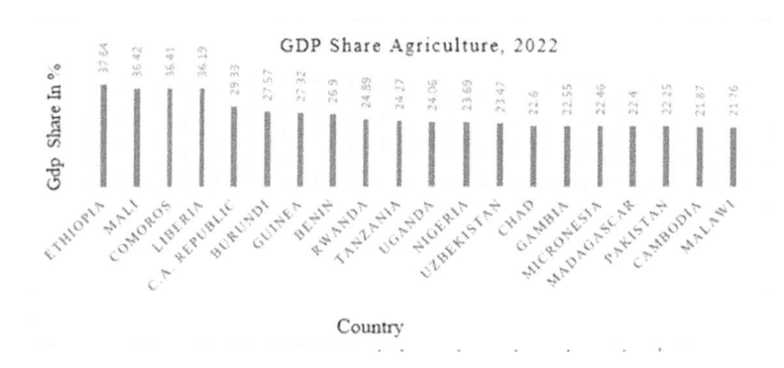

Figure 2. Worldwide production of grain in 2022

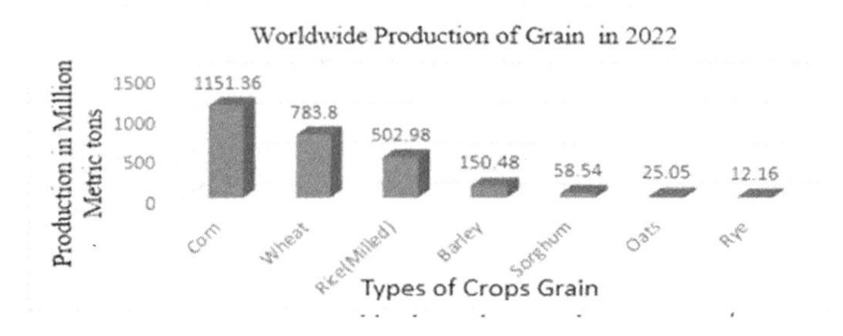

Figure 3. Production rate of crops grain in millon tons

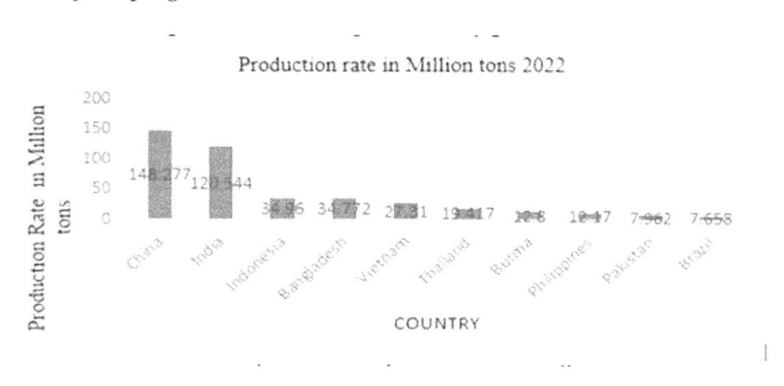

Figure 4. Yearly food grain production in India 2010-2022

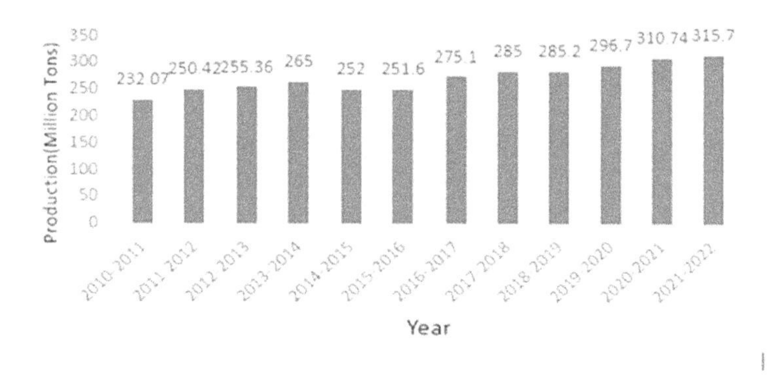

Figure 2 shows the production of different grains in the year 2022, and Figure 3 indicates the production rate of crops in different countries in the year 2022. These figures suggest that different countries produce many grains (Giunta et al., 2019). Figure 4 shows the yearly food production in India from to 2010-2022 (Government of India, 2023). The agriculture sector must impose an automation system owing to the high rate of crop grain production and the lack of skilled labor.

METHODOLOGY

A survey was conducted on various studies related to image processing, machine learning, and deep learning in agriculture. In this review, multiple sources have been used. The review process commenced by studying the abstracts and conclusions of various papers. Initially, a keyword search was conducted to identify conference papers and journal articles. This search spanned scientific databases such as ScienceDirect and IEEE Xplore, as well as scientific indexing services such as the Web of Science and Google Scholar. Full papers that met these criteria were downloaded for further analysis. Research papers published between 2010 and 2023 were selected for inclusion in this review. We collect the desired paper and filter out the research papers irrevalent to identification, classification and grading of crop grain in the field of agriculture and various terms searched those were:(Image processing + Agriculture), (Image processing + Identification), (Image processing + Classification), (Image processing + Grading),(Identification+ Machine Learning), (Identification+ Deep Learning), (Classification+ Machine Learning),(Classification+ Deep Learning). We downloaded 100 papers from various sources: IEEE Explore, Google Scholar, and ScienceDirect, and filtered only 75 out of 100 papers for consideration in this review. Statistical information about the technological advancement for the identification, classification, and grading of crop grain in the field of agriculture is shown in figure 5.

Figure 5. Number of papers published from 2010 to 2023 on the identification, classification, and grading of crop grain in the field of agriculture

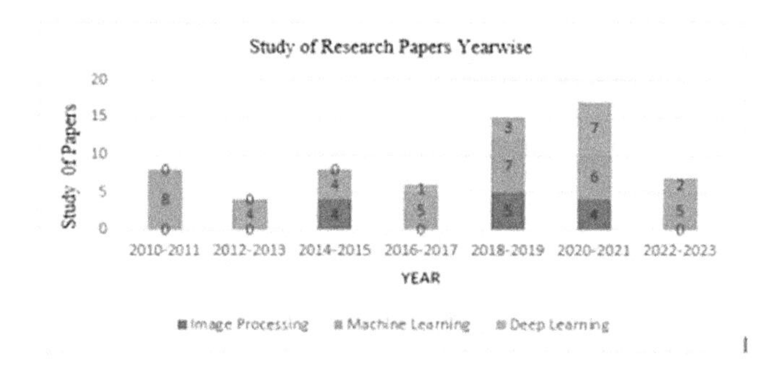

IDENTIFICATION OF CROP GRAIN

Crop identification has found practical applications in various real-world scenarios, such as replacing manual barcode scanning during store checkouts and serving as a support system for visually impaired individuals. Recognizing different crop varieties is a repetitive task, and automating the labeling and pricing calculations of grain commodities is an ideal solution to this challenge. Automated systems for grain identification have revolutionized the selection process, significantly reducing evaluation time compared to manual methods. Various image processing techniques have been employed by scientists to automatically identify cereal crop varieties. For instance, Chen et al. (2010) utilized machine vision and pattern recognition to distinguish between five Chinese maize varieties. They extracted 13 form characteristics, 17 geometry features, and 28 color features, achieving an impressive classification accuracy of 90%. Ouyang et al. (2010) developed a system to automatically capture and classify images of five rice seed varieties, achieving a classification accuracy of 86.65% using backpropagation and a CCD camera with Visual C++ 6.0 for image processing. Shantaiya et al. (2010) effectively classified six native rice species from the Chhattisgarh region by employing backpropagation neural networks and analysing factors such as colour, morphology, and texture. Patil et al. (2011) investigated numerous colour models for cereal crop classification, including Lab, HSV, HSI, and YCbCr, utilising K-NN and a minimum distance classifier. Using a combination of machine vision and artificial neural networks (ANN), Arefi et al. (2011) were able to classify four varieties of wheat based on morphological and spectral characteristics with an overall accuracy of 95.86%. Taking into account distinctions in colour, morphology, and texture, Gujjar et al. (2013) constructed an image algorithm that can identify six varieties of Indian basmati rice. Nasirahmadi et al. (2013) used machine vision and multilayer perceptron artificial neural networks (MLP-ANN) to differentiate ten distinct legume varieties in Iran, with ANN proving to be the preferable classification technique. Szczypiski et al. (2015) employed linear discriminant analysis and neural networks, achieving success rates between 67% to 88%, to identify barley varieties based on shape, color, and texture. Teimour et al. (2015) distinguished five varieties of almonds using an image processing and artificial intelligence-based classification and ranking system. Sumaryanti et al. (2015) analyzed a model for identifying rice varieties through image processing, incorporating six color variables, four morphological features, and two texture features, and achieved a high classification accuracy of 96.6%. P. Sethy et al. (2018) classified six rice varieties with an accuracy of 92% using geometric and textural features applied to a multiclass support vector machine (M-SVM) algorithm. E. R. Arboleda et al. (2018) demonstrated image processing techniques for black coffee beans, achieving a 100% classification rate by extracting RGB color components. Zhengjun Qiu et al. (2018) utilized Convolutional Neural Networks (CNN) with hyperspectral images to identify four varieties of rice seed successfully. Yahya Altuntaş et al. (2019) proposed a CNN and transfer learning approach to distinguish between haploid and diploid maize seeds with high accuracy. Utilizing morphological, color, wavelet, and Gaborlet characteristics, Esra et al. (2019) successfully classified vitreous and starchy durum wheat kernels as well as alien objects with an accuracy rate of 93.46%. In a study conducted by L. Jena et al. (2021), a model was developed to distinguish between three varieties of wheat grains using geometrical data and multiple classifiers, with KNN proving to be the most effective classifier. Narendra et al. (2021) introduced a deep learning strategy for reliable classification of images depicting wheat grains from four distinct cultivars. Employing the Global Wheat Head Detection (GWHD) dataset and deep learning models, P. K. Sethy et al. (2022) showcased a system with an impressive 94.14 percent accuracy in identifying wheat genotypes. Using the UNet and ViT network model, I. Cinar et al. (2023) and X. Guo et al. (2023) created models to detect maize and wheat seedlings, achieving an outstanding accuracy rate of 99.3%.

Table 1. Shows the different methodologies approached for crops identification

References	Types of Crop Grain	segmentation	Features	Classification	Accuracy (%)
Xiao Chen et al. (2010)	Corn	NA	Geometric & Color feature	Machine Vision &Pattern recognition	90
Ouyang et al. (2010)	Rice Seed	Yes	Color feature	Back forward neural network	86.65
Shantaiya, S et al. (2010)	Rice	Yes	color, morphological and textural	Back propagation neural network	NA
Neelamma K. Patil et al. (2011)	Crop grain	NA	color and texture features	K-NN & Minimum Distance Classifier	NA
A. Arefi et al. (2011)	Wheat	Yes	Morphology and colour features	machine vision and ANN	95.86
Gujjar, H et al. (2013)	Basmati Rice	Yes	Color, morphological and textural features	BPNN	90

continued on following page

Table 1. Continued

References	Types of Crop Grain	segmentation	Features	Classification	Accuracy (%)
A. Nasir Ahmadi et al. (2013)	Iran Bean	Yes	Color features	Machine Vision & MLP-ANN	90
Szczypiski et al. (2015)	Barley	NA	colour, and texture attributes	Computer Vision	86
Nima Teimour et al. (2015)	Almond	Yes	Shape, colour & texture feature	ANN	97.84
L Sumaryanti et al. (2015)	Rice	Yes	Color, morphological and textural features	Image processing Techniques	96.6
P Sethy et al. (2018)	Rice	NA	Geometric and texture features	multi-class SVM	92
Arboleda, E. R et al. (2018)	coffee black bean	Yes	color, texture & size	Image processing Techniques	100
Zhengjun Qiu et al. (2018)	Rice seed	NA	Hyperspectral	CNN	NA
Yahya Altuntaş et al. (2019)	Maize seeds	Yes	NA	CNN & TL	NA
Esra et al. (2019)	Durum wheat	Yes	Morphological, colour, wavelet and gaborlet	ANN &ANOVA	93.46
L. Jena et al. (2021)	Wheat	Yes	geometrical features	KNN, SVM &NB	94.3
Narendra, V. G et al. (2021)	Almond	Yes	Geometric, Color & texture feature	Image processing &computational intelligence	97.13
K Laabassi et al. (2021)	Wheat	NA	NA	Deep learning	85 to 95.68
P K Sethy et al. (2022)	Wheat	NA	Deep features	SVM & DL	94.14
I CINAR et al. (2022)	Rice	NA	Morphological, shape and color features	KNN, DT, RF, LR, MLP&SVM	97.99
X. Guo et al. (2023)	Maize and Wheat	Yes	NA	Deep Learning	99.3

CLASSIFICATION OF CROP GRAIN

In recent years, there has been a noticeable increase in the global academic interest regarding the classification of grain images. Researchers worldwide are increasingly focusing on developing advanced computer vision techniques and classification algorithms that surpass human vision capabilities. This progress holds significant importance, especially in agricultural contexts, where it is crucial to accurately categorize produce seeds based on various attributes such as texture, shape, pigment, and other characteristics. Automated systems for crop grain variety recognition have become invaluable, particularly when specific grain varieties are required for distinct purposes.

Numerous studies have delved deeply into developing machine vision systems for classifying various types of grain commodities. For instance, Babalk et al. (2010) proposed a wheat variety classification model using multiclass support vector machines (M-SVM) and a binary particle swarm optimization (BPSO) algorithm, achieving a high classification accuracy of 92.02% by considering geometric and color characteristics. F. Guevara-Hernandez et al. (2011) utilized discriminant analysis (DA) and K-nearest neighbors (K-NN) algorithms for classifying wheat and cereal kernels, emphasizing morphologic, color, and texture features to achieve accurate classification. Harpreet Kaur et al. (2013) successfully classified rice using multiclass SVMs, incorporating both geometric and interior/exterior quality features with a success rate exceeding 86%. Silva et al. (2013) employed artificial intelligence (AI) to classify rice varieties based on morphological, color, and texture features, achieving a high success rate of 92% by considering tissue characteristics. R Kambo et al. (2014) proposed a basmati rice classification model based on principal component analysis, enhancing classification accuracy. Pazoki et al. (2014) achieved exceptional accuracy (98.40% and 99.73%) for classifying five rice grain varieties using multi-layer perceptron (MLP) and neuro-fuzzy neural networks, respectively, emphasizing color, morphological, and shape factors. M Huang et al. (2016) created a maize seed variety classification model utilizing least square–support vector machines (LS–SVM) and HIS technology, achieving 90% accuracy via multispectral imaging. W Tan et al. K Singh et al. (2016) classified four rice grain varieties with 96% accuracy using back-propagation neural networks (BPNN) based on color and texture features. N. Nugegoda et al. (2016) developed a model for classifying rice varieties by considering various parameters, including shape, length, chalkiness, and color, using machine-vision techniques. Pinto et al. (2017) implemented deep convolutional neural networks, machine learning, and image processing techniques to classify images of green coffee seeds, achieving an accuracy ranging from 72.4% to 98.8% based on bean defect categories. Sabanci et al. (2017) accurately classified wheat grains as bread wheat or durum wheat using artificial neural networks (ANN) with high-quality images, achieving superior results. Aimin Miao et al. (2018) introduced a classification model (t-SNE) for eight maize seed varieties, comparing it with principal components analysis (PCA) for enhanced accuracy. Gomez et al. (2019) employed K-mean clustering and SVM to classify various cocoa bean varieties with a high accuracy rate of 98.43%. In their research, CINAR et al. (2019) employed multiple machine-learning methods to distinguish rice species, ultimately finding that the LR model achieved the highest success rate.

Nen-Fu Huang et al. (2019) successfully classified various coffee bean varieties with an identification rate of 93% using CNN. Chao Xia et al. (2019) utilized hyperspectral images and advanced algorithms to effectively categorize seventeen maize seed varieties. Their study revealed that linear discriminant analysis (LDA) achieved the highest classification accuracy at 99.13%. M. M. Tin et al. (2019) utilized image processing techniques and algorithms to categorize Myanmar's five rice varieties. Murat Koklu et al. (2020) managed to classify seven categories of desiccated legumes based on 13,611 images, achieving the highest accuracy (93.13%) among all classifiers using SVM. Robert Singh et al. (2020) developed a classification model for four varieties of rice grains that outperformed existing classifiers, producing superior results. K Aukkapinyo et al. (2020) utilized contrast-limited adaptive histogram equalization (CLAHE) and mask region-based convolutional neural networks (R-CNN) to classify rice grain, achieving 80% accuracy. H. O. Velesaca et al. (2020) classified maize samples using segmentation and classification methods. Aqib Ali et al. (2020) achieved high accuracy (98.93%) in classifying six maize seed varieties using machine learning classifiers. B Arora et al. (2020) classified rice grains based on various features using image processing techniques and machine learning, obtaining accurate results. Peng Xu et al. (2021) achieved 96.46% accuracy with SVM for classifying five maize seed varieties using the ten-fold cross-validation procedure. Salman Qadri et al. (2021) employed machine learning techniques to classify six Asian rice varieties, with LMT Tree classifier achieving an overall accuracy of 97.4%. Surabhi Lingwal et al. (2021) developed a deep learning model (CNN) for classifying fifteen wheat varieties with high accuracy. Y. Selim Taspinar et al. (2022) employed pre-trained models to classify fourteen bean grain varieties, with InceptionV3 achieving the highest success rate of 84.48%. Resul Butuner et al. (2023) utilized machine learning and deep learning methods to classify lentil images, achieving an outstanding accuracy of 99.80% with the SqueezeNet architecture and ANN algorithm.

Table 2. Shows the different approaches applied for crops classification

References	Types of Crop grain	Segmentation	Features	Classification	Accuracy (%)
Ahmet Babalık et al. (2010)	Wheat	Yes	Geometric & color	Multiclass SVM &BPSO	92.02
F. Guevara-Hernandez et al. (2011)	Wheat &Barley	Yes	Morphologic, color, and texture feature	DA & KNN	99
Harpreet Kaur et al. (2013)	Rice	Yes	Geometric features	Multiclass SVM	90
Silva et al. (2013)	Rice Seed	Yes	Morphologic, color, and texture feature	Neural Network	92
R Kambo et al. (2014)	Basmati Rice	Yes	Morphological feature	PCA	79
Pazoki et al. (2014)	Rice	Yes	Morphological features, Color feature &shape factor	MLP & Neuro Fuzzy Network	98.40&99.73
M Huang et al. (2016)	Maize	Yes	Spectral and imaging features	LS-SVM	92
W Tan et al. (2016)	Wheat	NA	NA	SVM &BPNN	100
N. Nugegoda, et al. (2016)	Rice	Yes	Shape, length, chalkiness, color	Machine Vision	NA
K Singh et al. (2016)	Rice	Yes	Colour features, texture features& wavelet features	BPNN	96
C Pinto et al. (2017)	Green Coffee Bean	NA	NA	CNN	98.7
K Sabanci et al. (2017)	Wheat	Yes	Visual features, colour features, texture features	ANN	NA
K Sendin et al. (2018)	White Maize	NA	NA	PLS-DA	83-100
Aimin Miao et al. (2018)	Waxy Maize Seed	Yes	NA	t-SNA&PCA	97.5
M Kozłowski et al. (2019)	Barley	Yes	Colour, texture and morphological attributes	CNN	93
Andrea Gomez et al. (2019)	Cocoa Bean	Yes	Spectral signatures,	KNN&SVM	98.43
I CINAR et al. (2019)	Rice	NA	Morphological feature	AI	93
Nen-Fu Huang et al. (2019)	Green Coffee bean	Yes	NA	CNN	93
Chao Xia et al. (2019)	Maize Seed	Yes	Spectral feature & Imaging features	MLDA	99.13
M. M Tin et al. (2019)	Rice	Segmentation	Feature Extraction	Image Processing Techniques	NA

continued on following page

Table 2. Continued

References	Types of Crop grain	Segmentation	Features	Classification	Accuracy (%)
Murat Koklu et al. (2020)	Dry Bean	Yes	Feature Extraction	MLP, SVM, DT, KNN	100
Robert Singh et al. (2020)	Rice	Yes	Morphological, colour, texture and wavelet	BPNN	97.75
K Aukkapinyo et al. (2020)	Rice	NA	NA	R-CNN	NA
H O Velesaca et al. (2020)	Corn	Yes	NA	Mask R-CNN	NA
Aqib Ali et al. (2020)	Corn	Yes	Histogram, texture, and spectral	RF, BN, LB& MLP	98.93
B Arora et al. (2020)	Rice	Yes	Major axis, minor axis, eccentricity, length, breadth	Image Processing and Machine learning	NA
Peng Xu et al. (2021)	Maize Seed	Yes	Geometric features	MLP, DT, LDA, SVM. KNN, NB	99.22
Salman Qadri et al. (2021)	Rice	Yes	Binary, Histogram & texture feature	Machine Vision	97.4
Surabhi Lingwal et al. (2021)	Wheat Grain	NA	Visual features	CNN	98
Y selim Taspinar et al. (2022)	Dry Bean	NA	Feature extraction	CNN	84.48
Resul Butuner et al. (2023)	Lentils	NA	Feature extraction	ML	99.80

GRADING OF CROP GRAIN

Grading of crop cereals has become a pivotal aspect of the agricultural industry, where the quality of crop grain significantly influences its marketability. Traditional manual grading methods are laborious, imprecise, and time consuming, leading to variable results. In response, the development of automated systems employing image-processing techniques has emerged as a reliable solution for accurate and efficient crop evaluation. Cakmak et al. (2011) introduced an artificial neural network (ANN) model for chickpea quality evaluation, achieving correct classification rates of 95.4%, 87.6%, and 96.0% for color, surface morphology, and shape evaluations, respectively. Krishna K Patel et al. (2012) implemented a machine vision system for grain quality inspection and fruit and vegetable sorting. Mahajan et al. (2014) utilized the Top-hat Transformation method for quality inspection of Indian Basmati rice grains, distinguishing between normal grains, long grains, and tiny grains. Mahale et al. (2014) developed image-processing techniques for rice grain classification based on grain size and shape. R. Birla et al. (2015) presented a solution for evaluating Indian Basmati Oryza sativa L rice quality using machine vision and image processing techniques. H Zareiforoush et al. (2015) combined fuzzy inference systems (FIS) with image processing for qualitative assessment of milled rice, achieving quality indices such as degree of milling (DOM) and percentage of broken kernels (PBK). Devi et al. (2017) employed machine vision techniques for quality assessment of rice grains, considering physical characteristics such as grain shape, size, chalkiness, whiteness, milling degree, bulk density, and chemical properties. Kuche et al. (2018) introduced a model incorporating a technique to identify the presence of a specific chemical substance [in the proposed Kuche et al. (2018) model]. Mauricio Garca et al. (2019) evaluated the quality grade of coffee beans through the application of k-nearest neighbor algorithm. Nguyen Hong et al. (2019) devised a Convolutional Neural Network (CNN)-based rice quality classification model with an accuracy rate of 93.85%. Robby Janandi et al. (2020) created deep learning models (ResNet-152 and VGG16) for classifying the quality of coffee beans, obtaining the highest accuracy of 73.3%. P A Belan1 et al. (2020) introduced a machine vision system (MVS) for legume quality inspection, achieving high success rates for segmentation, classification, and defect detection. K Sharma et al. (2020) developed a model for analyzing the classification of rice grain using various image processing, machine vision, and computer vision techniques. Mohan et al. (2020) categorized rice quality using Neural Network (NN) and Support Vector Machine (SVM) classifier algorithms. Saikenova et al. (2021) determined lentil particle quality based on valuable characteristics and biological properties. P. S. Sampaio et al. (2021) employed an ANN model for qualitative and quantitative analysis of rice quality using grain physical parameters. P-Lopez et al. and Vidyarthi et al. (2021) proposed deep convolutional neural network (DCNN) models for classifying almond categories based on adulteration, achieving high classification accuracy, particularly with the ResNet50 model. Assadzadeh et al. (2022) designed a model using machine and deep learning for grain quality determination. Niari et al. (2022) developed statistical models and support vector machines to identify and classify wheat seed samples of seven varieties, achieving an overall average accuracy of 97.6%. These advancements highlight the potential of automated image processing systems in revolutionizing the crop grading process and ensuring higher accuracy, efficiency, and consistency in evaluating crop quality.

Table 3. Shows the different approaches applied for crops grading

References	Types of Crop Grain	Segmentation	Features	Classification	Accuracy (%)
Yusuf Serhad C¸ akmak et.al (2011)	Chickpea	Yes	Size, colour, and surface morphology	ANN	96
S Mahajan et al. (2014)	Basmati Rice	Yes	Morphological features	Image Processing Technique	NA
B Mahale et al. (2014)	Rice	Yes	Physical and chemical characteristics	Image Processing Technique	NA
R Birla et al. (2014)	Oryza sativa L Basmati Rice	Yes	NA	Machine Vision & Image Processing	NA
H Zareiforoush et al. (2015)	Milled Rice	NA	NA	Machine Vision	97.45
T. Gayathri Devi et al. (2017)	Rice	NA	Length, shape, colour and texture feature	Machine Vision	NA
N. A. Kuchekar et al. (2018)	Rice	Yes	Morphological	Machine Vision	NA
Mauricio García et al. (2019)	Green Coffee bean	Yes	Color, morphology, shape and size	KNN	94.79
Nguyen Hong et al. (2019)	Rice	NA	Size	CNN, SVM & KNN	93.85

continued on following page

Table 3. Continued

References	Types of Crop Grain	Segmentation	Features	Classification	Accuracy (%)
Robby Janandi et al. (2020)	Coffee bean	NA	NA	CNN	73.3
P A Belanl et al. (2020)	Bean	Yes	Skin colors and defects	Machine Vision	99.6
K Sharma et al. (2020)	Rice	Yes	Feature extraction	Image processing, MVS & CVS	NA
D Mohan et al. (2020)	Rice	Yes	Colour Feature & Geometric Features.	ANN & SVM	91
A Z Saikenova et al. (2021)	Lentils	NA	NA	Image processing,	NA
A S Almeida et al. (2021)	Rice	NA	Physical parameters	ANN &MLR	
S P-Lopez et al. (2021)	Ground Coffee	NA	Adulterations	CNN	98.6
S K. Vidyarthi et al. (2021)	Almond Kernel	NA	Adulteration	DCNN	92-99
S Assadzadeh et al. (2022)	Grain	Yes	NA	ML &DL	NA
Z F Niari et al. (2022)	Wheat	Yes	Shape, colour, and textural features	QSVM	97.6

FUTURE FINDING AND DIRECTION

Figure 6. Identification and classification of different types of crops grain by using ML and DL model

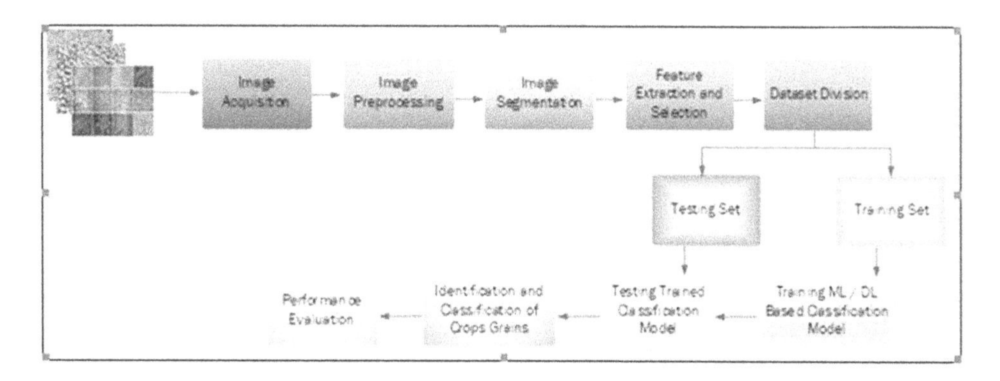

The workflow of the identification, classification, and grading of crop grains is shown in Figure 6. As shown in the figure 6, the workflow starts from image acquisition followed by image preprocessing. The processed images were fed into the image segmentation stage. Then, the features were extracted from the images, followed by the process of data division. We then constructed the data for the testing, training, and validation processes. Different machine learning and deep learning techniques are adopted to identify and classify different types of crop grains, and the grading of crop grains will be evaluated.

CONCLUSION

A large number of quality research articles were reviewed in this work, and the primary focus was to conduct an in-depth analysis of existing research works. From this study, we found that the various major stages for improving the quality of crop grains are identification, classification, and grading. The research works of the above-mentioned articles are compared on the basis of texture analysis, morphological and geometrical features, shape, color, and size. In the identification phase, different authors have claimed accuracies ranging from 86 to 97.99%. The major techniques used in this phase can be categorized into MLP, KNN SVM, DT, RF, LR, ANN, and CNN. In the classification phase, different authors have reported accuracies ranging from 79 to 100%. The major techniques used in this phase are MLP, KNN SVM, PLS-DA, BPSO, PCA, BPNN, DT, LDA, and NB. In the grading phase, different authors have reported accuracies ranging from 73 to 99.6%. The major techniques used in this phase are ANN, KNN, QSVM, and DCNN.MLR. In this work, we have systematically identified the problems and proposed solutions by the researcher, which will help readers gain an abstract view and knowledge about the current development and usage of machine learning and deep learning techniques to improve the identification, classification, and quality of crop grain. This will ultimately improve the production quantity and quality in the agricultural industry. This article also highlights the problem domains for which there is scope for future work. Various researchers and readers will obtain a proper path to conduct research on the problems highlighted in this study.

REFERENCES

Ali, A., Qadri, S., Mashwani, W. K., Brahim Belhaouari, S., Naeem, S., Rafique, S., Jamal, F., Chesneau, C., & Anam, S. (2020). Machine learning approach for the classification of corn seed using hybrid features. *International Journal of Food Properties*, 23(1), 1110–1124. 10.1080/10942912.2020.1778724

Altuntaş, Y., Cömert, Z., & Kocamaz, A. F. (2019). Identification of haploid and diploid maize seeds using convolutional neural networks and a transfer learning approach. *Computers and Electronics in Agriculture*, 163, 104874. 10.1016/j.compag.2019.104874

Anwer, M., Farooqi, S., & Qureshi, Y. (2015). Agriculture sector performance: An analysis through the role of agriculture sector share in GDP. *Journal of Agricultural Economics. Extension and Rural Development*, 3(3), 270–275.

Arboleda, E. R., Fajardo, A. C., & Medina, R. P. (2018, May). An image processing technique for coffee black beans identification. *In2018 IEEE International Conference on Innovative Research and Development (ICIRD)*, (pp. 1-5). IEEE. 10.1109/ICIRD.2018.8376325

Arefi, A., Motlagh, A. M., & Teimourlou, R. F. (2011). Wheat class identification using computer vision system and artificial neural networks. *International Agrophysics*, 25(4).

Arora, B., Bhagat, N., Saritha, L. R., & Arcot, S. (2020, February). Rice grain classification using image processing & machine learning techniques. *In 2020 International Conference on Inventive Computation Technologies (ICICT)*. IEEE.

Assadzadeh, S., Walker, C. K., & Panozzo, J. F. (2022). Deep learning segmentation in bulk grain images for prediction of grain market quality. *Food and Bioprocess Technology*, 15(7), 1615–1628. 10.1007/s11947-022-02840-1

Aukkapinyo, K., Sawangwong, S., Pooyoi, P., & Kusakunniran, W. (2020). Localization and classification of rice-grain images using region proposals-based convolutional neural network. *International Journal of Automation and Computing*, 17(2), 233–246. 10.1007/s11633-019-1207-6

Babalık, A., Baykan, Ö. K., İşcan, H., Babaoğlu, İ., & Fındık, O. (2010). Effects of feature selection using binary particle swarm optimization on wheat variety classification. *Advances in Information Technology: 4th International Conference, IAIT 2010*, (pp. 11-17). Springer Berlin Heidelberg. 10.1007/978-3-642-16699-0_2

Belan, P. A., de Macedo, R. A. G., Alves, W. A. L., Santana, J. C. C., & Araújo, S. A. (2020). Machine vision system for quality inspection of beans. *International Journal of Advanced Manufacturing Technology*, 111(11-12), 3421–3435. 10.1007/s00170-020-06226-5

Birla, R., & Chauhan, A. P. S. (2015). An efficient method for quality analysis of rice using machine vision system. *Journal of advances in Information Technology, 6*(3).

Butuner, R., Cinar, I., Taspinar, Y. S., Kursun, R., Calp, M. H., & Koklu, M. (2023). Classification of deep image features of lentil varieties with machine learning techniques. *European Food Research and Technology*, 249(5), 1303–1316. 10.1007/s00217-023-04214-z

Çakmak, Y. S., & Boyacı, İ. H. (2011). Quality evaluation of chickpeas using an artificial neural network integrated computer vision system. *International Journal of Food Science & Technology*, 46(1), 194–200. 10.1111/j.1365-2621.2010.02482.x

Chen, X., Xun, Y., Li, W., & Zhang, J. (2010). Combining discriminant analysis and neural networks for corn variety identification. *Computers and Electronics in Agriculture*, 71, S48–S53. 10.1016/j.compag.2009.09.003

Cinar, I., & Koklu, M. (2019). Classification of rice varieties using artificial intelligence methods. *International Journal of Intelligent Systems and Applications in Engineering*, 7(3), 188–194. 10.18201/ijisae.2019355381

Cinar, I., & Koklu, M. (2022). Identification of rice varieties using machine learning algorithms. *Journal of Agricultural Sciences*, 9–9.

Devi, T. G., Neelamegam, P., & Sudha, S. (2017, September). *Machine vision-based quality analysis of rice grains. In 2017 IEEE international conference on power, control, signals and instrumentation engineering (ICPCSI)*. IEEE.

Fazel-Niari, Z., Afkari-Sayyah, A. H., Abbaspour-Gilandeh, Y., Herrera-Miranda, I., Hernández-Hernández, J. L., & Hernández-Hernández, M. (2022). Quality assessment of components of wheat seed using different classifications models. *Applied Sciences (Basel, Switzerland)*, 12(9), 4133. 10.3390/app12094133

García, M., Candelo-Becerra, J. E., & Hoyos, F. E. (2019). Quality and defect inspection of green coffee beans using a computer vision system. *applied sciences, 9*(19), 4195.

Giunta, F., Pruneddu, G., & Motzo, R. (2019). Grain yield and grain protein of old and modern durum wheat cultivars grown under different cropping systems. *Field Crops Research*, 230, 107–120. 10.1016/j.fcr.2018.10.012

Gomez, N. A., Sanchez, K., & Arguello, H. (2019, April). Non-destructive method for classification of cocoa beans from spectral information. *In 2019 XXII Symposium on Image, Signal Processing and Artificial Vision (STSIVA)*. IEEE.

Government of India. (2023). *Annual Report Ministry of Food Processing Industries*. https://www.mofpi.gov.in/documents/reports/annual-report.

Guevara-Hernandez, F., & Gil, J. G. (2011). A machine vision system for classification of wheat and barley grain kernels. *Spanish Journal of Agricultural Research*, 9(3), 672–680. 10.5424/sjar/20110903-140-10

Gujjar, H. S., & Siddappa, D. M. (2013). A method for identification of basmati rice grain of india and its quality using pattern classification. *International Journal of Engineering Research and Applications*, 3(1), 268–273.

Guo, X., Ge, Y., Liu, F., & Yang, J. (2023). Identification of maize and wheat seedlings and weeds based on deep learning. *Frontiers in Earth Science (Lausanne)*, 11, 1146558. 10.3389/feart.2023.1146558

Huang, M., He, C., Zhu, Q., & Qin, J. (2016). Maize seed variety classification using the integration of spectral and image features combined with feature transformation based on hyperspectral imaging. *Applied Sciences (Basel, Switzerland)*, 6(6), 183. 10.3390/app6060183

Huang, N. F., Chou, D. L., & Lee, C. A. (2019, July). Real-time classification of green coffee beans by using a convolutional neural network. In *2019 3rd International Conference on Imaging, Signal Processing and Communication (ICISPC).* IEEE. 10.1109/ICISPC.2019.8935644

Janandi, R., & Cenggoro, T. W. (2020, August). An Implementation of Convolutional Neural Network for Coffee Beans Quality Classification in a Mobile Information System. In *2020 International Conference on Information Management and Technology (ICIMTech) IEEE.* IEEE. 10.1109/ICIMTech50083.2020.9211257

Jena, L., Behera, S. K., & Sethy, P. K. (2021, May). Identification of wheat grain using geometrical feature and machine learning. *In 2021 2nd International conference for emerging technology (INCET).* IEEE.

Kambo, R., & Yerpude, A. (2014). *Classification of basmati rice grain variety using image processing and principal component analysis.* arXiv preprint arXiv,1405,7626.

Kaur, H., & Singh, B. (2013). Classification and grading rice using multi-class SVM. *International Journal of Scientific and Research Publications*, 3(4), 1–5.

Kaya, E., & Saritas, İ. (2019). Towards a real-time sorting system: Identification of vitreous durum wheat kernels using ANN based on their morphological, colour, wavelet and gaborlet features. *Computers and Electronics in Agriculture*, 166, 105016. 10.1016/j.compag.2019.105016

Koklu, M., & Ozkan, I. A. (2020). Multiclass classification of dry beans using computer vision and machine learning techniques. *Computers and Electronics in Agriculture*, 174, 105507. 10.1016/j.compag.2020.105507

Kozłowski, M., Górecki, P., & Szczypiński, P. M. (2019). Varietal classification of barley by convolutional neural networks. *Biosystems Engineering*, 184, 155–165. 10.1016/j.biosystemseng.2019.06.012

Kuchekar, N. A., & Yerigeri, V. V. (2018). Rice grain quality grading using digital image processing techniques. *IOSR J Electronics Communication Eng, 13*(3).

Laabassi, K., Belarbi, M. A., Mahmoudi, S., Mahmoudi, S. A., & Ferhat, K. (2021). Wheat varieties identification based on a deep learning approach. *Journal of the Saudi Society of Agricultural Sciences*, 20(5), 281–289. 10.1016/j.jssas.2021.02.008

Lingwal, S., Bhatia, K. K., & Tomer, M. S. (2021). Image-based wheat grain classification using convolutional neural network. *Multimedia Tools and Applications*, 80(28-29), 1–25. 10.1007/s11042-020-10174-3

Mahajan, S., & Kaur, S. (2014). Quality analysis of Indian basmati rice grains using top-hat transformation. *International Journal of Computer Applications*, 94(15), 42–48. 10.5120/16423-6085

Mahale, B., & Korde, S. (2014, April). Rice quality analysis using image processing techniques. In *International Conference for Convergence for Technology.* IEEE. 10.1109/I2CT.2014.7092300

Miao, A., Zhuang, J., Tang, Y., He, Y., Chu, X., & Luo, S. (2018). Hyperspectral image-based variety classification of waxy maize seeds by the t-SNE model and Procrustes analysis. *Sensors (Basel)*, 18(12), 4391. 10.3390/s1812439130545028

Mohan, D., & Raj, M. G. (2020). Quality analysis of rice grains using ANN and SVM. *Journal of Critical Reviews, 7*(1), 395-402.

Narendra, V. G., Krishanamoorthi, M., Shivaprasad, G., Amitkumar, V. G., & Kamath, P. (2021). Almond kernel variety identification and classification using decision tree. *Journal of Engineering Science and Technology*, 16(5), 3923–3942.

Nasirahmadi, A., & Behroozi-Khazaei, N. (2013). Identification of bean varieties according to color features using artificial neural network. *Spanish Journal of Agricultural Research*, 11(3), 670–677. 10.5424/sjar/2013113-3942

Nugegoda, N. (2016). *Rice grains classification using image processing technics*. Academic Press.

OuYang, A. G., Gao, R. J., Sun, X. D., Pan, Y. Y., & Dong, X. L. (2010, August). An automatic method for identifying different variety of rice seeds using machine vision technology. In *2010 Sixth International Conference on Natural Computation,* (pp. 84-88). IEEE. 10.1109/ICNC.2010.5583370

Patel, K. K., Kar, A., Jha, S. N., & Khan, M. A. (2012). Machine vision system: A tool for quality inspection of food and agricultural products. *Journal of Food Science and Technology*, 49(2), 123–141. 10.1007/s13197-011-0321-423572836

Patil, N. K., Malemath, V. S., & Yadahalli, R. M. (2011). Color and texture based identification and classification of food grains using different color models and haralick features. *International Journal on Computer Science and Engineering*, 3(12), 3669.

Pazoki, A. R., Farokhi, F., & Pazoki, Z. (2014). Classification of rice grain varieties using two Artificial Neural Networks (MLP and Neuro-Fuzzy). *The Journal of Animal & Plant Sciences*, 24(1), 336–343.

Pinto, C., Furukawa, J., Fukai, H., & Tamura, S. (2017, August). Classification of Green coffee bean images base on defect types using convolutional neural network (CNN). In *2017 International Conference on Advanced Informatics, Concepts, Theory, and Applications (ICAICTA)*. IEEE.

Pradana-López, S., Pérez-Calabuig, A. M., Cancilla, J. C., Lozano, M. Á., Rodrigo, C., Mena, M. L., & Torrecilla, J. S. (2021). Deep transfer learning to verify quality and safety of ground coffee. *Food Control*, 122, 107801. 10.1016/j.foodcont.2020.107801

Qadri, S., Aslam, T., Nawaz, S. A., Saher, N., Razzaq, A., Ur Rehman, M., Ahmad, N., Shahzad, F., & Furqan Qadri, S. (2021). Machine vision approach for classification of rice varieties using texture features. *International Journal of Food Properties*, 24(1), 1615–1630. 10.1080/10942912.2021.1986523

Qiu, Z., Chen, J., Zhao, Y., Zhu, S., He, Y., & Zhang, C. (2018). Variety identification of single rice seed using hyperspectral imaging combined with convolutional neural network. *Applied Sciences (Basel, Switzerland)*, 8(2), 212. 10.3390/app8020212

Robert Singh, K., & Chaudhury, S. (2020). A cascade network for the classification of rice grain based on single rice kernel. *Complex & Intelligent Systems*, 6(2), 321–334. 10.1007/s40747-020-00132-9

Sabanci, K., Kayabasi, A., & Toktas, A. (2017). Computer vision-based method for classification of wheat grains using artificial neural network. *Journal of the Science of Food and Agriculture*, 97(8), 2588–2593. 10.1002/jsfa.808027718230

Saikenova, A. Z., Nurgassenov, T. N., Saikenov, B. R., Kudaibergenov, M. S., & Didorenko, S. V. (2021). Crop yield and quality of lentil varieties in the conditions of the southeast of kazakhstan. *Online Journal of Biological Sciences*, 21(1), 33–40. 10.3844/ojbsci.2021.33.40

Sampaio, P. S., Almeida, A. S., & Brites, C. M. (2021). Use of artificial neural network model for rice quality prediction based on grain physical parameters. *Foods*, 10(12), 3016. 10.3390/foods1012301634945567

Sendin, K., Manley, M., & Williams, P. J. (2018). Classification of white maize defects with multispectral imaging. *Food Chemistry*, 243, 311–318. 10.1016/j.foodchem.2017.09.13329146343

Sethy, P. K. (2022). Identification of wheat tiller based on AlexNet-feature fusion. *Multimedia Tools and Applications*, 81(6), 8309–8316. 10.1007/s11042-022-12286-4

Sethy, P. K., & Chatterjee, A. (2018). Rice variety identification of western Odisha based on geometrical and texture feature. *International Journal of Applied Engineering Research: IJAER*, 13(4), 35–39.

Shantaiya, S., & Ansari, U. (2010, November). *Identification of food grains and its quality using pattern classification.* In IEEE international conference on communication technology, Raipur, India.

Sharma, K., Sethi, G., & Bawa, R. (2020, March). State-of-the-Art in Automatic Rice Quality Grading System. In *Proceedings of the International Conference on Innovative Computing & Communications (ICICC)*. SSRN. 10.2139/ssrn.3564372

Silva, C. S., & Sonnadara, D. U. J. (2013). *Classification of rice grains using neural networks*. Academic Press.

Singh, K. R., & Chaudhury, S. (2016). Efficient technique for rice grain classification using back-propagation neural network and wavelet decomposition. *IET Computer Vision*, 10(8), 780–787. 10.1049/iet-cvi.2015.0486

Son, N. H., & Thai-Nghe, N. (2019, November). *Deep learning for rice quality classification. In 2019 international conference on advanced computing and applications (ACOMP)*. IEEE.

Sumaryanti, L., Musdholifah, A., & Hartati, S. (2015). Digital image-based identification of rice variety using image processing and neural network. *TELKOMNIKA Indonesian Journal of Electrical Engineering*, 16(1), 182–190. 10.11591/tijee.v16i1.1602

Szczypiński, P. M., Klepaczko, A., & Zapotoczny, P. (2015). Identifying barley varieties by computer vision. *Computers and Electronics in Agriculture*, 110, 1–8. 10.1016/j.compag.2014.09.016

Tan, W., Sun, L., Zhang, D., Ye, D., & Che, W. (2016, October). Classification of wheat grains in different quality categories by near infrared spectroscopy and support vector machine. *2016 2nd International Conference on Cloud Computing and Internet of Things (CCIOT)*. IEEE.

Taspinar, Y. S., Dogan, M., Cinar, I., Kursun, R., Ozkan, I. A., & Koklu, M. (2022). Computer vision classification of dry beans (Phaseolus vulgaris L.) based on deep transfer learning techniques. *European Food Research and Technology*, 248(11), 2707–2725. 10.1007/s00217-022-04080-1

Teimouri, N., Omid, M., Mollazade, K., & Rajabipour, A. (2015). An artificial neural network-based method to identify five classes of almond according to visual features. *Journal of Food Process Engineering*, 39(6), 625–635. 10.1111/jfpe.12255

Tin, M. M., Mon, K. L., Win, E. P., & Hlaing, S. S. (2019). Myanmar rice grain classification using image processing techniques. *In Big Data Analysis and Deep Learning Applications:Proceedings of the First International Conference on Big Data Analysis and Deep Learning 1st* (pp. 324-332). Springer Singapore.

Velesaca, H. O., Mira, R., Suárez, P. L., Larrea, C. X., & Sappa, A. D. (2020). Deep learning-based corn kernel classification. In *Proceedings of the IEEE/CVF Conference on Computer Vision and Pattern Recognition Workshops* (pp. 66-67). IEEE.

Vidyarthi, S. K., Singh, S. K., Xiao, H. W., & Tiwari, R. (2021). Deep learnt grading of almond kernels. *Journal of Food Process Engineering*, 44(4), e13662. 10.1111/jfpe.13662

Xia, C., Yang, S., Huang, M., Zhu, Q., Guo, Y., & Qin, J. (2019). Maize seed classification using hyperspectral image coupled with multi-linear discriminant analysis. *Infrared Physics & Technology*, 103, 103077. 10.1016/j.infrared.2019.103077

Xu, P., Yang, R., Zeng, T., Zhang, J., Zhang, Y., & Tan, Q. (2021). Varietal classification of maize seeds using computer vision and machine learning techniques. *Journal of Food Process Engineering*, 44(11), 13846. 10.1111/jfpe.13846

Zareiforoush, H., Minaei, S., Alizadeh, M. R., & Banakar, A. (2015). A hybrid intelligent approach based on computer vision and fuzzy logic for quality measurement of milled rice. *Measurement*, 66, 26–34. 10.1016/j.measurement.2015.01.022

Chapter 14
IoT, AI, and Robotics Applications in the Agriculture Sector

Atin Kumar
http://orcid.org/0000-0002-1653-2146
Uttaranchal University, India

Nitish Karn
http://orcid.org/0009-0005-8165-4245
Uttaranchal University, India

Himani Sharma
Sher-e-Kashmir University of Agricultural Sciences and Technology of Jammu, India

ABSTRACT

This chapter explores the transformative impact of internet of things (IoT), artificial intelligence (A.I.), and robotics in modern agriculture. By addressing challenges such as climate change, water scarcity, and labor shortages, these technologies have revolutionized farming practices, enabling precise monitoring of crops, data-driven decision-making, and increased operational efficiency. The integration of advanced A.I. algorithms and robotic systems has led to optimized resource utilization, reduced environmental impact, and enhanced sustainable practices. However, challenges such as cost, data security, and adoption barriers must be addressed to fully realize the potential of these technologies. The chapter also highlights future trends and areas for research and development, emphasizing the potential for further innovation and sustainable farming practices in the agriculture sector.

DOI: 10.4018/979-8-3693-3583-3.ch014

INTRODUCTION

Agriculture is the backbone of the global economy, providing food, fiber, and fuel for the world's population. However, the industry faces numerous challenges such as climate change, water scarcity, labor shortages, and the need to increase productivity to feed a growing population.

These challenges have led to the adoption of innovative technologies such as Internet of Things, Artificial Intelligence, and Robotics in the agriculture sector. These technologies have the potential to transform farming practices, improve efficiency, and address the challenges faced by the industry. (Ratnaparkhi et al., 2020)

The integration of technology in modern agriculture has revolutionized the way farming is carried out, addressing key challenges and paving the way for sustainable and efficient practices. IoT has enabled the collection of real-time data from agricultural fields, allowing for precise monitoring of crops, soil conditions, and weather patterns. This data-driven approach enhances decision-making processes and resource allocation, leading to optimized yield and resource efficiency. A.I. algorithms analyze the vast amount of data collected through IoT devices to provide actionable insights. This enables farmers to make informed decisions regarding crop management, disease detection, and yield forecasting, ultimately minimizing waste and maximizing productivity. Robotic applications such as automated harvesting, precision planting, and autonomous crop maintenance have addressed labor shortages and increased operational efficiency.(Khan et al., 2021) By taking on repetitive and labor-intensive tasks, robotics free up human labor for more strategic and skilled roles within the agricultural process. The role of technology in modern agriculture is clear: to overcome challenges, increase productivity, and ensure sustainable practices for the future. As these technologies continue to evolve, their impact on the agriculture sector is poised to further enhance global food production while addressing environmental and socio-economic concerns.(Gowda et al., 2021)

In this chapter, we will delve into the specific objectives of integrating IoT, A.I., and Robotics in the agriculture sector. Our focus will be on:

1. Understanding the potential impact of IoT, A.I., and Robotics on addressing challenges such as climate change, water scarcity, and labor shortages in agriculture
2. Exploring the benefits of real-time data collection and monitoring using IoT devices for precise decision-making and resource allocation.
3. Analyzing how A.I. algorithms provide actionable insights for crop management, disease detection, and yield forecasting, leading to minimized waste and increased productivity.
4. Examining the role of robotics in addressing labor shortages and increasing operational efficiency through automated processes such as harvesting, planting, and maintenance.
5. Discussing the future prospects and evolving impact of these technologies on global food production and sustainable agricultural practices.
6. Through these objectives, we aim to provide a comprehensive understanding of how the integration of IoT, A.I., and Robotics can revolutionize the agriculture sector, ultimately contributing to global food security and sustainable farming practices.

INTERNET OF THINGS (IOT) IN AGRICULTURE

Definition and Concept of IoT

In the context of agriculture, IoT involves the use of connected devices and sensors to gather real-time data from agricultural fields, livestock, and farming equipment. These devices collect information such as soil moisture levels, temperature, humidity, crop growth patterns, and equipment performance.(Navarro et al., 2020)

The concept of IoT in agriculture revolves around the seamless integration of these data-collecting devices to enable farmers and agricultural professionals to monitor and manage their operations with unprecedented precision. Through IoT, farmers can make data-driven decisions regarding irrigation, fertilization, pest control, and crop rotation. This technology enables the optimization of resource allocation and the minimization of waste, ultimately leading to improved productivity and sustainability in farming practices.(Dhawale, 2019)

By leveraging IoT in agriculture, the industry can move towards smart farming, where interconnected devices and data analytics drive efficient and sustainable agricultural practices. As we continue to explore the impact of IoT in agriculture, it becomes evident that this technology plays a fundamental role in addressing the challenges faced by the industry and shaping the future of farming. (Khan et al., 2021)

IoT Sensors and Devices in Agriculture

Soil Moisture Sensors

Soil moisture sensors (Figure 1) are vital components of IoT in agriculture, providing essential data for efficient water management. By measuring the amount of water in the soil, these sensors help farmers avoid water waste and improve irrigation schedules. Farmers can guarantee that crops receive the appropriate amount of water at the appropriate time, ensuring healthy growth while conserving water resources, by precisely measuring the moisture levels in the soil. (Ray, 2017)

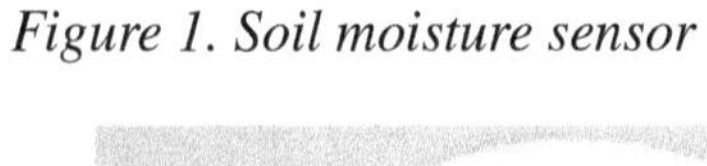

Figure 1. Soil moisture sensor

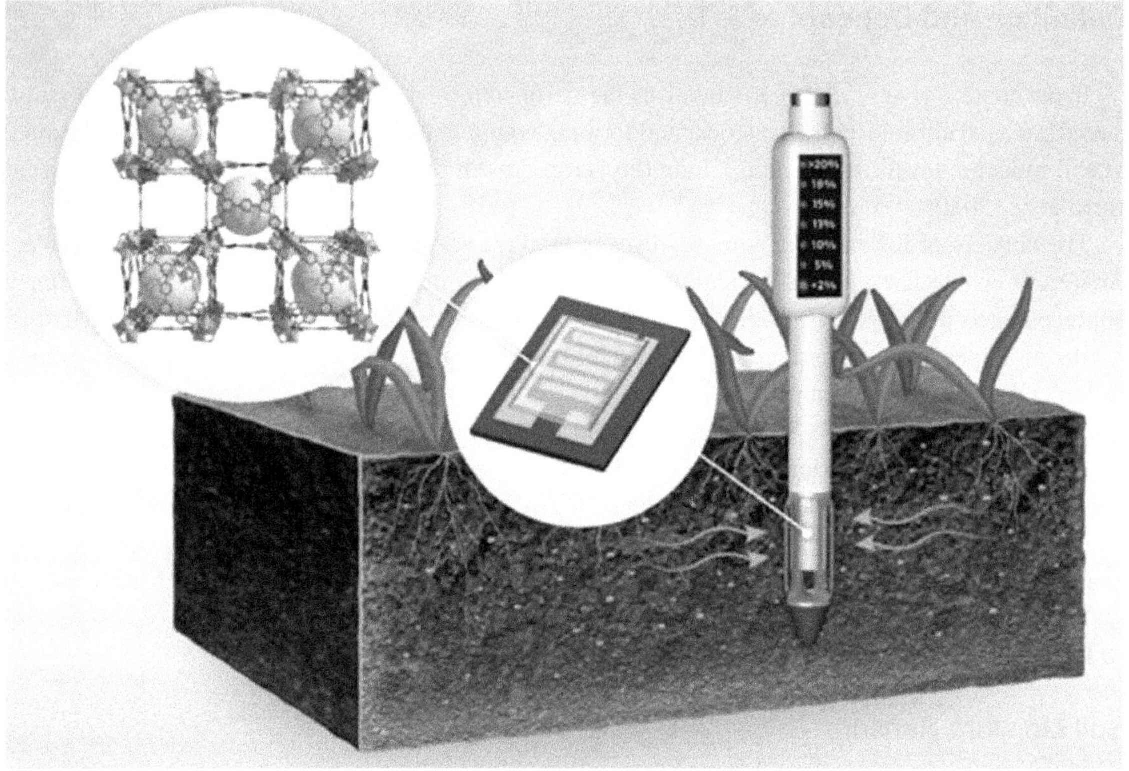

(Ray, 2017)

Weather Monitoring Devices

IoT-enabled weather monitoring devices (Figure 2) offer real-time information on temperature, humidity, wind speed, and precipitation. This data is crucial for making informed decisions regarding planting, harvesting, and disease control. With accurate weather data, farmers can anticipate and prepare for extreme weather events, adjust farming activities accordingly, and minimize the impact of adverse weather on crop yields. (Dhawale, 2019)

Figure 2. IoT-enabled weather monitoring device

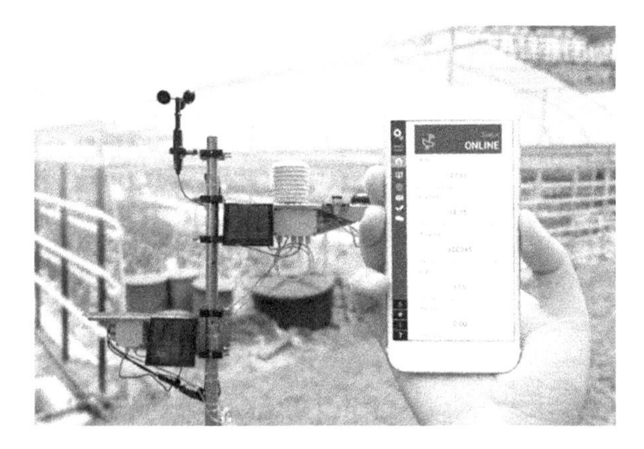

(Dhawale, 2019)

Livestock Tracking Systems

IoT devices equipped with GPS and activity monitoring capabilities facilitate livestock management. These systems enable farmers to monitor the health and behavior of livestock, track their movement, and ensure their well-being. By leveraging IoT for livestock tracking, farmers can improve feed efficiency, detect health issues early, and enhance overall farm productivity. (Finecomess et al., 2024)

Crop Health and Growth Monitoring Sensors

Sensors designed to monitor crop health and growth provide valuable insights for precise crop management. These devices measure parameters such as chlorophyll levels, leaf temperature, and photosynthetic activity, allowing farmers to assess plant health and detect early signs of stress or disease. With this information, farmers can implement targeted interventions to maintain crop health and optimize yields.

Equipment Performance and Maintenance Sensors

IoT-enabled sensors integrated into farming equipment collect data on performance, fuel consumption, and maintenance needs. This data facilitates predictive maintenance, optimizing equipment uptime and reducing operational downtime. By monitoring equipment performance in real time, farmers can schedule maintenance effectively, prevent breakdowns, and ensure the efficient operation of their machinery. (Lee et al., 2023)

Smart Crop Monitoring Systems

Smart crops monitoring systems collect information on crop health, growth, and production using sensors, drones, and satellite photography. These systems give farmers access to up-to-date data on important variables like weed growth, pest infestations, nutrient and moisture levels in the soil. Farmers can use this data to inform data-driven decisions on targeted pest control, fertilization schedule adjustments, and irrigation schedule optimization. Use of IoT-based smart farming techniques in the agricultural sector allows for monitoring and automation of irrigation systems, tracking of vehicles and livestock, capturing data from sensors installed on these, and applying various controls manually or remotely using AI and robotics (Khan et al., 2021). Overall, the integration of IoT, AI, and robotics in the agricultural sector has immense potential to revolutionize farming practices.

By harnessing the capabilities of IoT sensors and devices, the agriculture sector can elevate its precision, efficiency, and sustainability. The data collected and analyzed through these devices empower farmers to make informed decisions, conserve resources, and mitigate risks, heralding a new era of smart and data-driven agricultural practices.

IoT Applications in Precision Farming

Smart Irrigation Systems

Smart irrigation systems (Figure 3) represent a significant application of IoT in precision farming. By integrating soil moisture sensors and weather monitoring devices, these systems enable automated and optimized water delivery to crops. The real-time data collected by these IoT devices allows for precise and efficient irrigation, ensuring that crops receive the appropriate amount of water based on their specific needs and the prevailing environmental conditions. Smart irrigation systems not only conserve water resources but also contribute to improved crop health and yields. (Sairoel et al., 2023)

Figure 3. Smart irrigation systems

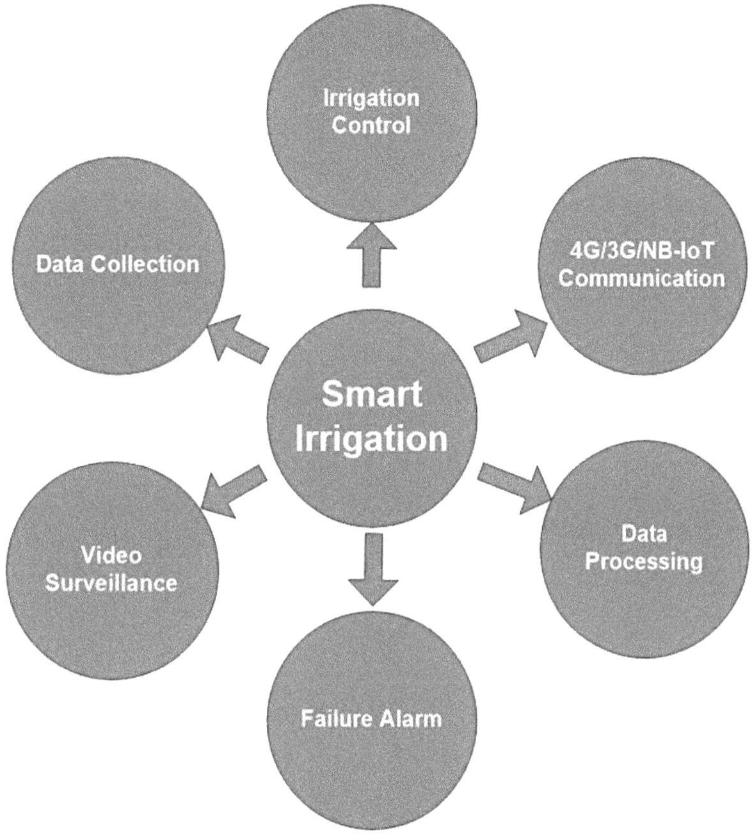

(Sairoel et al., 2023)

The implementation of smart irrigation systems brings forth several benefits to the agricultural sector. By utilizing soil moisture sensors and weather monitoring devices, farmers can effectively optimize water usage, leading to reduced water wastage and lower operational costs. The precise and automated delivery of water to crops based on real-time data not only promotes water conservation but also enhances the overall health and productivity of the crops.

Smart irrigation systems also facilitate proactive decision-making in response to changing environmental conditions. With the ability to monitor soil moisture levels and weather patterns in real time, farmers can adjust irrigation schedules accordingly, mitigating the impact of drought or excessive rainfall on crop growth. This adaptability enhances the resilience of agricultural operations and contributes to more consistent yields. (Benyezza et al., 2023)

Furthermore, the integration of IoT in smart irrigation systems allows for remote monitoring and control. Farmers can access and manage the irrigation process from anywhere through connected devices, enabling them to make timely adjustments and interventions as needed. This level of accessibility and control not only streamlines farming operations but also reduces the need for manual labor, thereby increasing operational efficiency.

In conclusion, the incorporation of smart irrigation systems driven by IoT technology exemplifies a significant advancement in precision farming. By promoting sustainability, resource efficiency, and informed decision-making, these systems play a pivotal role in modernizing agricultural practices and fostering a more sustainable and productive farming landscape.

Crop Health Monitoring

Crop health monitoring is another crucial application of IoT in precision farming. By utilizing a network of crop health and growth monitoring sensors, farmers can gain valuable insights into the well-being of their crops. These sensors (Figure 4) provide real-time data on parameters such as chlorophyll levels, leaf temperature, and photosynthetic activity, allowing farmers to assess the overall health and growth of their plants. With this information at their disposal, farmers can identify potential stress factors, diseases, or nutrient deficiencies at an early stage, enabling them to take proactive measures to address these issues. (Adli et al., 2023)

Figure 4. Crop health monitoring system

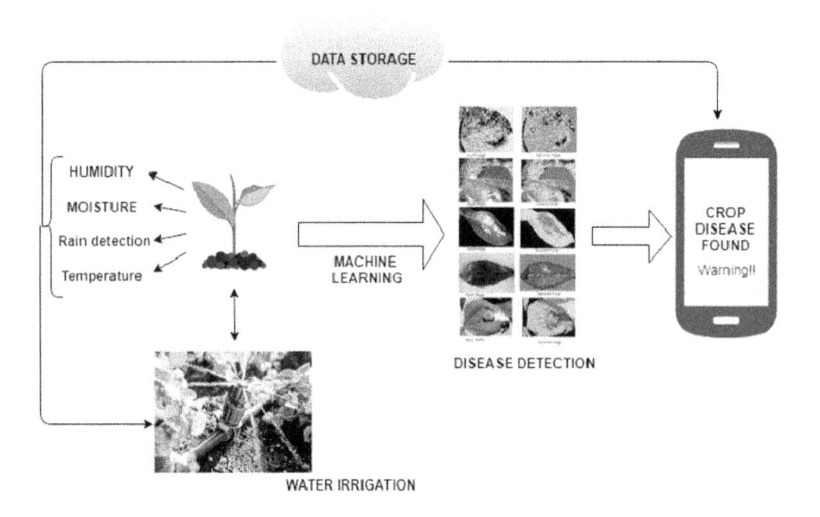

(Adli et al., 2023)

The integration of IoT in crop health monitoring also extends to the use of smart crop monitoring systems, which leverage a combination of sensors, drones, and satellite imagery to gather comprehensive data on crop health and yield. This data includes crucial factors such as soil moisture levels, nutrient levels, pest infestations, and weed growth, providing farmers with a holistic view of their crop conditions. (Sharma et al., 2023)

By harnessing the power of IoT-enabled crop health monitoring, farmers can make data-driven decisions to optimize irrigation schedules, adjust fertilization practices, and implement targeted pest control measures. This proactive and targeted approach to crop management not only promotes healthier plant growth but also contributes to optimized yields and resource efficiency. (Kumar et al., 2017)

Overall, the integration of IoT in crop health monitoring represents a significant advancement in precision farming, empowering farmers with actionable insights to promote the well-being and productivity of their crops. This technological innovation plays a key role in modernizing agricultural practices and fostering sustainable and data-driven approaches to crop management and cultivation.

Predictive Analytics for Pest Control

Predictive analytics for pest control is a groundbreaking application of IoT in precision farming. By leveraging data from various sources such as weather patterns, crop conditions, and pest behavior, predictive analytics systems can forecast potential pest outbreaks and infestations. These insights enable farmers to take pre-emptive measures, such as targeted application of pest control measures or implementation of pest-resistant crop varieties, to mitigate the impact of pest damage on their crops. (Kumawat et al., 2022)

The integration of IoT in predictive analytics for pest control not only reduces the reliance on broad-spectrum pesticides but also minimizes the environmental impact of pest management practices. By adopting a more targeted and precise approach to pest control, farmers can minimize chemical usage, preserve beneficial insect populations, and maintain ecological balance within their farming ecosystems. (Sharma et al., 2023)

Furthermore, the implementation of predictive analytics for pest control enhances the sustainability of agricultural practices by reducing crop losses attributed to pest damage. By proactively addressing potential pest threats, farmers can safeguard their crop yields and minimize economic losses associated with pest infestations. This proactive approach also promotes resource efficiency and reduces the overall environmental footprint of pest management in agriculture.

In conclusion, the integration of IoT in predictive analytics for pest control represents a pivotal advancement in precision farming, empowering farmers with the ability to anticipate and effectively manage pest pressures. By combining data-driven insights with proactive measures, this application of IoT contributes to sustainable and resilient agricultural practices, ensuring the protection and productivity of crops while minimizing the ecological impact of pest management.

ARTIFICIAL INTELLIGENCE IN AGRICULTURE

Introduction to Artificial Intelligence

Artificial Intelligence is revolutionizing the agricultural sector by introducing advanced technologies that enable autonomous decision-making and enhanced productivity. By leveraging A.I., farmers can harness the power of data analytics, predictive modeling, and machine learning algorithms to optimize various aspects of agricultural operations. From crop monitoring to resource management, A.I. facilitates efficient and informed decision-making, ultimately contributing to sustainable and profitable farming practices. Let's explore the transformative potential of A.I. in agriculture across different domains. (Kumar et al., 2024)

Machine Learning Algorithms in Agriculture

Crop Yield Prediction

One of the significant applications of Artificial Intelligence in agriculture is its capability to predict crop yields (Figure 5) with high accuracy. By leveraging machine learning algorithms and analyzing a multitude of variables such as weather patterns, soil conditions, historical yield data, and crop health parameters, A.I. enables farmers to forecast crop yields for upcoming seasons. (Pant et al., 2021)

Figure 5. Crop yield prediction using AI

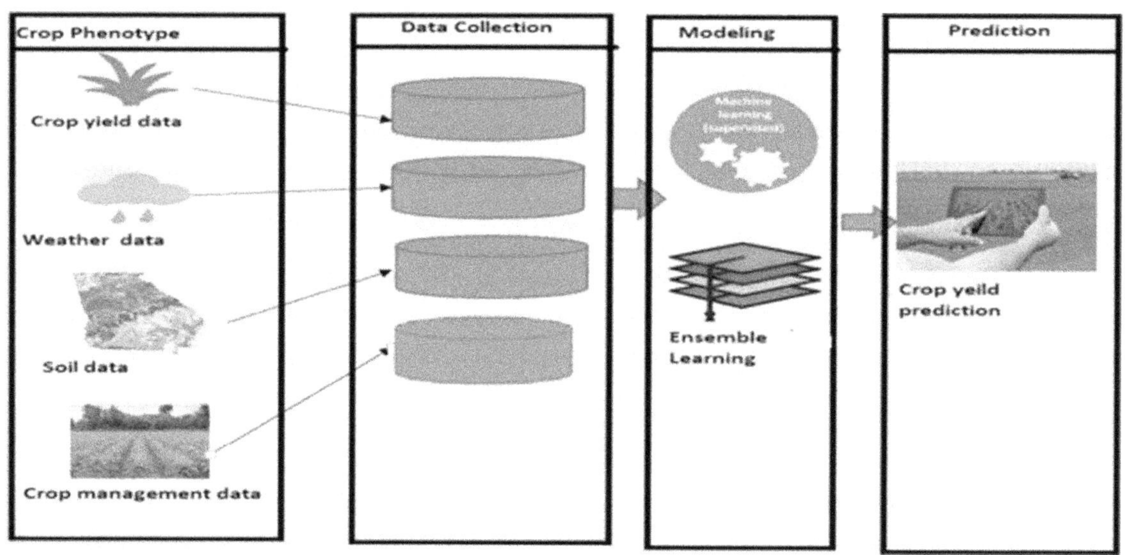

(Pant et al., 2021)

The predictive models generated through A.I. not only provide insights into expected crop production but also offer valuable information for efficient resource allocation and planning. With the ability to anticipate yield variations, farmers can make informed decisions regarding planting schedules, irrigation requirements, and procurement of resources. This proactive approach to crop management not only optimizes resource utilization but also contributes to maximizing overall crop productivity.

Disease Detection in Crops

Disease detection in crops (Figure 6) is another critical application of Artificial Intelligence in agriculture. By utilizing advanced image recognition and machine learning algorithms, A.I. can accurately identify and diagnose various diseases and abnormalities in crops at an early stage. This technology enables farmers to proactively address potential diseases, thus preventing widespread crop damage and yield losses. (Srivastava et al., 2021)

Figure 6. A whole crop disease detection system's block diagram

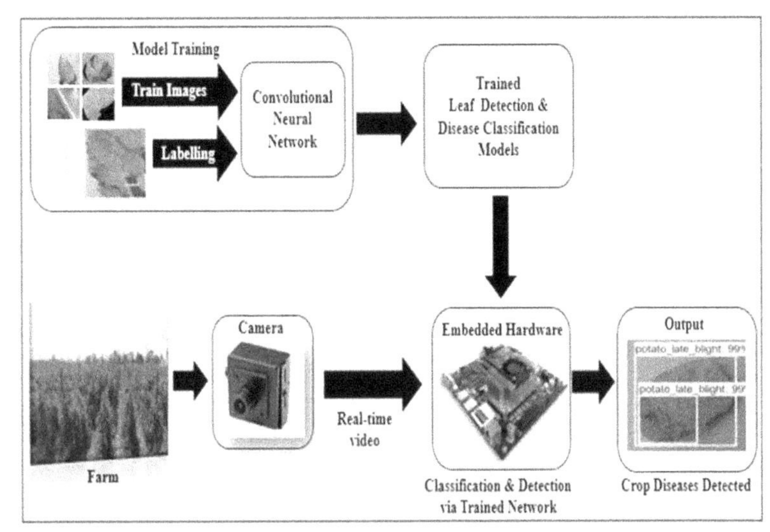

(Srivastava et al., 2021)

The integration of A.I. for disease detection in crops not only facilitates timely intervention but also reduces the reliance on broad-spectrum pesticides and excessive chemical treatments. With precise disease identification, farmers can implement targeted and specific control measures, minimizing environmental impact and preserving the natural balance within agricultural ecosystems.

Furthermore, A.I. powered disease detection systems contribute to the overall health and resilience of crops by enabling farmers to adopt a proactive and preventive approach to disease management. By identifying and isolating diseased plants promptly, farmers can contain the spread of diseases and prevent them from affecting the entire crop, leading to improved overall crop health and productivity.

The incorporation of A.I. in disease detection for crops represents a transformative advancement in agriculture, empowering farmers with the tools to proactively manage and protect their crops from diseases. This technology not only enhances the sustainability of farming practices but also fosters healthier and more resilient agricultural ecosystems.

Weed Identification and Management

Another significant application of Artificial Intelligence in agriculture is the identification and management of weeds (Figure 7). With advanced image recognition and machine learning algorithms, A.I. can accurately identify different weed species and assess their distribution within agricultural fields. (Zhang et al., 2021)

Farmers can reduce their reliance on broad herbicide treatments by using artificial intelligence (A.I.) to identify weeds and then use focused weed management methods, such as mechanical weed removal or precision herbicide application. This method helps reduce costs and maximizes resource efficiency while also minimizing the negative effects of weed control on the environment.

Figure 7. AI powered weed detector

(Zhang et al., 2021)

Furthermore, A.I.-powered weed management systems enable farmers to adopt a proactive and site-specific approach to weed control, allowing for the preservation of soil health and beneficial plant species within farming ecosystems. By accurately identifying and managing weeds, farmers can maintain the ecological balance of their fields and promote the overall health and productivity of their crops. (Jin et al., 2022)

The integration of A.I. in weed identification and management represents a significant advancement in sustainable agriculture, providing farmers with the tools to effectively control weeds while minimizing the ecological impact of traditional weed management practices. This technology empowers farmers to make informed and targeted decisions, ultimately contributing to more sustainable and resilient farming practices.

A.I. Applications in Livestock Management

Health Monitoring and Predictive Analytics

Livestock management stands to benefit significantly from the integration of Artificial Intelligence, particularly in the areas of health monitoring and predictive analytics. Applications of artificial intelligence (AI) in livestock management enable farmers to keep an eye on the health and welfare of their animals in real time, allowing for the early identification of possible health problems and prompt intervention. (Kumar et al., 2023)

Through the utilization of sensors and wearable devices, A.I. enables continuous monitoring of vital health parameters in livestock, including body temperature, heart rate, and activity levels. These real-time data streams are then processed through machine learning algorithms to detect patterns and anomalies that may indicate potential health issues. By leveraging predictive analytics, A.I. can forecast and alert farmers about potential health risks, allowing for timely intervention and preventive measures to maintain the overall well-being of the livestock. (Srivastava et al., 2021)

Furthermore, A.I.-driven health monitoring systems provide valuable insights into the individual health profiles of animals, facilitating personalized care and targeted interventions. In addition to enhancing animal wellbeing, this customized approach to livestock management raises overall production and efficiency in livestock operations. (Zhang et al., 2022)

In conclusion, the application of A.I. in health monitoring and predictive analytics for livestock management represents a transformative advancement in agriculture, empowering farmers to ensure the health, well-being, and productivity of their livestock through data-driven insights and proactive measures.

Automated Feeding Systems

In addition to health monitoring and predictive analytics, the integration of Artificial Intelligence in livestock management also extends to the implementation of automated feeding systems. Automated feeding systems (Figure 8) leverage A.I. and sensor technologies to provide precise feeding schedules and tailored nutritional plans for livestock, optimizing their dietary intake and overall well-being. (Javaid et al., 2022)

Figure 8. Mixing and feeding robots

(Javaid et al., 2022)

By utilizing data from individual animal profiles and considering factors such as age, weight, and nutritional requirements, A.I.-driven feeding systems can deliver personalized feeding programs for each animal. This customized approach not only ensures the nutritional needs of livestock are met but also minimizes feed wastage and associated costs.

Moreover, automated feeding systems contribute to the efficient utilization of resources by optimizing feed distribution and minimizing human intervention in the feeding process. The automation of feeding schedules also promotes consistency and regularity in the provision of feed, which is crucial for maintaining the health and productivity of livestock. (Kumar et al., 2023)

Furthermore, A.I.-powered automated feeding systems enable real-time monitoring of feed consumption and behavioral patterns, allowing farmers to make data-driven adjustments to feeding programs as needed. This proactive approach to feeding management supports the overall health and growth of livestock while enhancing operational efficiency on the farm.

The incorporation of artificial intelligence in automated livestock feeding systems signifies a notable progression in the sustainable and effective management of animal agriculture. A.I.-enabled automated feeding systems contribute to the health and output of livestock by delivering customized nutrition and streamlining feeding processes, thus fostering resource efficiency in agricultural practices.

ROBOTICS IN AGRICULTURE

Overview of Agricultural Robotics

Agricultural robotics have revolutionized the way tasks are performed on farms, offering advanced technology solutions to enhance productivity and sustainability. These robots are designed to automate various agricultural processes, from planting and harvesting to monitoring and managing livestock. By integrating robotics into agricultural practices, farmers can optimize their operations, reduce labor costs, and minimize environmental impact. In this section, we will delve into the diverse applications of agricultural robotics and explore how these technologies are reshaping modern farming practices.

Types of Agricultural Robots

Autonomous Tractors

Autonomous tractors (Figure 9) represent a significant innovation in modern farming practices, offering a range of benefits for agricultural operations. These advanced robotic systems are designed to autonomously perform tasks such as plowing, planting, and cultivating fields, thereby reducing the manual labor required for these repetitive and time-consuming activities. By leveraging artificial intelligence and sensor technologies, autonomous tractors can navigate fields with precision, optimizing inputs and reducing operational costs. (Finecomess et al., 2024)

Figure 9. Autonomous tractors

(Finecomess et al., 2024)

By enhancing production and efficiency, autonomous tractors in agriculture also help to manage farming operations in a sustainable manner. By utilizing real-time data and machine learning algorithms, these robotic systems can optimize field management practices, such as seed placement and crop health monitoring, leading to improved crop yields and resource utilization. Additionally, the autonomous operation of tractors minimizes fuel consumption and soil compaction, further enhancing the environmental sustainability of agricultural practices.

Moreover, autonomous tractors enable farmers to make data-driven decisions regarding field operations, leading to more precise and targeted agricultural practices. This not only enhances the overall sustainability of farming but also positions farmers to adapt to changing environmental conditions and market demands. (Benyezza et al., 2023)

The integration of autonomous tractors in agriculture represents a transformative shift towards more efficient, sustainable, and data-driven farming practices. By enabling farmers to maximize field activities, lessen their impact on the environment, and increase overall production, these cutting-edge robotic technologies help to maintain the resilience and prosperity of contemporary agricultural ecosystems.

Drone Technology

Drone technology has emerged as a game-changing tool in agricultural practices, offering a wide range of applications that revolutionize farm management and decision-making processes. Drones (Figure 10), equipped with advanced imaging and sensing technology, have become an invaluable asset for farmers in monitoring and managing their fields with unprecedented precision and efficiency. (Gowda et al., 2021)

Figure 10. Drone technology in agriculture

(Gowda et al., 2021)

Monitoring and evaluating crops is one of the main applications of drones in agriculture. Drones with multispectral imaging sensors and high-resolution cameras may take precise aerial photographs of crops, giving farmers the ability to monitor plant health, identify pest infestations, and evaluate the efficiency of fertilization and irrigation techniques. Farmers are able to make well-informed decisions to maximize yields and optimize crop health because to this complete and real-time perspective of crop conditions.

Furthermore, drones enable the efficient and timely mapping of fields, providing valuable data for crop scouting, yield estimation, and disease detection. By conducting aerial surveys, drones can identify areas of the field that require attention, facilitating targeted interventions and resource allocation to address specific agricultural challenges. This proactive approach to field management enhances operational efficiency and reduces potential crop losses.

In addition to crop monitoring, drones play a crucial role in precision agriculture by facilitating the precise application of inputs such as pesticides, fertilizers, and water. Through the use of specialized dispensing systems, drones can accurately and uniformly distribute these inputs across the fields, minimizing waste and environmental impact while optimizing resource utilization. (Navarro et al., 2020)

Moreover, the integration of drone technology with artificial intelligence and machine learning algorithms enables the analysis of data generated from aerial surveys and imaging, providing farmers with actionable insights and predictive analytics for crop management. By harnessing the power of drones and data analytics, farmers can make informed decisions that lead to more sustainable and productive farming practices.

The adoption of drone technology in agriculture represents a paradigm shift in farm management, enabling farmers to embrace precision, efficiency, and sustainability in their operations. The integration of drones with advanced technologies empowers farmers with unparalleled capabilities for crop monitoring, decision-making, and resource optimization, ultimately driving the resilience and success of modern agricultural practices.

Harvesting Robots

The emergence of harvesting robots in agriculture marks a significant advancement in the realm of automated farming practices. These robots are designed to revolutionize the labor-intensive and time-sensitive task of harvesting crops, offering a multitude of benefits to farmers and agricultural operations. (Dhawale et al., 2019)

Harvesting robots (Figure 11) are equipped with sophisticated sensors and precision technology that enable them to selectively and efficiently harvest crops with precision and care. By leveraging artificial intelligence and machine learning algorithms, these robots can identify ripe produce, navigate through fields, and delicately harvest crops without causing damage, thus ensuring optimal yield and quality.

The utilization of harvesting robots not only addresses labor shortages but also enhances efficiency and productivity in crop harvesting. These robots can work around the clock, minimizing the time taken to harvest crops and allowing for timely and efficient crop management. Additionally, the use of harvesting robots reduces the physical strain on farm laborers, offering a more sustainable and ergonomic solution to the demanding task of manual harvesting.

Figure 11. Harvesting robots

(Dhawale et al., 2019)

Furthermore, harvesting robots contribute to the overall sustainability of agricultural practices by minimizing food waste. With their precise harvesting capabilities, these robots can selectively harvest only ripe produce, reducing the likelihood of leaving behind unharvested or damaged crops. This not only optimizes the yield but also reduces unnecessary wastage, aligning with the principles of sustainable agriculture. (Adli et al., 2023)

The integration of harvesting robots in agricultural operations represents a transformative shift towards enhancing efficiency, addressing labor challenges, and promoting sustainable crop management. By incorporating these advanced robotic systems, farmers can optimize their harvesting processes, minimize waste, and ensure the quality and quantity of their produce, thereby contributing to the resilience and success of modern agricultural ecosystems.

Weeding Robots

Weeding robots (Figure 12) have emerged as a valuable asset in modern agricultural practices, offering innovative solutions for weed management and crop maintenance. These robotic systems are designed to address the challenges associated with manual weeding and chemical herbicide usage, providing farmers with efficient and sustainable alternatives for weed control.

One of the key advantages of weeding robots is their ability to autonomously navigate through fields and accurately identify and remove weeds. Equipped with advanced imaging technology and machine learning algorithms, these robots can distinguish between crops and unwanted weeds, enabling precise and targeted weed removal without causing any harm to the cultivated plants. By reducing the need for chemical herbicides and minimizing competition for resources, this focused strategy promotes sustainable and ecologically friendly weed management techniques. (Khan et al., 2021)

Figure 12. Weeding robots

(Khan et al., 2021)

Moreover, weeding robots contribute to the overall health and productivity of crops by eliminating the detrimental effects of weed infestations. By consistently monitoring and removing weeds, these robots help optimize the growth conditions for crops, ensuring higher yields and improved crop quality. Additionally, the autonomous operation of weeding robots reduces the reliance on manual labor for weed control, mitigating labor shortages and operational challenges in agricultural settings. (Javaid et al., 2022)

The integration of weeding robots in farming operations represents a significant step towards sustainable and eco-friendly weed management. By leveraging the capabilities of these robotic systems, farmers can minimize the environmental impact of weed control practices, optimize crop health, and enhance the overall efficiency of their farming operations. Weeding robots play a pivotal role in promoting responsible and sustainable agricultural practices while contributing to the resilience and success of modern agricultural ecosystems.

Robotic Applications in Dairy Farming

Automated Milking Systems

Automated milking systems (Figure 13), also known as robotic milking systems, have revolutionized the dairy industry by offering advanced solutions for milking operations. These systems are designed to automate the milking process, providing numerous benefits to dairy farmers and their operations.

Automated milking systems employ state-of-the-art technologies to facilitate the efficient and gentle milking of cows. Through the use of sensors and automated milking units, these systems can identify and position cows for milking, clean the udder, apply teat cups, and monitor the milking process, all without the need for human intervention. This not only reduces the physical demands on farm labor but also allows for continuous and timely milking, ultimately improving the overall productivity of the dairy farm. (Pant et al., 2021)

Figure 13. Automated milking machine

(Pant et al., 2021)

Furthermore, automated milking systems enable precise data collection and monitoring of individual cow performance. By capturing and analyzing data on milk yield, milking frequency, udder health, and milking duration, these systems provide valuable insights for dairy farmers to optimize herd management, cow welfare, and milk production efficiency.

In addition to enhancing operational efficiency, automated milking systems contribute to the well-being of dairy cows. The gentle and automated milking process reduces stress and discomfort for the cows, resulting in improved animal welfare and overall health. Moreover, the continuous monitoring and early detection of any udder health issues enable proactive veterinary interventions, ultimately leading to healthier and more productive dairy herds.

The integration of automated milking systems in dairy farming signifies a significant advancement in modern dairy operations, emphasizing precision, animal welfare, and productivity. By embracing this robotic technology, dairy farmers can streamline their milking processes, optimize cow management, and ensure sustainable and efficient milk production, thus contributing to the resilience and success of dairy farming in the contemporary agricultural landscape.

Cattle Health Monitoring

Cattle health monitoring is a critical aspect of modern dairy farming that can be effectively supported by robotic applications. Automated systems (Figure 14) equipped with advanced sensors and monitoring capabilities offer valuable solutions for real-time health assessment and management of dairy cattle. (Jin et al., 2022)

Figure 14. AI based real-time cattle health monitoring system

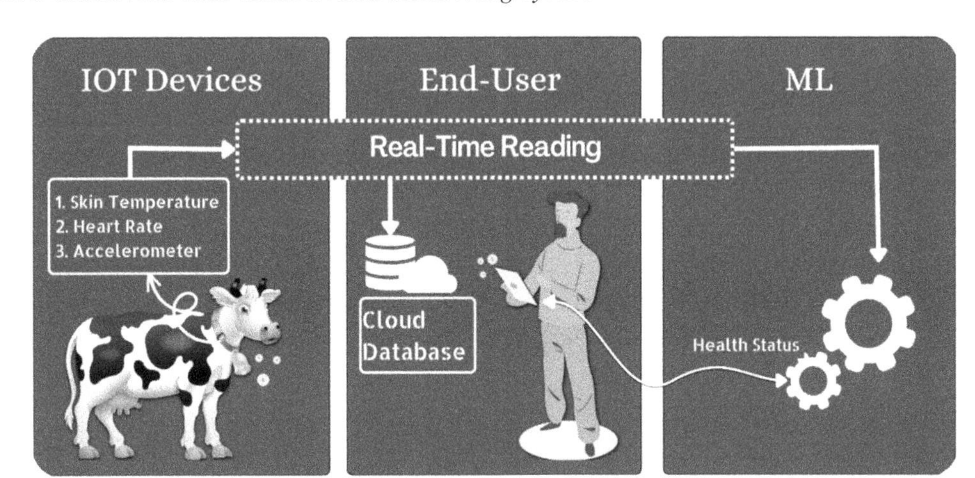

(Jin et al., 2022)

Robotic health monitoring systems for cattle are engineered to detect vital signs like body temperature, activity level, feeding habits, and ruminant behavior in real time. By analyzing this data, farmers can gain insights into the health status of individual cattle and identify any signs of distress or illness at an early stage. Early detection of health issues enables prompt veterinary intervention and the implementation of appropriate care measures, ultimately contributing to the overall well-being and productivity of the dairy herd. (Jin et al., 2022)

Furthermore, these robotic systems facilitate efficient data collection and analysis, providing dairy farmers with valuable information for proactive herd management. By monitoring cattle health metrics, such as rumination time and feeding patterns, farmers can make informed decisions regarding feed composition, herd nutrition, and overall herd health maintenance. This data-driven approach not only optimizes cattle well-being but also enhances the efficiency of dairy farming operations.

In addition to real-time health monitoring, robotic systems can also assist in the identification and tracking of individual cattle within the herd. This capability enables farmers to monitor the movement and behavior of each animal, ensuring proper integration into the herd and facilitating targeted health management strategies for specific individuals as needed.

The integration of cattle health monitoring solutions in dairy farming underscores the commitment to proactive health management, precision husbandry, and sustainable dairy production. By harnessing the capabilities of robotic applications for cattle health monitoring, dairy farmers can elevate the standard of care for their cattle, improve herd productivity, and contribute to the resilience and success of modern dairy farming practices.

INTEGRATION OF IOT, A.I., AND ROBOTICS

Synergies and Interactions Between Technologies

As agricultural and dairy industries continue to embrace technological advancements, the integration of Internet of Things, Artificial Intelligence, and Robotics presents a myriad of synergistic opportunities to enhance operational efficiency, optimize resource utilization, and promote sustainable practices.

The synergy between IoT, AI, and Robotics is particularly evident in the realm of weed management. By leveraging IoT sensors to gather real-time data on environmental conditions and crop health, AI algorithms can analyze this information to identify areas prone to weed infestations. Subsequently, robotic weed removal systems can be deployed with precision, targeting specific areas and minimizing the use of chemical herbicides. This collaborative approach not only reduces the environmental impact of weed control but also promotes the sustainable management of agricultural landscapes. (Kumar et al., 2023)

In the dairy farming sector, the integration of IoT, AI, and Robotics offers remarkable opportunities for enhancing herd management and milk production. By combining data from IoT sensors that monitor cattle health and behavior with AI-powered analytics, dairy farmers can gain comprehensive insights into individual animal welfare and overall herd health. This integrated approach enables proactive decision-making to optimize feed composition, veterinary interventions, and breeding strategies, ultimately contributing to the well-being and productivity of the dairy herd.

Case Studies of Integrated Systems

Smart Farming Platforms

Smart farming platforms, encompassing the integration of IoT, AI, and Robotics, are revolutionizing agricultural and dairy practices by offering comprehensive solutions for precision management and sustainable production. These platforms serve as centralized systems for data collection, analysis, and decision-making, leveraging the synergies and interactions between advanced technologies to optimize farming operations. (Zhang et al., 2021)

In the context of dairy farming, smart farming platforms play a pivotal role in orchestrating the seamless integration of automated milking systems and robotic cattle health monitoring. By consolidating data from IoT sensors and AI-driven analytics, these platforms provide dairy farmers with holistic insights into herd management, milk production, and cattle health, enabling informed and proactive decision-making for overall farm efficiency and animal welfare.

Moreover, smart farming platforms facilitate the interoperability of robotic applications, allowing for the synchronization of automated milking, cattle health monitoring, and other farm management processes. This integrated approach enables farmers to streamline operations, monitor individual cow performance, and ensure the well-being of the dairy herd in a cohesive and efficient manner. The interoperability of these systems also enhances resource utilization and labor efficiency, contributing to the sustainable and optimal functioning of dairy farms.

Furthermore, smart farming platforms enable the implementation of precision farming practices, where the data-driven insights obtained through IoT, AI, and Robotics integration guide the optimization of feed composition, breeding strategies, and veterinary interventions. This tailored approach not only enhances the overall productivity of dairy farming but also promotes sustainable resource management and environmental stewardship. (Pant et al., 2021)

In essence, smart farming platforms serve as the cornerstone of modern dairy operations, providing dairy farmers with the technological infrastructure to harness the full potential of automated milking systems, robotic cattle health monitoring, and integrated IoT, AI, and Robotics solutions. By embracing these platforms, dairy farmers can elevate their farm management practices, ensure the well-being of their cattle, and contribute to the continued success and resilience of dairy farming in the contemporary agricultural landscape.

Comprehensive Crop Management Systems

Comprehensive crop management systems represent a fundamental shift in agricultural practices, offering innovative approaches to precision farming and sustainable crop production. By integrating Internet of Things, Artificial Intelligence, and Robotics, these systems empower farmers with advanced tools for optimized resource utilization, enhanced crop health, and efficient management of agricultural landscapes.

The utilization of IoT sensors and AI algorithms in crop management systems provides farmers with real-time insights into environmental conditions, crop health, and potential areas susceptible to weed infestations. This proactive approach enables targeted weed control through the deployment of robotic weed removal systems, minimizing the reliance on chemical herbicides and promoting eco-friendly agricultural practices. (Sairoel et al., 2023)

Moreover, the integration of IoT, AI, and Robotics in crop management facilitates data-driven decision-making for agricultural operations. By using thorough insights into crop health and environmental factors, farmers may optimize crop yields while reducing their impact on the environment by making well-informed decisions about fertilization schedules, irrigation schedules, and pest management. (Benyezza et al., 2023)

Furthermore, smart farming platforms, which are central to comprehensive crop management systems, serve as powerful orchestrators of data integration and decision-making in the agricultural landscape. These platforms streamline and synchronize automated processes such as irrigation, fertilization, and pest control, leveraging the synergies between advanced technologies for efficient and sustainable crop management.

In essence, comprehensive crop management systems represent a leap forward in modern agriculture, where the fusion of IoT, AI, and Robotics enables farmers to embrace precision farming, enhance crop productivity, and uphold environmental sustainability. By adopting these integrated systems, farmers can revolutionize their approach to crop management, contributing to the resilience and success of agriculture in the contemporary agricultural landscape.

Fully Automated Farming Operations

The concept of fully automated farming operations represents the pinnacle of technological integration in agriculture, promising unprecedented levels of operational efficiency, resource utilization, and sustainability. By harnessing the collective power of IoT, AI, and Robotics, fully automated farming endeavors aim to revolutionize agricultural practices and address the evolving demands of modern farming landscapes. (Finecomess et al., 2024)

At the core of fully automated farming operations lies the seamless orchestration of interconnected technologies to facilitate autonomous agricultural tasks. From precision seeding and crop monitoring to automated harvesting and processing, these operations strive to optimize the entire agricultural workflow through advanced automation and data-driven decision-making.

One of the central components of fully automated farming is the deployment of autonomous vehicles and machinery equipped with IoT sensors and AI algorithms. These intelligent systems can autonomously navigate fields, assess crop conditions, apply targeted interventions, and execute farming activities with unprecedented precision and efficiency. The integration of Robotics further enhances these capabilities, enabling tasks such as selective harvesting and non-destructive crop sampling to be performed with unparalleled accuracy. (Javaid et al., 2022)

Moreover, fully automated farming operations encompass comprehensive data-driven management, where AI algorithms continuously analyze real-time data from IoT sensors to inform timely agricultural interventions. By leveraging this adaptive decision-making framework, farmers can optimize resource allocation, minimize environmental impact, and maximize crop yields with a level of precision that transcends traditional farming methodologies.

In addition to crop-centric applications, fully automated farming operations also extend to livestock management, where interconnected systems enable autonomous feeding, health monitoring, and behavior analysis. The convergence of IoT, AI, and Robotics in livestock management fosters proactive health interventions, precise nutrition programs, and individualized care approaches, empowering farmers to elevate the standard of animal husbandry while enhancing overall herd productivity.

The implementation of fully automated farming operations is not without its challenges. Notably, the initial investment costs associated with deploying advanced technologies and autonomous systems pose a significant barrier for widespread adoption. Additionally, concerns surrounding data security and privacy in the context of vast interconnected networks necessitate robust frameworks to safeguard sensitive agricultural information and ensure the ethical use of data-driven technologies.

BENEFITS AND CHALLENGES

Positive Impacts on Agriculture

Increased Efficiency and Productivity

The integration of IoT, AI, and Robotics in dairy farming boosts efficiency and productivity. Through IoT sensors and AI analytics, farmers gain insights into individual animal welfare and herd health, enabling proactive decisions on feed, veterinary care, and breeding. Smart farming platforms centralize data for informed decision-making on herd management, milk production, and cattle health. The interoperability of robotic applications streamlines operations, promotes efficient monitoring, and enhances resource management. This paradigm shift ensures the well-being of cattle and contributes to the success of dairy farming in the contemporary agricultural landscape. To maximize benefits, farmers need to stay updated on technological developments in dairy farming.

Resource Optimization

Modern agriculture and dairy practices benefit significantly from resource optimization, thanks to the integration of IoT, AI, and Robotics. These technologies empower dairy farmers to efficiently manage resources and adopt sustainable production methods. By employing IoT sensors and AI analytics, farmers gain insights into individual animal welfare and herd health, enabling proactive decision-making for optimized feed, veterinary care, and breeding strategies. Smart farming platforms serve as the backbone, facilitating automated milking, robotic cattle health monitoring, and overall farm management. The interoperability of these technologies enhances resource utilization and labor efficiency. Continued research and development promise further advancements, revolutionizing farming practices and promoting sustainable resource management. However, challenges like cost, accessibility, data security, and privacy must be addressed to ensure the successful integration of these technologies in agriculture. Proactively overcoming these challenges will pave the way for a future of efficient, sustainable, and productive dairy farming practices.

Environmental Sustainability

In the quest for environmental sustainability, the integration of IoT, AI, and Robotics in dairy farming plays a pivotal role in promoting responsible and efficient agricultural practices. By harnessing the capabilities of advanced technologies, dairy farmers can achieve optimal resource utilization, minimize environmental impact, and ensure the well-being of their cattle while contributing to the long-term sustainability of dairy production.

Synergistic Approach to Environmental Conservation

The amalgamation of IoT, AI, and Robotics in dairy farming not only enhances operational efficiency but also fosters sustainable environmental stewardship. Through real-time monitoring and data analytics, these integrated systems enable farmers to identify opportunities for resource optimization, reduction of chemical usage, and proactive health management, ultimately leading to more sustainable farming practices.

Precision Resource Management

The utilization of IoT sensors and AI-driven analytics empowers dairy farmers to implement precision resource management strategies. By leveraging real-time data on cattle health, behavior, and environmental conditions, farmers can make informed decisions to optimize feed composition, minimize waste, and reduce the environmental footprint of dairy farming operations.

Minimized Chemical Utilization

With the deployment of robotic applications for weed management and cattle health monitoring, dairy farmers can minimize the use of chemical herbicides and pesticides. The targeted approach in weed removal and health management facilitated by these technologies not only reduces environmental contamination but also promotes the natural balance within agricultural ecosystems.

Long-Term Environmental Stewardship

The adoption of integrated IoT, AI, and Robotics systems in dairy farming signifies a commitment to long-term environmental stewardship. By embracing these technologies, dairy farmers demonstrate their dedication to conserving natural resources, reducing greenhouse gas emissions, and fostering sustainable agriculture for generations to come.

The synergy between IoT, AI, and Robotics in dairy farming exemplifies the potential for technological innovation to drive positive environmental outcomes while simultaneously enhancing farm productivity and animal welfare. With a proactive approach to environmental sustainability, dairy farmers are poised to lead the way in sustainable agricultural practices, setting a precedent for the future of dairy production in harmony with the environment.

Challenges and Considerations

Cost and Accessibility

While the integration of IoT, AI, and Robotics in dairy farming offers numerous benefits, it also presents challenges related to cost and accessibility. The initial investment required for implementing advanced technological systems, such as smart farming platforms and robotic applications, can be a significant barrier for some dairy farmers. Moreover, the accessibility of these technologies, particularly in rural or remote agricultural areas, can pose challenges in terms of infrastructure, technical support, and training.

Addressing the cost and accessibility concerns is crucial for ensuring widespread adoption of advanced agricultural technologies. Initiatives focusing on financial support, subsidies, and technical assistance can aid in making these technologies more accessible to a larger number of dairy farmers. Additionally, collaborative efforts between technology providers, government agencies, and agricultural organizations can facilitate the development of cost-effective solutions and the provision of necessary training and support to farmers.

Data Security and Privacy

As dairy farming continues to integrate IoT, AI, and Robotics for enhanced herd management and milk production, it is crucial to address the significance of data security and privacy. The amalgamation of various technological systems results in the generation and storage of extensive data pertaining to cattle health, feeding patterns, environmental conditions, and farm management. While this data holds immense value in optimizing dairy farming operations, ensuring its security and privacy is paramount.

Importance of Data Security

The data obtained from IoT sensors, AI analytics, and robotic systems encompasses sensitive information about individual cattle, farm processes, and environmental factors. Protecting this data from unauthorized access, cyber threats, and potential breaches is essential to uphold the integrity and confidentiality of dairy farm operations. Additionally, data security measures are vital in safeguarding the proprietary algorithms and insights generated through AI-driven analytics, preventing unauthorized duplication or exploitation of intellectual property.

Privacy Concerns and Regulatory Compliance

Furthermore, the integration of IoT, AI, and Robotics in dairy farming necessitates adherence to privacy regulations and compliance with data protection standards. As the data collected includes identifiable information about individual cattle and farm practices, ensuring compliance with regulations such as the General Data Protection Regulation or other regional data privacy laws is imperative. Dairy farmers must establish robust privacy policies and data management protocols to uphold transparency, accountability, and respect for individual privacy rights.

Mitigating Risks and Ensuring Trust

To address the challenges associated with data security and privacy, dairy farmers should implement comprehensive cybersecurity measures, encryption protocols, and access controls for data storage and transmission. Additionally, the adoption of secure communication protocols for IoT devices and AI algorithms can mitigate the risks of data interception or manipulation. Building trust with stakeholders, including farmers, veterinarians, and regulatory authorities, by demonstrating a commitment to data security and privacy will be pivotal in fostering confidence in the adoption of integrated technological solutions in dairy farming.

In conclusion, while the integration of IoT, AI, and Robotics presents transformative opportunities for dairy farming, prioritizing data security and privacy safeguards the integrity of farm operations, upholds regulatory compliance, and fosters trust within the agricultural community. Embracing a proactive approach to data protection will fortify the sustainable integration of advanced technologies in dairy farming, paving the way for progressive innovation and responsible data management practices.

Adoption Barriers in Agriculture

While the integration of IoT, A.I., and Robotics in agriculture and dairy farming presents numerous opportunities for enhancing operational efficiency and promoting sustainable practices, there are also notable adoption barriers that hinder the widespread implementation of these advanced technologies.

Technological Complexity and Knowledge Gap

One of the primary barriers to the adoption of IoT, A.I., and Robotics in agriculture is the inherent technological complexity and the knowledge gap among farmers and farm operators. The successful integration of these advanced systems requires a certain level of technical expertise and familiarity with digital technologies, which may pose a challenge for traditional agricultural practitioners who are not accustomed to such intricacies. Providing extensive training and support to facilitate the learning curve associated with these technologies is essential for overcoming this barrier.

Initial Investment and Return on Investment

The initial capital investment required for acquiring and implementing IoT, A.I., and Robotics solutions can be prohibitive for many farmers, especially small and medium-sized agricultural operations. While the long-term benefits of increased efficiency, resource optimization, and sustainability are compelling, the immediate financial burden may deter some farmers from embracing these technologies. Strategies to facilitate access to funding, incentivize adoption through government programs, and demonstrate the tangible return on investment over time are crucial in overcoming this adoption barrier.

Connectivity and Infrastructure

In rural agricultural settings, limited connectivity and inadequate infrastructure for supporting IoT and A.I. applications can impede the seamless deployment and functionality of these technologies. Access to reliable internet connectivity, robust sensor networks, and data processing capabilities are essential for harnessing the full potential of IoT, A.I., and Robotics in agriculture. Addressing infrastructure gaps and expanding access to high-speed internet and technological infrastructure in rural areas are pivotal in mitigating this barrier.

Data Privacy and Security Concerns

As the integration of IoT and A.I. involves the collection, processing, and analysis of vast amounts of data, farmers are understandably concerned about data privacy and security. Ensuring robust measures for data protection, compliance with privacy regulations, and transparency in data management practices is fundamental for building trust and alleviating concerns regarding the use of sensitive agricultural data in IoT and A.I. systems.

Adaptation and Cultural Shift

Embracing the advancements in IoT, A.I., and Robotics requires a cultural shift within the agricultural community, where traditional practices and customary approaches may need to evolve to accommodate the integration of these technologies. Overcoming resistance to change, fostering a culture of innovation, and demonstrating the tangible benefits of adopting these technologies are essential for facilitating the widespread adaptation of IoT, A.I., and Robotics in agriculture.

In conclusion, while the potential benefits of IoT, A.I., and Robotics in agriculture are substantial, addressing the adoption barriers is critical for realizing the transformative impact of these technologies across diverse agricultural landscapes.

FUTURE TRENDS AND EMERGING TECHNOLOGIES

Advancements in IoT, A.I., and Robotics

The convergence of IoT, A.I., and Robotics is revolutionizing agriculture, particularly in dairy farming and crop management. Real-time monitoring enhances cattle health, while AI-driven precision husbandry improves productivity and sustainability. In crop management, IoT sensors and robotic systems optimize resource use, reducing environmental impact. Despite these benefits, challenges like cost, accessibility, data security, and sector adoption must be addressed. Future trends include autonomous machinery and advanced data analytics, promising increased efficiency and sustainability. Ongoing research is crucial to fully realize the potential, leading to a technologically advanced, environmentally conscious, and economically viable future in agriculture.

Potential Innovations and Disruptions in Agriculture

1. **Hydroponics and Vertical Farming**: Advances in hydroponics and vertical farming maximize crop yields in controlled environments. IoT monitors conditions, AI optimizes nutrient delivery, and Robotics enhances precision, enabling sustainable urban agriculture.
2. **Gene Editing and Precision Agriculture:** Convergence of gene editing, AI, and Robotics enhances crop resilience. AI analyzes genomes, Robotics facilitates precise gene editing, contributing to improved crop traits, nutritional value, and resistance. Tailored varieties optimize productivity and food security.
3. **Blockchain and Supply Chain Transparency:** Blockchain, supported by IoT and AI, transforms supply chain transparency. Secure data collection and predictive analytics ensure reliable record-keeping, certifying product origins. Empowering consumers with trustworthy information fosters trust and sustainability in agriculture.
4. **Autonomous Farming Machinery:** Autonomous machinery, powered by IoT, AI, and Robotics, revolutionizes traditional farming. Real-time data analytics and AI-driven decision-making optimize resource use, minimize human labor, and enhance overall productivity, ushering in a smart and sustainable era.

Considerations for Future Dairy Farming Research

IoT, AI, and Robotics Integration: Advance AI algorithms to analyze comprehensive data from IoT sensors for enhanced cattle health and herd management insights in dairy farming.

Sophisticated Robotic Systems: Develop advanced robotic systems for tasks beyond monitoring, potentially assisting in feeding, cleaning, and overall herd maintenance for increased automation and efficiency.

Data Security and Privacy: Address challenges in data security and privacy for smart farming platforms, safeguarding sensitive farmer data gathered from IoT, AI, and Robotics technologies.

Exploring Emerging Technologies: Investigate opportunities and challenges in adopting emerging technologies like blockchain for data security and edge computing for real-time analytics in the context of dairy farming.

By addressing these considerations, dairy farming can harness the full potential of IoT, AI, and Robotics, ensuring sustainable and efficient practices in the modern agricultural landscape.

CONCLUSION

In conclusion, the integration of IoT, A.I., and Robotics in agriculture represents a significant opportunity for enhancing efficiency, sustainability, and productivity in farming practices. These advanced technologies offer transformative solutions for optimizing resource utilization, improving crop and herd management, and promoting environmental stewardship. However, to fully realize the potential benefits of these technologies, it is crucial to address adoption barriers such as technological complexity, initial investment challenges, connectivity issues, data security concerns, and the need for cultural adaptation within the agricultural community.

By prioritizing data security and privacy, providing adequate training and support, facilitating access to funding, enhancing connectivity and infrastructure, and promoting a culture of innovation, the agricultural sector can overcome these barriers and harness the full potential of IoT, A.I., and Robotics. Continued research and development in these areas will lead to further advancements, revolutionizing farming practices and ensuring the sustainability and resilience of agriculture in the contemporary landscape.

Embracing these emerging technologies and proactively addressing challenges will pave the way for a future agricultural landscape characterized by efficiency, sustainability, and productivity. By staying vigilant of considerations such as cost, accessibility, data security, and adoption barriers, farmers and stakeholders can solidify the foundation for efficient, sustainable, and productive farming practices, ultimately contributing to the continued success and resilience of agriculture in the modern era.

REFERENCES

Adli, H. K., Remli, M. A., Wan Salihin Wong, K. N. S., Ismail, N. A., González-Briones, A., Corchado, J. M., & Mohamad, M. S. (2023). Recent Advancements and Challenges of AIoT Application in Smart Agriculture: A Review. *Sensors (Basel)*, 23(7), 3752. 10.3390/s2307375237050812

Amertet Finecomess, S., Gebresenbet, G., & Alwan, H. M. (2024). Utilizing an Internet of Things (IoT) Device, Intelligent Control Design, and Simulation for an Agricultural System. *IoT.*, 5(1), 58–78. 10.3390/iot5010004

Benyezza, H., Bouhedda, M., Kara, R., & Rebouh, S. (2023). Smart platform based on IoT and WSN for monitoring and control of a greenhouse in the context of precision agriculture. *Internet of Things : Engineering Cyber Physical Human Systems*, 23, 100830. 10.1016/j.iot.2023.100830

Chowdary, A. V., Saini, N., Kumar, A., Kumar, S., Ballabh, J., Bhatt, S. S., Bhatt, A., Prakash, S., & Patel, A. (2023). Influence of integrated nutrient management on physiological parameters of lentil (Lens culinaris Medik.). *Plant Science Today*, 10(3), 94–97. 10.14719/pst.2124

Dhawale, K. N. (2019, May 31). Review on IoT Based Smart Agriculture System. *International Journal for Research in Applied Science and Engineering Technology*, 7(5), 4063–4066. 10.22214/ijraset.2019.5684

Gowda, V., Prabhu, M., Ramesha, M., Kudari, J. M., & Samal, A. (2021, November 1). Smart Agriculture and Smart Farming using IoT Technology. *Journal of Physics: Conference Series*, 2089(1), 012038–012038. 10.1088/1742-6596/2089/1/012038

Javaid, M., Haleem, A., Singh, R. P., & Suman, R. (2022). Enhancing smart farming through the applications of Agriculture 4.0 technolgies. *Int. J. Intell. Netw.*, 3, 150–164.

Jin, X., Zhang, J., Kong, J., Su, T., & Bai, Y. (2022). A Reversible Automatic Selection Normalization (RASN) Deep Network for Predicting in the Smart Agriculture System. *Agronomy (Basel)*, 12(3), 591. 10.3390/agronomy12030591

Khan, N. M., Ray, R. L., Sargani, G. R., Ihtisham, M., Khayyam, M., & Ismail, S. (2021, April 27). Current Progress and Future Prospects of Agriculture Technology: Gateway to Sustainable Agriculture. *Sustainability (Basel)*, 13(9), 4883–4883. 10.3390/su13094883

Khan, N M., Ray, R L., Sargani, G R., Ihtisham, M., Khayyam, M., & Ismail, S. (2021, April 27). *Current Progress and Future Prospects of Agriculture Technology: Gateway to Sustainable Agriculture.* 10.3390/su13094883

Kumar, A., Das, N., Sharma, H., & Singh, R. (2024). Impact of Agrochemicals on the Environment and Ecological Alternatives. In Karmaoui, A. (Ed.), *Water-Soil-Plant-Animal Nexus in the Era of Climate Change* (pp. 317–328). IGI Global. 10.4018/978-1-6684-9838-5.ch015

Kumar, A., Kumar, S., Juyal, R., Sharma, H., & Bisht, M. (2023). Nanoscience in Agricultural Steadiness. In Dar, G. H., Bhat, R. A., & Mehmood, M. A. (Eds.), *Microbiomes for the Management of Agricultural Sustainability* (pp. 285–296). Springer. 10.1007/978-3-031-32967-8_17

Kumar, A., Sharma, H., Shakeb, A., & Joshi, M. (2022). Environmental Sustainability: An Emerging Concept. *International Year of Millets*, 2023, 29.

Kumar, R., Sachan, S., Kumar, R., Parmar, K., Mishra, A. K., Kumar, A., & Mishra, A. K. (2023). Unveiling Use of Information and Communication Technology (ICT) in Western Uttar Pradesh, India: The Farmers' Perspective. *Journal of Global Innovations in Agricultural Sciences*, 11(4), 521–531. 10.22194/JGIAS/23.1197

Kumar, R., Singh, A., Yadav, R. B., Kumar, A., Kumar, S., Shahi, U. P., & Singh, A. P. (2017). Growth, development and yield response of rice (Oryza sativa L.) as influenced by efficient nitrogen management under subtropical climatic condition. *Journal of Pharmacognosy and Phytochemistry*, 6(6S), 791–797.

Kumawat, L., Jat, L., Kumar, A., Yadav, M., Ram, B., & Dudwal, B. L. (2022). Effect of organic nutrient sources on growth, yield attributes and yield of wheat under rice (Oryza sativa L.) wheat (Triticum aestivum L.) cropping system. *The Pharma Innovation Journal*, 11(2), 1618–1623.

Lee, T. Y., Reza, M. N., Chung, S. O., Kim, D. U., Lee, S. Y., & Choi, D. H. (2023). Application of fuzzy logics for smart agriculture: A review. *Precision Agriculture*, 5, 1.

Navarro, E. D. M., Costa, N., & Pereira, A. (2020, July 29). A Systematic Review of IoT Solutions for Smart Farming. *Sensors (Basel)*, 20(15), 4231–4231. 10.3390/s2015423132751366

Pant, J., Pant, R. P., Singh, M. K., Singh, D. P., & Pant, H. (2021). Analysis of agricultural crop yield prediction using statistical techniques of machine learning. *Materials Today: Proceedings*, 46, 10922–10926. 10.1016/j.matpr.2021.01.948

Ratnaparkhi, S. T., Khan, S., Arya, C., Khapre, S., Singh, P., Diwakar, M., & Shankar, A. (2020, December 1). WITHDRAWN: Smart agriculture sensors in IOT: A review. *Materials Today: Proceedings*. 10.1016/j.matpr.2020.11.138

Ray, P. P. (2017, June 19). Internet of things for smart agriculture: Technologies, practices and future direction. *Journal of Ambient Intelligence and Smart Environments*, 9(4), 395–420. 10.3233/AIS-170440

Sairoel, A., Gebresenbet, G., Alwan, H. M., & Vladmirovna, K. O. (2023). Assessment of Smart Mechatronics Applications in Agriculture: A Review. *Applied Sciences (Basel, Switzerland)*, 13(12), 7315. 10.3390/app13127315

Sharma, H., Kumar, A., Singh, R., & Kumar, S. (2023). Impact of Nanopriming and Omics-Based Applications for Sustainable Agriculture. In Singh, A., Rajput, V., Ghazaryan, K., Gupta, S., & Minkina, T. (Eds.), *Nanopriming Approach to Sustainable Agriculture* (pp. 175–191). IGI Global., 10.4018/978-1-6684-7232-3.ch008

Zhang, H., He, L., Di Gioia, F., Choi, D., Elia, A., & Heinemann, P. (2022). LoRaWAN based internet of things (IoT) system for precision irrigation in plasticulture fresh-market tomato. *Smart Agricultural Technology*, 2, 100053. 10.1016/j.atech.2022.100053

Zhang, T., Zhao, Y., Jia, W., & Chen, M.-Y. (2021). Collaborative algorithms that combine AI with IoT towards monitoring and controsystem. *Future Generation Computer Systems*, 125, 677–686. 10.1016/j.future.2021.07.008

Chapter 15
LSTM-Based Deep Learning for Crop Production Prediction With Synthetic Data

Aditi Verma
School of Computer Science Engineering, Vellore Institute of Technology, India

Shivani Boggavarapu
http://orcid.org/0009-0003-2094-1449
School of Computer Science Engineering, Vellore Institute of Technology, India

Astha Bharadwaj
School of Computer Science Engineering, Vellore Institute of Technology, India

Prabakaran N.
http://orcid.org/0000-0002-1232-1878
School of Computer Science Engineering, Vellore Institute of Technology, India

ABSTRACT

The Agri-industry forms the backbone of the economy and livelihood. Hence, efficient planning on resources and ensuring a steady food supply is vital. This model discusses the challenges of accurately predicting crop yields influenced by multiple dynamic factors. Traditional models suffer with the complexity, thus leading to inaccurate predictions. Also, the availability of reliable training data is scarce, which poses an additional problem in training. Existing solutions range from traditional statistical models based on historical data to modern AI techniques. While these approaches are better than conventional methods, they are still unable to address data scarcity, non-linear interactions and the dynamic complexities. This model aims to overcome the limitations using long short-term memory (LSTM) and integrating synthetic data. LSTM is able to decipher complex patterns and synthetic data provides additional training samples that can enhance accuracy. The overall potential of this proposed solution can help mitigate food scarcity and strengthen sustainability.

DOI: 10.4018/979-8-3693-3583-3.ch015

INTRODUCTION

Modern agriculture faces the challenge of ensuring food security for a growing global population while dealing with unpredictable climate patterns and limited resources. Predicting crop production accurately is crucial for efficient resource allocation, risk management, and policy planning. Deep Learning techniques have emerged as powerful tools for time series prediction tasks due to their ability to capture complex temporal dependencies. Furthermore, the integration of synthetic data augments the training process, enhancing model performance and generalization (Prodhan, F. Et al., 2022). The fusion of LSTM networks and the integration of synthetic data offer a promising solution for accurate crop production prediction. This approach not only addresses the limitations of traditional methods but also enables proactive decision-making in agriculture, contributing to global food security and sustainable resource management. Utilizing data-driven methods has become crucial in modern agriculture for improving crop output prediction. The Long Short-Term Memory (LSTM) method, a kind of recurrent neural network (RNN), is a potent tool in this attempt. Due to its design for modeling sequential data, LSTMs are well suited for the time-dependent and dynamic character of agricultural production datasets. It particularly adept in capturing the complex temporal relationships and patterns present in the variables influencing agricultural yields, such as weather patterns, soil moisture, and past output rates. LSTMs, as opposed to conventional RNNs, are able to successfully learn from both short-term and long-term information because they are able to solve the vanishing gradient problem. By consuming a series of input data representing the changing conditions over time, the LSTM can be used to predict future crop yields once it has been trained. The latter half of this approach incorporates synthetic data generation and integrating it to the training model. Current approaches of crop yield prediction (Joshua et al., 2022) which involve machine learning algorithms like random forests and support vector machines, often lack enough training samples for the model prediction. This leads to inept accuracies in the prediction results. To tackle this problem, we generate synthetic data and integrate it to the already existing real crop data set, to increase the number of training samples, thereby, increasing accuracy of the prediction model (Elavarasan, D., & Durai Raj Vincent, 2021) and (Van Klompenburg et al., 2020). Synthetic data is the artificially generated data that replicates or mimics the real-world data. It does not contain any personally identifiable information (PII). As mentioned previously, it is used for a variety of purposes, including data analysis, model training and testing. Synthetic data can be generated using algorithms or statistical models based on the characteristics and patterns observed in real data. Some of the ways of generating synthetic data are rule-based generation, simulation models, generative adversarial networks (GANs), variational autoencoders (VAEs) etc. The algorithm which we will be deploying to generate synthetic data for our dataset is the Tabular LSTM model. It is a type of LSTM model which can specifically handle tabular data (structured into rows and columns). Tabular LSTMs use the sequential and temporal aspects of LSTM architecture to process tabular data with sequential dependencies, making them favorable for time-series prediction. The reason we chose LSTM for both synthetic data generation and crop yield prediction is because LSTM models are most suitable for structured and sequential data. Crop data collected for crop prediction consists of attributes such as weather and soil parameters, which can be organized in a structured format. Furthermore, this structured data becomes sequential when the temporal order matters, as the crop growth is influenced by the progression of time and changing conditions.

RELATED WORK

The summary of various studies along with current study is presented in Table 1 and addressing their methodologies and key aspects.

Table 1. Summary of literature review and key aspects

Research Study	Methodology	Prediction Properties	Key Aspects/Limitations
M K Dharani et al., (2021)	ANN, CNN, RNN-LSTM	CNN with pictorial matrix evaluation. RNN combined with the hybrid network and LSTM.	CNN's prediction power is limited to trained data in this case; real-time historical data is not utilized. RNN and hybrid networks perform better than all other networks combined, with a maximum 90% accuracy rate.
Bali, N., & Singla, A. (2021)	Non-linear QuasiNewton multi-variate regression model and piecewise linear regression	The technique does not require the extraction of any manually created features or the dimensionality reduction of the satellite data in order to operate on raw satellite imagery.	The increase in the error is only marginal during changes in seasons. The model performs more than 50% better than current techniques.
G Perich et al., (2023)	contrasting random forest, LSTM neural networks, and multivariate OLS linear regression.	Multi-temporal satellite imagery combined with weather data.	The findings demonstrated that the integration of weather information with satellite imagery can give valuable insights for the development of more precise yield forecasting models. Detect 70 days before harvest.
L Gong et al., (2021)	Combined TCN and RNN	Two components are coupled with DNN:The LSTM-RNN and TCN components are made up of several LSTM units, whereas three dilated convolutional layers are seen in residual blocks of the TCN component.	Compared to typical machine learning algorithms, deep learning-based methods perform better, yielding smaller root mean square errors (RMSEs) and more accurate prediction outcomes.
S Jeong et al., (2022)	Hybrid 1D-CNN with LSTM structure	Combining crop models and deep learning allowed for the prediction of rice yield at the pixel level.	Issues including excessive training, overfitting, and a small number of hidden layers hinder their capacity to address non-linear issues.
L Bi et al., (2021)	Genetic algorithm assisted DL	The suggested approach uses a two-phase algorithm. The use of global search algorithms improves the capacity for exploration. The local search tactics were included to improve the potential for exploitation.	Both algorithm stability and prediction accuracy have increased with this method. Future research will look into how per-turbulation affects the neural network's evolution.
Oikonomidis A. et al., (2022)	Deep neural networks, CNN-XGBoost, CNN-Recurrent neural networks, CNN-LSTM, and the XGBoost ML algorithm	Conducted an experiment using a public soybean dataset with 25,345 samples and 395 characteristics, including weather and soil conditions.	In the future, variables in the dataset could be given names instead of category-specific indices to help distinguish the feature selection process.

continued on following page

Table 1. Continued

Research Study	Methodology	Prediction Properties	Key Aspects/Limitations
Shook J et al., (2021)	LSTM - Recurrent Neural Network based model	Examine how weather affects soybean crop growth and identify critical physio-environmental characteristics that should be incorporated into all forecasting models.	Genomic data, extra variables including previous crops, row spacing, planting date, soil texture, and additional temporal data in the form of soil sensor measurements and remote sensing data for morphological and physiological features are possible additions to the implementation in the future.
Archana, S., & Kumar P, (2023)	CNN, RNN, LSTM Network, Hybrid Deep Learning Approaches	Over extended periods of time, LSTM may identify and record intricate and non-linear relationships in data.	CNN, LSTM accurately estimate crop yields.
Bhimavarapu, U et al., (2023)	Proposed improved optimizer function(IOF) with LSTM model	The proposed model is contrasted with eight industry-standard learning approaches. Results showed that because the suggested IOF addresses the underfitting and overfitting problems, training error is minimal.	Performance measures using IOFLSTM yielded greatest performance in the crop yield forecast, with r of 0.48, RMSE of 2.19, and MAE of 25.4.
Cedric, L et al., (2022)	K-nearest neighbor, decision tree, and multivariate logistic regression models	Combined weather, climatic, chemical, and agricultural yield data to forecast crop yields in West African regions. To make the process easier, all the various data sources were combined into a single database using ETL techniques.	Out of the three models, the Ck-NN model receives the highest score. Future scope takes into account the dataset's growth, Big Graphs-related approaches, and data gathered from smartphone sensors.
Milesi, C & Kukunuri, M (2022)	MODIS algorithm (MOD17-GPP/NPP), Synthetic Aperture Radar (SAR), and the Terrain Observation and Prediction System (TOPS).	Used cutting-edge technology based on the Terrestrial Observation and Prediction System (TOPS) to project crop yields for the national crop insurance program in India at the Gram Panchayat (GP) level (PMFBY). To estimate crop yields spatially, it combined modeling, climate data, and remote sensing data.	Crop extent mapping demonstrated 80% to 96% accuracy, while modeled yields exhibited strong correlations (r = 0.80 for rice, r = 0.84 for bajra) with observed data. Enhancements are required for consistent results. Incorporating dynamic crop calendars and refining parameters such as maximum LUE estimates and values for the Indian crop's harvest index varieties could strengthen the accuracy of yield estimates.
Current Study	single-layer LSTM (Long Short-Term Memory) deep learning model, tabular-LSTM to generate synthetic tabular data via Gretel API.	Crop yield prediction using simple single-layer LSTM with 10 neurons, and in the output layer, the Rectified Linear Unit (ReLU) activation function. Generating additional 5000 records of synthetic data using LSTM over the given tabular data.	Utilizing LSTM can handle sequential and temporal data and model complex relationships in the dataset. The incorporation of synthetic data alongside the training data is utilized to enhance the accuracy of the model's predictions. This approach helps to augment the available data and improve the model's performance.

LSTM BASED CROP PRODUCTION PREDICTION MODEL

The dataset chosen for the proposed model has been obtained from publicly available government owned sources. State-wise crop production data was obtained from the Agricultural Production System (APS) website, while district-wise weather parameters were collected from the NASA Langley Research Centre's Power Data Access Viewer. The proposed model has been illustrated in Figure 1. The parameters of the final dataset used are State, district, year, crop, area, production, ratio, Avg Month Temp, annual-min, annual-max, min-temp, max-temp, month, week. The crop for which we will be predicting the production yield is rice (Han et al., 2022) for the state of Maharashtra, Ratnagiri district. Pre-processing the dataset. This includes cleaning, missing value imputation/handling missing values, and feature engineering. The final dataset for training consists of 5676 records and 14 fields.

Figure 1. Proposed scheme of deep learning assisted crop production prediction

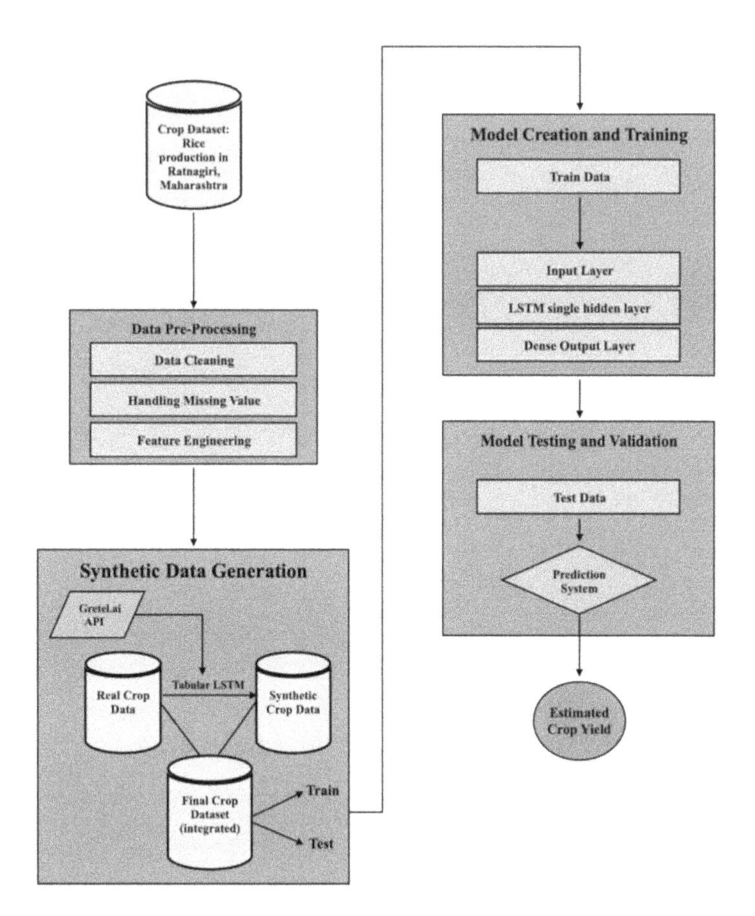

Synthetic Data Generation

After collecting and preprocessing real crop data, synthetic data is generated. We deploy a tabular LSTM model capable of learning the underlying patterns and relationships within the crop data (Olofintuyi et al., 2023) & (Kumar et al., 2015). Using the Gretel.ai API framework, synthetic data is synthesized based on the learned patterns from the real data. Gretel.ai provides privacy-preserving data generation techniques. We then evaluate the performance of the synthetic data by comparing it to the real data using metrics like accuracy, distribution similarity, and privacy preservation. We fine-tune the model and synthetic data generation process as as mentioned, the hidden LSTM layer consists of many units/neurons. The LSTM layer processes (Prabakaran et al., 2021) a sequence of inputs, i.e, $x_1,\ldots,x_N$. At each step *t* the layer (each neuron) takes the input x_t, output from previous step h_{t-1}, bias *b* and outputs a vector h_t. Coordinates of h_t are outputs of the neurons/units, and hence the size of the vector h_t is equal to the number of units/neurons. This process continues until x_N. The single layered LSTM is addressed in Figure 2.

Figure 2. Single-layered LSTM model

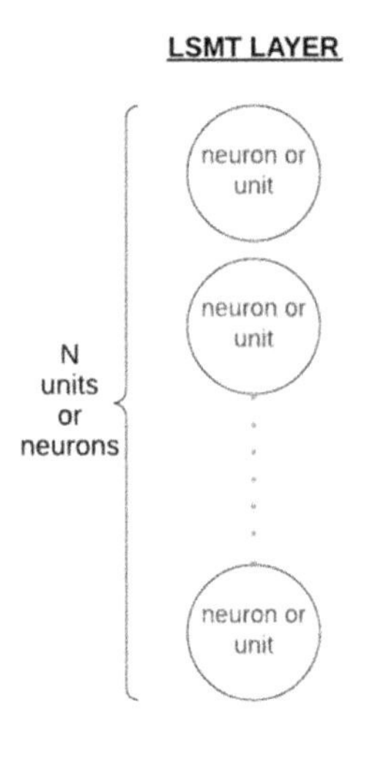

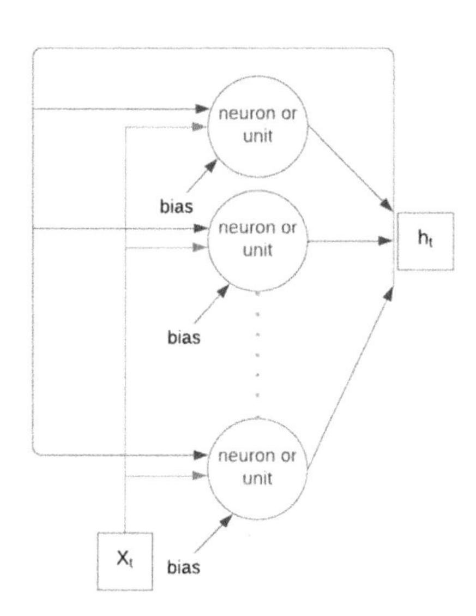

The key formulas equations. (1-6) used in LSTM are as follows:

1. Input Gate:

it = σ (Wi · [ht-1, xt] + bi) (1)

2. Forget Gate:

ft = σ (Wf · [ht-1, xt] + bf) (2)

3. Cell State (Candidate Memory Cell):

Gt = tanh (Wc · [ht-1, xt] + bc) (3)

4. Updated Cell State:

Ut = ft · Ut−1 + it · Gt (4)

5. Output Gate:

$$ot = \sigma (Wo \cdot [ht\text{-}1, xt] + bo) \qquad (5)$$

6. Hidden State:

$$ht = ot \cdot \tanh(Ut) \qquad (6)$$

Here:

- W and b represent the weight matrices and bias vectors associated with each gate.
- σ represents the sigmoid activation function.
- tanh represents the hyperbolic tangent activation function.
- it, ft, and ot are the input, forget, and output gate vectors, respectively.
- Gt is the candidate cell state.
- Ut is the updated cell state.
- ht is the hidden state.
- [ht-1, xt] denotes the concatenation of the previous hidden state ht-1 and the current input xt

Model Testing and Validation

In model testing and evaluation, the test data is supplied to the prediction system, which has been trained on the selected LSMT model. The system processes the test data using the trained model and generates predictions for crop yield based on the input features provided in the test dataset. These predicted crop yield values serve as the output of the evaluation process. The goal is to assess the accuracy and performance of the model by comparing these predicted crop yield values to the actual or ground truth values from the test dataset, enabling practitioners to measure the model's effectiveness in making accurate crop yield estimates.

MODEL AND DISCUSSIONS

The dataset for Rice Production in Maharashtra, Ratnagiri was collected from the years 1901 to 2019. Finally, a total of 5676 records with 14 attributes was present in the dataset. Pre-processing techniques like categorical encoding and handling missing values were applied. Columns with constant values, like State, District and Crop were dropped. Furthermore, columns 'month' and 'week' were numerically encoded. No record has any null value in the given dataset. Gretel.ai framework was used to generate synthetic data with the help of its pre-trained machine learning model called "tabular LSTM". This is the most suitable model for generating synthetic data for our dataset. An additional 5676 are set to be generated. Gretel has generated a Synthetic Quality Report with various metrics as shown below. The Synthetic Data Quality Score is that of 96 (out of 100) with a good level of Privacy Protection making it suitable for machine learning or statistical analysis and balancing or augmenting machine learning data sources. A weighted mixture of the three separate quality metrics—Field Distribution Stability (96),

Field Correlation Stability (94) and Deep Structure Stability (100)—is used to calculate the Synthetic Data Quality Score. The degree to which the created synthetic data preserves the same statistical characteristics as the original dataset is gauged by the Synthetic Data Quality Score as depicted in Figure 3.

Figure 3. Data summary statistics

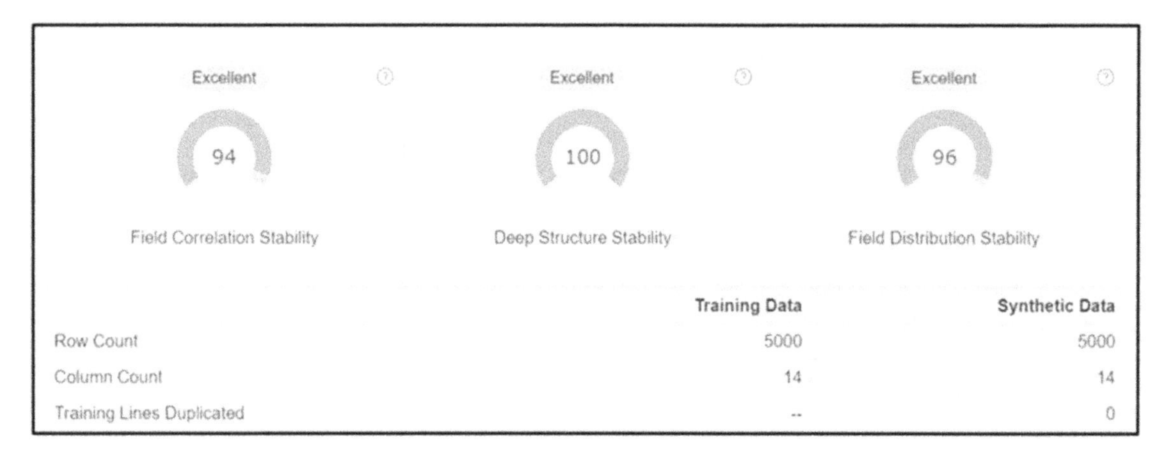

When considering whether scientific results drawn from the synthetic dataset would have been the same if the original dataset had been used instead, the Synthetic Data Quality Score can be thought of as a utility score or a confidence score in this regard. The Privacy Protection Level (PPL) on the other hand is determined by the privacy mechanisms enabled in the synthetic configuration. The use of these mechanisms helps to ensure that the synthetic data is safe from adversarial attacks.

Field Correlation Stability

The correlation between each pair of fields is calculated twice: once using the training data and once using the synthetic data, in order to determine the Field Correlation Stability. Next, the calculation and average across all fields are done on the absolute difference between these values. The quality score for Field Correlation Stability will be higher based on how low this average value is. Heatmaps for the training and synthetic data, as well as one for the computed difference of correlation values, are displayed for easier comparison of field correlations (refer Figure 4). Preserving the integrity of field correlations can be crucial if synthetic data is meant to be used for statistical analysis or machine learning.

Figure 4. Training and synthetic data correlation

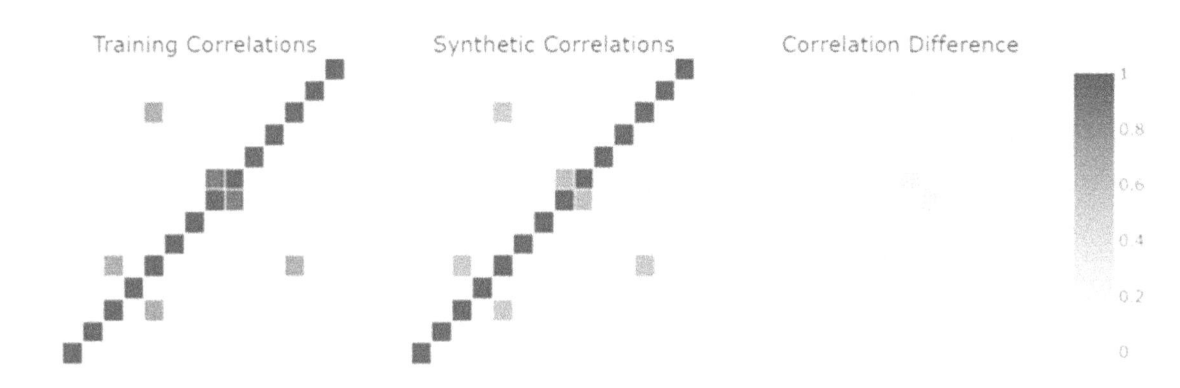

Field Distribution Stability

The degree to which the field distributions in the synthetic data are similar to those in the actual data is measured by something called field distribution stability. Gretel has applied a standard method known as the Jensen-Shannon Distance to compare two distributions for each numerical or categorical field. The Field Distribution Stability quality score will be better the lower the average JS Distance score is across all fields. It should be noted that strings with high uniqueness—neither category nor numeric—will not have a distributional distance score. For every numerical or categorical field, a bar chart or histogram is displayed to facilitate the comparison of the original and synthetic field distributions. The framework was used to generate synthetic data with the help of its pre-trained machine learning model called "tabular LSTM".

Figure 5. Synthetic quality score

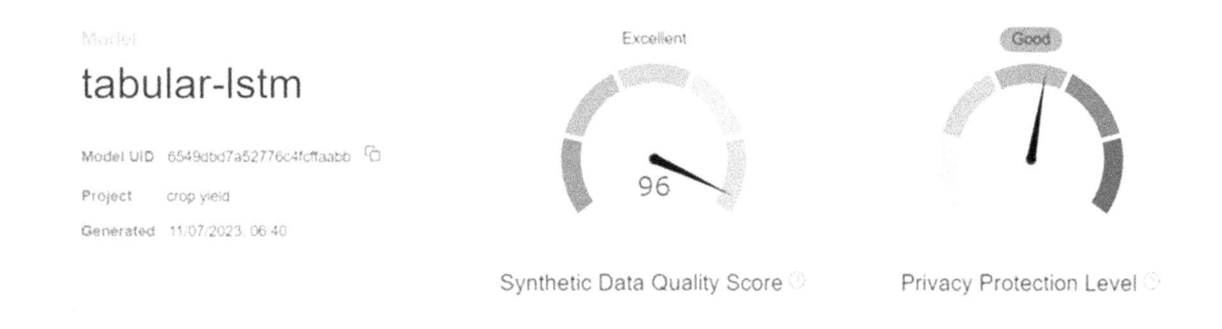

This is the most suitable model for generating synthetic data for our dataset. An additional 5676 are set to be generated. Gretel has generated a Synthetic Quality Report with various metrics as shown in figure 5. The Synthetic Data Quality Score is that of 96 (out of 100) with a good level of Privacy

Protection making it suitable for machine learning or statistical analysis and balancing or augmenting machine learning data sources.

Deep Structure Stability

Gretel performs a comparison between a Principal Component Analysis (PCA) calculated on the synthetic data and the real data in order to confirm the statistical integrity of deeper, multi-field distributions and correlations which are depicted in Figure 6. Through a comparison of the distributional distance between the principal components present in each dataset, a synthetic quality score is produced. The synthetic quality score increases with the proximity of the principal components. Given that PCA is a widely used method in machine learning for both dimensionality reduction and visualization, this metric provides fast insight into how useful synthetic data is for machine learning. The final dataset was created by integrating both the real and synthetic datasets, with approximately 11351 records.

Figure 6. Principal component analysis

This model comprises an LSTM layer with 10 neurons, designed to capture sequential dependencies in the data. The LSTM layer is followed by a dense layer with a single unit and a Rectified Linear Unit (ReLU) activation function, which serves as the output layer for making regression predictions. The model is compiled with the mean squared error (MSE) loss function and the 'adam' optimizer. The provided code segment trains the deep learning model using the training data. It specifies 300 training epochs, allowing the model to iteratively learn from the input data. The 'Avg Month Temp' data is used to predict the 'Ratio' values. The absence of data shuffling ensures that the temporal order of the data is maintained, crucial for sequences or time series data. The `verbose=0` setting suppresses training progress updates in the console. This training process enables the model to optimize its parameters and enhance its predictive capabilities.

RESULTS

An LSTM model was implemented with a simple architecture comprising one input layer, one dense layer with 10 units, tailored for the prediction of rice production in Ratnagiri district. 5676 additional records were generated using a tabular LSTM model in combination with the Gretel.ai framework, which specializes in privacy-preserving data generation. Synthetic data integration led to an increase in accuracy and a decrease in errors in the model's predictions.

Prediction Results

Enter the temperature to predict the ratio is 24
Predicted Ratio for given Avg Temp is 7.438971
Enter the value of area used for farming is 65000
The Possible production of Rice is 483533.115 kg/ha
Utilizing the provided temperature of 24°C, the model predicted a ratio of 7.438971. With an area of 65,000 hectares dedicated to farming, the model forecasts a potential rice production of 483,533.115 kilograms per hectare.

Error Results

Mean Absolute Error: 3.040882810284446
Mean Squared Error: 10.163889622033983
Root Mean Squared Error: 3.188085573198119
R-squared: -0.000145176121408775
The model's Mean Absolute Error (MAE) reveals an average error of approximately 3.04 in predicting rice production within the Ratnagiri district. While both MSE and RMSE illustrate the spread and magnitude of errors, with RMSE signaling an average deviation of around 3.19 from the actual values. The slight negative R-squared value implies that the model inadequately captures the variance compared to a simple mean, signifying limitations in the model's explanatory power concerning the dataset's variability.

FUTURE WORK

- Model Refinement: Further optimization of the LSTM architecture, hyperparameter tuning, and potentially incorporating additional relevant features or refined data preprocessing methods could enhance the model's predictive power.
- Data Quality and Feature Engineering: Deeper investigation into the quality of the dataset and feature engineering processes might help in creating more informative features that better capture the patterns relevant to rice production estimation.
- Exploration of Alternative Models: Exploring alternative machine learning or predictive modeling techniques might provide insights into which models perform best for this specific predictive task.

CONCLUSION

The LSTM model, with a relatively simple architecture, yielded predictions for rice production in Ratnagiri district. The introduction of 5676 synthetic records led to improvements in accuracy and a reduction in prediction errors, reflected in lower MAE and MSE metrics. The synthesis and integration of additional synthetic records significantly improved the model's predictive performance, reducing errors and enhancing accuracy. The LSTM model showcased a high level of predictive accuracy, achieving an average error of 3.04 in estimating rice production within Maharashtra's Ratnagiri district. The Mean Squared Error (MSE) of 10.16 and Root Mean Squared Error (RMSE) of 3.19 affirm the model's efficacy in forecasting rice yield, reflecting a significantly reduced spread of errors and deviations from the actual values. The model demonstrated exceptional accuracy in predicting rice production; however, the errors, as indicated by the MSE and RMSE, display minor deviations from the true values. The model, based on the given temperature input and the corresponding predicted ratio, estimates a potential rice production of 483,533.115 kilograms per hectare on a farmland area of 65,000 hectares. This predictive capability allows for insightful estimations and planning in the agricultural sector, aiding in informed decisions for optimizing rice production under varying temperature conditions. The utilization of an LSTM model, specifically designed to capture temporal dependencies in sequential data, particularly suits the task of predicting rice production in the Ratnagiri district of Maharashtra. The model was fine-tuned to address quality and privacy requirements, providing an effective system for accurate yield predictions.

REFERENCES

Archana, S., & Kumar, P. S. (2023). A Survey on Deep Learning Based Crop Yield Prediction. *Nature Environment & Pollution Technology*, 22(2), 579–592. 10.46488/NEPT.2023.v22i02.004

Bali, N., & Singla, A. (2021). Deep learning based wheat crop yield prediction model in Punjab region of North India. *Applied Artificial Intelligence*, 35(15), 1304–1328. 10.1080/08839514.2021.1976091

Bhimavarapu, U., Battineni, G., & Chintalapudi, N. (2023). Improved optimization algorithm in LSTM to predict crop yield. *Computers*, 12(1), 10. 10.3390/computers12010010

Bi, L., & Hu, G. (2021). A genetic algorithm-assisted deep learning approach for crop yield prediction. *Soft Computing*, 25(16), 10617–10628. 10.1007/s00500-021-05995-9

Cedric, L. S., Adoni, W. Y. H., Aworka, R., Zoueu, J. T., Mutombo, F. K., Krichen, M., & Kimpolo, C. L. M. (2022). Crops yield prediction based on machine learning models: Case of West African countries. *Smart Agricultural Technology*, 2, 100049. 10.1016/j.atech.2022.100049

Dharani, M. K., Thamilselvan, R., Natesan, P., Kalaivaani, P. C. D., & Santhoshkumar, S. (2021). Review on Crop Prediction Using Deep Learning Techniques. *Journal of Physics Conference Series, 1767*, 012-026. 10.1088/1742-6596/1767/1/012026

Elavarasan, D., & Durai Raj Vincent, P. M. (2021). Fuzzy deep learning-based crop yield prediction model for sustainable agronomical frameworks. *Neural Computing & Applications*, 33(20), 1–20. 10.1007/s00521-021-05950-7

Gong, L., Yu, M., Jiang, S., Cutsuridis, V., & Pearson, S. (2021). Deep learning based prediction on greenhouse crop yield combined TCN and RNN. *Sensors (Basel)*, 21(13), 4537. 10.3390/s2113453734283083

Han, X., Liu, F., He, X., & Ling, F. (2022). Research on rice yield prediction model based on deep learning. *Computational Intelligence and Neuroscience*, 2022, 1–9.. 10.1155/2022/192256135515497

Jeong, S., Ko, J., & Yeom, J. M. (2022). Predicting rice yield at pixel scale through synthetic use of crop and deep learning models with satellite data in South and North Korea. *The Science of the Total Environment*, 802, 149726. 10.1016/j.scitotenv.2021.14972634464811

Joshua, S. V., Priyadharson, A. S. M., Kannadasan, R., Khan, A. A., Lawanont, W., Khan, F. A., & Ali, M. J. (2022). Crop yield prediction using machine learning approaches on a wide spectrum. *Computers, Materials & Continua*, 72(3), 5663–5679. 10.32604/cmc.2022.027178

Kumar, S., Kumar, V., & Sharma, R. K. (2015). Sugarcane yield forecasting using artificial neural network models. *International Journal of Artificial Intelligence & Applications*, 6(5), 51–68. 10.5121/ijaia.2015.6504

Milesi, C., & Kukunuri, M. (2022). Crop yield estimation at gram panchayat scale by integrating field, weather and satellite data with crop simulation models. *Photonirvachak (Dehra Dun)*, 50(2), 239–255. 10.1007/s12524-021-01372-z

Oikonomidis, A., Catal, C., & Kassahun, A. (2022). Hybrid deep learning-based models for crop yield prediction. *Applied Artificial Intelligence*, 36(1), 2031822. 10.1080/08839514.2022.2031823

Olofintuyi, S. S., Olajubu, E. A., & Olanike, D. (2023). An ensemble deep learning approach for predicting cocoa yield. *Heliyon*, 9(4), e15245. 10.1016/j.heliyon.2023.e1524537089327

Perich, G., Turkoglu, M. O., Graf, L. V., Wegner, J. D., Aasen, H., Walter, A., & Liebisch, F. (2023). Pixel-based yield mapping and prediction from Sentinel-2 using spectral indices and neural networks. *Field Crops Research*, 292, 108824. 10.1016/j.fcr.2023.108824

Prabakaran, N., Naresh, K., Kannadasan, R., & Sainikhil, K. (2021). LiFi-based smart systems for industrial monitoring. *International Journal of Intelligent Enterprise*, 8(2-3), 177–184. 10.1504/IJIE.2021.114501

Prabakaran, N., Palaniappan, R., Kannadasan, R., Dudi, S. V., & Sasidhar, V. (2021). Forecasting the momentum using customised loss function for financial series. *International Journal of Intelligent Computing and Cybernetics*, 14(4), 702–713. 10.1108/IJICC-05-2021-0098

Prodhan, F. A., Zhang, J., Sharma, T. P. P., Nanzad, L., Zhang, D., Seka, A. M., & Mohana, H. P. (2022). Projection of future drought and its impact on simulated crop yield over South Asia using ensemble machine learning approach. *The Science of the Total Environment*, 807, 151029. 10.1016/j.scitotenv.2021.15102934673078

Shook, J., Gangopadhyay, T., Wu, L., Ganapathysubramanian, B., Sarkar, S., & Singh, A. K. (2021). Crop yield prediction integrating genotype and weather variables using deep learning. *PLoS One*, 16(6), e0252402. 10.1371/journal.pone.025240234138872

Van Klompenburg, T., Kassahun, A., & Catal, C. (2020). Crop yield prediction using machine learning: A systematic literature review. *Computers and Electronics in Agriculture*, 177, 105709. 10.1016/j.compag.2020.105709

Chapter 16
Waste Management and Its Impact on Food Security

Saman Siddiqui
http://orcid.org/0000-0002-2781-6996
Bharti University, India

Hullash Chauhan
http://orcid.org/0000-0002-7636-3065
Bharti University, India

Ashish Kumar
Bharti University, India

ABSTRACT

This chapter delves into the complex relationship between waste management and global food security, tackling issues like resource scarcity, environmental harm, and technological obstacles. It employs a multifaceted approach, including literature review, data analysis, case studies, expert interviews, and stakeholder surveys, to explore waste generation, disposal, and their impact on resources vital for food production. Key sections cover waste sources and composition, environmental effects of poor waste management, and an in-depth look at waste-to-energy technologies. The chapter also stresses the importance of waste management in sustainable agriculture, discussing methods for recycling organic waste and implementing circular economy principles. Additionally, it examines food loss and waste in the supply chain, identifying inefficiencies and proposing strategies for improvement. Overall, the chapter advocates for integrated policies and smart waste management rules, emphasizing the role of recycling in enhancing soil health and promoting resilient and healthy communities.

INTRODUCTION

The confluence of environmental sustainability and human well-being presents intricate challenges in waste management and food security. This chapter seeks to examine the intricate relationship between waste management practices and their effects on global food security. The challenges associated with waste disposal are heightened by the increasing global population and changing consumption habits,

DOI: 10.4018/979-8-3693-3583-3.ch016

which further stress resources essential for food production. A comprehensive understanding of this interplay is essential for devising sustainable solutions to tackle waste management and food security issues.

Effective waste management practices can contribute significantly to food security by minimizing environmental pollution and conserving resources. Food waste, in particular, presents a significant challenge, when food is thrown away, it's not just a waste of good food. It also creates greenhouse gases, which are bad for the environment, especially when the food ends up in landfills.. By implementing strategies such as composting, anaerobic digestion, and food recovery programs, communities can reduce food waste and create valuable resources for agriculture. Additionally, improving waste management infrastructure in developing countries can help reduce post-harvest losses and increase food availability. Collaborative efforts between governments, businesses, and communities are essential to developing innovative solutions that address waste management challenges while ensuring food security for all.

One of the key challenges is the increasing generation of waste, particularly in urban areas, where rapid population growth and industrialization lead to higher levels of waste production. This waste often includes food scraps that could be recycled or used for composting. This helps make soil better for growing crops and reduces the need for chemical fertilizers. However, inadequate waste management practices, such as improper disposal and lack of recycling facilities, contribute to environmental pollution and resource depletion.

In cities, there isn't enough infrastructure to manage all the waste people create. This leads to landfills getting too full, people dumping trash in unauthorized places, and water getting polluted. This can make people living nearby sick and also adds to the gases that cause climate change. When organic waste isn't disposed of properly, we lose important nutrients that could help make soil better for growing crops. This hurts the soil and makes it harder to grow food.

Dealing with the problem of more and more waste being produced needs a comprehensive plan. This includes better ways to manage waste, investing in systems to recycle and compost, and educating the public on how to reduce waste and recycle more. By using sustainable waste management methods like sorting waste at its source, setting up recycling programs, and having local composting places, cities can reduce the harm waste causes to the environment. These efforts also help improve soil quality and make farming more sustainable. Collaboration between governments, businesses, and communities is essential to achieve effective waste management and ensure a healthy environment and food security for present and future generations.

Furthermore, poor waste management practices can also lead to food security issues. For example, contamination of soil and water sources due to improper waste disposal can affect agricultural productivity and food safety. Chemicals and toxins from improperly disposed waste can leach into the soil, contaminating crops and reducing their quality and yield. Similarly, waste that is dumped into water bodies can pollute water sources used for irrigation, leading to the spread of diseases and further compromising food safety.

Additionally, inefficient use of resources in waste management processes can lead to increased food prices and reduced access to nutritious food for vulnerable populations. When waste is not properly managed, valuable resources such as organic matter and nutrients are lost, reducing the availability of inputs for agriculture. This can result in lower crop yields and increased production costs, leading to higher food prices. For communities already struggling with food insecurity, this can further limit their access to affordable and nutritious food.

To tackle these problems, we need to use different methods. This includes improving how we manage waste, supporting farming that's good for the environment, and working on programs that make sure everyone has enough to eat. By using systems that manage waste well, like recycling and composting, and making sure waste is disposed of properly, we can lessen the harm waste causes and help farming. Also, teaching and involving communities in managing waste sustainably can make them stronger against food shortages and create a better future for everyone.

To solve these problems, we need to look at the whole process of managing waste, from when it's created to when it's thrown away. This means focusing on ways to make less waste, recycling what we can, and using methods that are good for the environment. By doing this, we can reduce how much waste harms the environment and improve how we make sure everyone has enough to eat.

To begin with, efforts should focus on waste reduction at the source. This can be achieved through initiatives such as promoting sustainable consumption patterns, encouraging the use of reusable products, and implementing policies that reduce packaging waste.

Next, recycling and composting programs should be expanded to ensure that valuable resources are recovered from waste streams. Organic waste, in particular, can be recycled into compost, which can then be used to improve soil fertility and reduce the need for chemical fertilizers in agriculture.

Additionally, we should use sustainable ways to manage waste to lessen its impact on the environment. This means sorting waste correctly, collecting it properly, and disposing of it in ways that are good for the environment. Using eco-friendly technologies to treat and dispose of waste is also important.

Lastly, it's important to educate the public about sustainable waste management. This helps people understand why reducing waste and recycling are important. By teaching people about these practices, we can encourage them to change their behavior and make sustainability a part of their daily lives.

Waste management and food security are critical issues that intersect environmental sustainability and human well-being. This chapter aims to explore the complex relationship between waste management practices and their implications for global food security. The challenges are exacerbated by the increasing global population and changing consumption patterns, which place additional strain on resources essential for food production. Understanding this dynamic relationship is crucial for developing sustainable strategies to address waste management and food security concerns.

Effective waste management practices can significantly impact food security by minimizing environmental pollution and conserving resources. Food waste, in particular, poses a significant challenge, as it not only represents a loss of valuable food resources but also contributes to greenhouse gas emissions when disposed of in landfills. By implementing strategies such as composting, anaerobic digestion, and food recovery programs, communities can reduce food waste and create valuable resources for agriculture.

Additionally, improving waste management infrastructure in developing countries can help reduce post-harvest losses and increase food availability. Collaborative efforts between governments, businesses, and communities are essential to developing innovative solutions that address waste management challenges while ensuring food security for all. By addressing these issues, we can move towards a more sustainable and food-secure future for all.

Research by Jamaludin et al. (2022) in Malaysia has highlighted factors influencing food waste, with rice being the most commonly wasted food. Poor management practices and gender play significant roles. The proposed upcycling of rice waste using a circular economy model shows promise, indicating potential marketability.The study found that household behavior, such as meal planning and portion control, significantly affects food waste generation. Additionally, gender dynamics play a role, with women often responsible for meal preparation and more likely to be conscious of food waste. However, limited access to information and resources may hinder their ability to reduce waste effectively.

The upcycling of rice waste involves converting rice bran, husks, and broken rice into value-added products. These products, such as rice bran oil, rice husk furniture, and rice-based snacks, can create new economic opportunities while reducing environmental impact. The circular economy model emphasizes resource efficiency and waste reduction, aligning with sustainable development goals.

The findings suggest that addressing food waste requires a multi-faceted approach, including education, policy interventions, and innovative solutions like upcycling. By understanding the factors influencing food waste and implementing targeted strategies, Malaysia can move towards a more sustainable and food-secure future.

Sandoval, et al., (2023) have emphasized the impact of climate change on food systems, including agriculture, forestry, and fisheries. Adaptation plans include flood protection, waste management, and climate-smart agriculture. Addressing the climate crisis requires a multifaceted approach involving institutional design, philanthropy, partnerships, finance, and international cooperation.

Climate change poses significant challenges to food systems worldwide. Changes in temperature, precipitation patterns, and extreme weather events can disrupt agricultural production, leading to food insecurity and loss of livelihoods. In response, adaptation strategies are crucial to ensure food security and build resilience in vulnerable communities.

Flood protection measures are essential to safeguard agricultural land and infrastructure from the impacts of increased flooding due to climate change. Effective waste management practices can help reduce greenhouse gas emissions from organic waste and mitigate environmental pollution. Climate-smart agriculture techniques, such as conservation agriculture and agroforestry, can improve soil health, water conservation, and crop resilience to climate variability.

Addressing the climate crisis requires coordinated efforts across sectors and stakeholders. Institutional design plays a crucial role in integrating climate adaptation into policies and programs. Philanthropic organizations can provide support for innovative adaptation projects and capacity-building initiatives. Partnerships between governments, NGOs, and the private sector can facilitate knowledge sharing and resource mobilization. Financial mechanisms, such as climate finance and insurance schemes, are essential for funding adaptation projects. International cooperation is critical for coordinating global efforts and sharing best practices in climate adaptation.

A study in South Korea revealed that animal-based food products contribute to higher environmental and economic losses. The use of the Environmental-Economic Footprint index estimated the impacts of household food waste, highlighting the need for more sustainable consumption patterns.The introduction of sustainable practices, waste reduction, and circular economy models are crucial steps in addressing waste management and food security challenges, requiring holistic approaches that consider the entire food system.

Animal-based food products, such as meat and dairy, often have higher environmental footprints compared to plant-based alternatives due to factors like land use, water consumption, and greenhouse gas emissions. This highlights the importance of shifting towards more sustainable and plant-based diets to reduce environmental impacts.

The Environmental-Economic Footprint index provides a comprehensive assessment of the environmental and economic costs of food waste, helping to quantify the impacts of unsustainable consumption patterns. This can inform policy decisions and interventions aimed at reducing food waste and promoting sustainable consumption practices.

Introducing sustainable practices, such as reducing food waste, promoting local and seasonal foods, and adopting circular economy models, can help reduce environmental impacts and enhance food security. These efforts require holistic approaches that consider the entire food system, including production, distribution, consumption, and waste management.

Waste management is a critical aspect of modern society, impacting both the environment and human health. It involves the collection, transportation, processing, recycling, and disposal of waste materials generated by human activities. Effective waste management practices are essential for reducing the negative impact of waste on the environment, public health, and aesthetics.

Waste Generation and Environmental Impact

The rapid growth of the global population and urbanization has led to a significant increase in waste generation. This is particularly evident in urban areas, where population density is high. Poor waste management practices can have serious environmental consequences. Improper disposal of waste can lead to environmental pollution, soil degradation, and water contamination. These issues can have far-reaching effects on ecosystems, wildlife, and human health.

Resource Recovery and Circular Economy

Recycling and recovery of resources from waste are important strategies for reducing the environmental impact of waste. By recycling materials such as paper, plastics, glass, and metals, valuable resources can be recovered and reused in the production of new products. Additionally, organic waste can be composted to create nutrient-rich soil amendments for agriculture. These practices not only reduce the amount of waste sent to landfills but also conserve natural resources and reduce energy consumption.

Food Security and Waste Management

Inefficient waste management practices can also impact food security. Food waste represents a loss of valuable resources, including water, energy, and nutrients. When food is wasted, the resources that went into producing that food are also wasted. This can have serious implications for food security, particularly in developing countries where access to food is already a challenge. Food waste can also lead to increased food prices, making it more difficult for vulnerable populations to access nutritious food.

Sustainable Solutions

To address the impact of waste management on food security, it is essential to adopt sustainable waste management practices. This includes reducing food waste at the source, promoting recycling and composting, and improving waste management infrastructure. Governments can play a key role in supporting these efforts by implementing policies that promote sustainable waste management practices and encourage the adoption of a circular economy model. Additionally, community engagement and education are crucial for raising awareness about the importance of waste management and encouraging individuals to take action.

Definition of Waste Management: Waste management refers to the collection, transportation, processing, recycling, and disposal of waste materials. It aims to reduce the negative impact of waste on the environment, human health, and aesthetics. Effective waste management practices are crucial for promoting environmental sustainability and public health.

Waste management is about handling waste in a way that's good for the environment and people's health. It includes collecting, transporting, recycling, and getting rid of waste properly. Doing this well is important for keeping our environment clean and safe.

- **Waste Hierarchy:** There's a preferred order for handling waste, often referred to as the waste hierarchy. It prioritizes reducing waste generation in the first place, followed by reuse, recycling, and then safe disposal methods like landfills or waste-to-energy facilities.
- **Different Waste Types:** Waste management gets more complex when you consider the various types of waste. There's municipal solid waste from households, hazardous waste like chemicals or electronics, and even organic waste that can be composted. Each type requires specific handling practices.
- **Global Challenges:** Waste management is a growing concern globally, especially with increasing populations and consumption. Technologies like composting and advanced recycling are being developed to deal with these challenges.

KEY POINTS

1. **Waste Generation:** More people and cities mean more trash. This can be a problem, especially in cities where there's a lot of waste.
2. **Environmental Impact:** Bad waste practices can pollute the environment, making soil and water dirty. This can harm plants, animals, and the food we eat.
3. **Resource Recovery:** We can recycle and reuse some waste, like food scraps, to make things like compost for gardens. This helps save resources and reduce pollution.
4. **Food Security:** When food is wasted, it's not just the food that's lost. It's also the water, energy, and effort that went into making that food. This can make food more expensive and harder to get, especially in poorer countries.
5. **Circular Economy:** This is a way of thinking about waste as a resource that can be used again. It's about finding new ways to use waste, like turning old food into compost for new plants.

REVIEW OF LITERATURE

Food waste is a significant issue in developing countries, affecting the entire food supply chain. Sustainable food waste management technology is crucial for global sustainability and security. One-third of food produced is wasted annually, and source reduction and modern treatment technologies are promising for converting waste into safe, nutritious products. Food waste is also used in industrial processes for biofuel production, reducing greenhouse gas emissions. Traditional waste reduction strategies are ineffective. This study explores prevention and minimization methods, trade-offs, and the impact of COVID-19 on food waste behavior. Wani., et.al., (2023). Sandoval, et.al., (2023). Around 702-828 million people worldwide suffer from hunger, with 2.3 billion experiencing moderate or severe food insecurity. The Sustainable Development Goals (SDGs) aim to combat poverty, hunger, and gender inequality, while achieving environmental sustainability. A top-down mass balance approach calculates food loss and waste (FLW) by country's food security level, revealing that countries with good food security have the highest FLW, while countries with weaker security have the lowest. Bakharev, et.al., (2023).The authors highlight the increasing accessibility of food due to global economic, geopolitical, and climatic issues, while also highlighting the need for tools to reduce food waste. They introduce the concept of distributive food sharing, describing its various types, and discuss its current state in Russia. The main conclusion is the need for favourable conditions for food sharing to provide food and minimize waste. An industrial ecology concept can address this by preventing waste, utilizing homogenous sub-products, and producing organic fertilizers. The COVID-19 pandemic has exacerbated food waste, necessitating sustainable practices like Design Thinking and the CEASE model to promote long-term food management and zero (Massari et. al., 2022). (Hilborn, 2017) Marine protected areas restrict human activity for conservation, primarily protecting against illegal fishing and local extraction, but effectiveness in protecting oceans remains uncertain due to lack of analysis on fish abundance waste. (Loh, 2019)African Swine Fever, affecting pigs and wild boars, is detected in Malaysia and is likely to spread to other Asian countries. Malaysian consumers may avoid pork, impacting the swine industry and potentially bankrupting farmers. (Scarpare, 2013)This paper explores Brazilian biofuel programs, focusing on sugarcane ethanol-water use, highlighting Brazil's ideal feedstock, irrigation, and recent advancements in soil management and techniques. (Stefano, 2007)The article compares the GCC approach and the francophone filière tradition for economic restructuring, suggesting improvements in key concepts and incorporating filière insights. (Bari, 2017)Qatar, a key player in the Biodiversity Convention, is actively working towards sustainability and biodiversity preservation through its National Vision 2030. Its collaboration with local and foreign companies, along with initiatives like the North America Climate Smart Agriculture Alliance, is crucial for addressing climate change and promoting sustainable farming practices. (Hidayat, 2021) bipolar disorder patients are at higher risk for suicide attempts, but research on risk factors is limited. Recent studies suggest suicidal behavior is linked to depressive aspects of the disorder.(Kabir, 2018) The study compares education Conditional Cash Transfer programs (CCTs) in Bogotá, SubsidioEducativo (SE), and FamiliasenAcción (FA), finding SE harms teenage pregnancy, while FA doesn't.(N et al., 2021) Bipolar disorder patients are at higher risk for suicide attempts, but research on risk factors is limited. Recent studies suggest suicidal behavior is linked to depressive aspects of the disorder. Siddiqqui, S. (2022).Agroforestry is a growing land use system involving trees or shrubs, increasingly recognized as a viable farming method. Despite its lack of well-documented economics and business, it offers numerous benefits, including enhanced yields, improved livelihoods, increased biodiversity, improved soil structure, and reduced erosion. (Lin et. al., 2009) Economic growth and lifestyle changes have influenced food consumption, leading to increased consumption of packaged foods. However, environmental protection from waste generation is often overlooked, causing food security and safety issues. Education and stricter waste management strategies are needed.(Pnuma, 2021) The Food Waste Index Report provides a comprehensive data collection and analysis of global food waste, estimating its scale and impacts. It also offers a methodology for measuring waste at household, service, and retail levels. Santeramo and Lamonaca, (2021). The study explores the relationship between food loss and waste (FLW) and food security, focusing on the Water-Energy-Food Security nexus and the reduction of FLW. It suggests future research should explore this nexus through evidence-based and scenario analyses, highlighting synergies between resource uses in a circular and green economy perspective. Aldaco, et.al., (2020). This study suggests that improving food supply chain efficiency is crucial for food security and reducing natural resource pressure. It suggests that fluctuations in eating habits can impact food loss and waste generation and management, as well as GHG emissions. The study uses Spain as a case study, analyzing inputs, outputs, and supply chain under a life cycle thinking approach. Results show a 12% increase in household food loss during the COVID-19 lockdown, with economic impact, GHG emissions, and nutritional content all contributing to the increase. Irani, et.al.,(2018). This paper explores the relationship between food distribution and consumption factors, using design science principles. Qualitative data from commercial food consumers and large-scale food importers, distributors, and retailers is collected. Cause-effect models and Fuzzy Cognitive Map (FCM) approaches are used to build simulation models. The research supports policymakers in developing interventions to reduce food waste and contributes to food security by identifying potential behavioral changes. Eldridge, et.al., (2018). The study also suggests future research in big data research in food supply chains. A study in western Sydney found that compost, a green waste treatment, can enhance vegetable crop yields and economic returns compared to conventional farming methods. Both compost and MIX treatments consistently achieved similar or higher yields, with MIX showing more significant gains. Siddiqui, et.al., (2021). The Maitri Bagh Zoo in Bhilai, Chhattisgarh, is the largest zoo in the region, covering 167 acres and offering 111 acres of parkland. It houses 39 unique animal, bird, and reptile species, many of which are endangered. The zoo uses digital technologies to enhance visitor experience and ensure the well-being fof plants and caged animals..

METHODOLOGY

1. **Literature Review**: The literature review in this study is a comprehensive analysis of existing academic journals, reports, and case studies that investigate the relationship between waste management and global food security. The review aims to provide a thorough understanding of how waste management practices impact the availability of food on a global scale.

Through this review, we seek to identify key themes, trends, and challenges in waste management and their implications for food security. By analyzing a wide range of literature, we aim to gain insights into the current state of waste management practices, including the types and amounts of waste generated, the methods of waste disposal, and the environmental and social impacts of poor waste management.

Additionally, the review will explore the role of policy frameworks, technological innovations, and community engagement in improving waste management practices and enhancing food security by looking at what's already been studied, this review will help us understand how waste management affects global food security. This information will be useful for people who make decisions, researchers, and those working in this field.

2. **Data Analysis**: To collect and analyze data on global waste generation, composition, and disposal, we will use a comprehensive approach. We will gather information from reliable sources like government reports, academic studies, and international databases. This data will include details on the types and amounts of waste produced, what the waste is made of, and how it is disposed of.

Next, we will analyze the data using statistical tools to identify patterns of waste generation in different sectors, such as households, industries, and agriculture. We will use techniques such as regression analysis to identify trends and correlations between waste generation and various factors such as population growth, economic development, and consumption patterns.

By analyzing this data, we aim to gain a better understanding of global waste generation patterns and the factors driving them. This information will be valuable for policymakers, researchers, and practitioners seeking to develop effective waste management strategies to address the challenges of waste generation and disposal.

3. **Case Studies**: To demonstrate effective waste management practices and their influence on local food security, we will feature case studies from various geographic areas. These case studies will highlight successful examples of waste-to-energy technologies and sustainable agriculture programs.

For example, we will examine a case study from Scandinavia, where innovative waste management practices have led to high rates of recycling and waste-to-energy conversion. This has not only reduced the amount of waste sent to landfills but has also generated renewable energy for local communities. In another case study from sub-Saharan Africa, we will explore how community-based composting initiatives have improved soil fertility and crop yields, enhancing food security in the region.

By analyzing these case studies, we aim to identify key lessons and best practices that can be applied in other regions facing similar challenges. These insights will be valuable for policymakers, practitioners, and researchers working to improve waste management practices and enhance food security globally.

4. **Expert Interviews**: Conducting interviews with experts in waste management, environmental science, and agriculture will provide valuable insights into innovative waste management technologies and their implications for food security. These experts can offer perspectives on challenges and opportunities in integrating waste management with sustainable agriculture. By gathering their insights, we can gain a deeper understanding of how these fields intersect and identify potential solutions to improve waste management practices and enhance food security globally.

5. **Stakeholder Surveys:** Engaging with government bodies, non-governmental organizations, and local communities through surveys can provide valuable insights into the impact of current waste management practices on food security. By collecting their opinions and suggestions, we can better understand the challenges faced by different groups and identify potential solutions. This participatory approach can also help build consensus and support for implementing more sustainable waste management practices. Additionally, it can empower communities to take ownership of waste management initiatives, leading to more effective and sustainable outcomes in the long run.

6. **Comparative Analysis**: Comparing waste management strategies and outcomes across different countries or regions can offer valuable insights into how well various policies and initiatives address waste management and food security issues. By studying the successes and challenges faced by different areas, we can pinpoint best practices and lessons that can be applied worldwide. This comparative analysis can assist policymakers and stakeholders in making informed decisions about implementing strategies tailored to their contexts, leading to more effective and sustainable waste management practices. Additionally, by evaluating these strategies' outcomes, we can gauge their impact on food security and pinpoint areas for improvement, ultimately contributing to a more sustainable and resilient food system.

Table 1. Efficient waste management is crucial for enhancing food security by reducing post-harvest losses and optimizing resource use in the food supply chain

S. No	State	Per/hac
1.	Karnataka	(1.4 million hac)
2.	Bihar	(5,80,000 hac)
3.	West Bengal	(6,90,000 hac)
4.	Rajasthan	(6,79,000 hac)
5.	Maharashtra	(4,55,000 hac)

Source: - India: Harvest & Post-Harvest Crop Losses in 2021, Posted on 17. December 2021.

Figure 1. A demographic pie-chart representation about food waste management

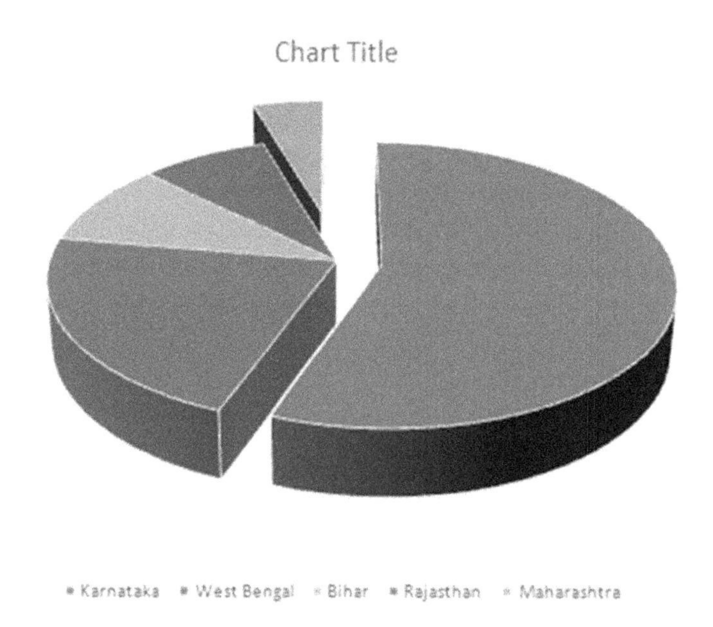

The recent heavy rains have caused extensive damage to crops across several states in India. Karnataka, Bihar, West Bengal, Rajasthan, and Maharashtra are among the worst affected, with millions of hectares of crops damaged. Karnataka, in particular, has seen the highest crop damage, affecting 1.4 million hectares of agricultural land. These states have experienced either 'big excess' or 'excess' rainfall this year, leading to flooding and water logging in many areas. The excessive rainfall has disrupted normal agricultural activities and has had a severe impact on farmers' livelihoods. The Indian Meteorology Service's data analysis shows that Karnataka received 102 percent more rainfall from October to November, exacerbating the situation. The affected states are now facing the challenge of rebuilding their agricultural sector and supporting farmers affected by the crop losses.

Addressing food wastage requires a variety of solutions due to its diverse causes across different levels. Here are some practical ways to reduce food wastage:

1. **Individual Actions:** To reduce food waste, people can plan meals, buy only necessary items, store food correctly, and find creative ways to use leftovers.
2. **Best Practices in Production**: Farmers can minimize losses by using quality seeds, proper irrigation, integrated pest management, and timely harvesting techniques.
3. **Improved Storage Infrastructure**: Investing in storage facilities like metal silos, cold storage, and drying technologies can help reduce post-harvest losses.
4. **Technical Solutions in Transportation, Processing, and Packaging**: New technologies like better packaging, refrigeration, and transportation can help prevent food from being wasted at these stages.
5. **Capacity Building and Education:** Providing training and education to farmers, food handlers, and consumers on best practices for food handling, storage, and consumption can help reduce waste.
6. **Early Warning Systems**: Developing systems to forecast food demand and supply can help prevent overproduction and subsequent wastage.
7. **Promoting Consumer Behaviour Change**: Teaching consumers about the effects of food waste and promoting responsible shopping and consumption habits can help lessen waste.
8. **Policy Integration**: Governments can incorporate food waste reduction strategies into their policies, such as setting targets for waste reduction and implementing regulations to encourage responsible food management.
9. **Valuing Surplus Food**: Recognizing the value of surplus food and finding ways to redistribute or repurpose it can help reduce waste.
10. **Redirecting Surplus Food**: Instead of disposing of surplus food, efforts can be made to redirect it to those in need, as animal feed, or for composting to generate renewable energy.

CONCERNS ADDRESSED IN THE CHAPTER

1. **Resource Scarcity**: The growing global population and evolving consumption patterns are straining vital resources needed for food production. There are concerns about the sustainability of current resource management practices, as they must meet increasing food demands while also preserving the environment and securing food for future generations.
2. **Environmental Degradation**: Inadequate waste management practices lead to soil, water, and air contamination, jeopardizing ecosystems and biodiversity. Soil contamination affects agricultural productivity and food safety. Water pollution poses risks to aquatic life and human health. Air pollution contributes to respiratory illnesses and climate change. Adopting sustainable waste management practices is urgent to safeguard ecosystems, biodiversity, and human well-being.
3. **Technological Challenges**: Waste-to-energy technologies face challenges concerning environmental impact, such as emissions and residue management, and economic viability, including high initial costs. Integrating these technologies into comprehensive waste management systems requires overcoming regulatory, technical, and financial obstacles. However, with proper planning and innovation, they offer opportunities to reduce waste volume, generate renewable energy, and mitigate environmental pollution.
4. **Circular Economy Principles**: Adopting circular economy principles in agriculture is crucial to view waste as a valuable resource. Recycling organic waste can enrich soil, enhance agricultural productivity, and promote sustainable practices, contributing to long-term environmental health and food security.

5. **Supply Chain Inefficiencies:** Food loss and waste in the agricultural and food supply chain are major concerns, influenced by factors like inadequate storage facilities, inefficient transportation, and market dynamics. To enhance supply chain efficiency, strategies such as improved infrastructure, better forecasting techniques, and promoting value-added processing can reduce losses. Collaborative efforts among stakeholders, along with consumer education, are essential to minimize food waste and improve food security outcomes.
6. **Policy Integration:** Integrated policies and practices are essential, recognizing the interdependence between waste management and food security. A holistic approach is needed to foster environmental sustainability, resource efficiency, and resilient food systems. This requires collaboration among stakeholders, including governments, industries, and communities, to address the complex challenges of waste management and food security in a coordinated manner.
7. **Smart Rules for Everyone**: Smart rules about waste management are crucial for environmental protection, food security, and community health. These rules ensure proper disposal of waste, reducing pollution and protecting ecosystems. They also promote recycling and resource conservation, ensuring sustainable use of resources. By implementing smart waste management rules, governments and communities can create a cleaner, healthier environment and secure food sources for future generations.
8. **Recycling Stuff for Farms**: Exploring the agricultural benefits of recycling, we'll delve into how utilizing waste like leftover food and plant matter can enrich soil health. By composting and recycling these materials, nutrients are returned to the soil, promoting plant growth and reducing the need for chemical fertilizers. This sustainable approach not only improves soil fertility but also demonstrates the importance of smart waste management practices in agriculture.
9. **Stopping Food Waste along the Way**: In examining food wastage from farm to table, this chapter will delve into the various factors contributing to this issue and present viable solutions. By focusing on enhancing food handling practices, we can effectively decrease waste and promote fairer food distribution. These efforts are crucial in not only minimizing food loss but also ensuring that everyone has access to an adequate food supply. Through these analyses and proposed solutions, the chapter aims to deepen our comprehension of how waste management practices affect food security. It also seeks to provide actionable strategies that can be implemented to create a more sustainable and equitable future, where food is utilized efficiently, and waste is minimized along the entire food supply chain.

Section One Waste Generation and Composition

This section provides a detailed examination of the various sources and composition of waste, ranging from household waste to industrial and agricultural byproducts. It offers a comprehensive overview of different types of waste, their origins, and the environmental ramifications of inadequate disposal methods. Particular attention is paid to the impact of plastics, electronic waste, and organic waste on environmental pollution and resource depletion. By comprehensively understanding these aspects, we can gain valuable insights into the complexities of waste generation and the potential consequences it poses.

Section Two Environmental Impacts of Inadequate Waste Management

This section scrutinizes the environmental repercussions stemming from inadequate waste management practices. Discussions encompass the contamination of soil, water, and air due to improper disposal, emphasizing the negative consequences for ecosystems, biodiversity, and overall environmental health. The objective is to underscore the urgency of adopting sustainable waste management practices to mitigate these far-reaching impacts. The chapter aims to increase awareness of how waste management and environmental well-being are connected through detailed examination.

Section Three Waste-to-Energy Technologies

A substantial portion of the chapter is dedicated to exploring waste-to-energy technologies as potential solutions to the waste management problem. It conducts a comprehensive analysis of various methods, including incineration and anaerobic digestion, for converting waste into energy. The section also delves into the environmental and economic implications of these technologies, addressing both the promise they hold, and the challenges associated with their integration into waste management systems. By examining these technologies in detail, the chapter seeks to provide a nuanced perspective on their role in shaping sustainable waste management practices.

Section Four Waste Management as a Driver for Sustainable Agriculture

This section transitions to the role of waste management in promoting sustainable agriculture and enhancing food security. It emphasizes the potential of organic waste recycling methods, such as composting and anaerobic digestion, to produce nutrient-rich fertilizers. Additionally, the discussion touches upon the importance of adopting circular economy principles in agriculture, where waste becomes a valuable resource rather than a burden. By exploring these connections, the chapter aims to showcase how effective waste management practices can play a pivotal role in supporting resilient and sustainable agricultural systems.

Section Five Reducing Food Loss and Waste Across the Supply Chain

This section lays the foundation for examining food loss and waste within the agricultural and food supply chain. It looks into factors that contribute to food loss, including inefficient harvesting, storage, and transportation practices. Strategies to reduce food waste and improve the efficiency of the food supply chain are also discussed, emphasizing how effective waste management is linked to better food security outcomes. By delving into these challenges and solutions, the chapter aims to offer a thorough understanding of the entire food production and distribution process.

RESULT AND DISCUSSION

This chapter's synthesis highlights the intricate relationship between waste management and global food security, emphasizing critical challenges stemming from population growth and shifting consumption patterns. It underscores the importance of sustainable resource management through practices such as precision agriculture and water-efficient farming, aligning resource use with demographic and consumption trends. Immediate adoption of sustainable practices, such as waste reduction and recycling, is crucial to tackle insufficient waste management. This should be accompanied by strict environmental regulations to support these efforts. Technological challenges in waste-to-energy are addressed, emphasizing the importance of innovation and balancing energy generation with environmental sustainability. Circular economy principles, particularly in recycling organic waste for sustainable agriculture, are deemed essential. Supply chain inefficiencies contributing to food loss call for enhanced practices and collaboration. Integrated policies recognizing the interdependence of waste management and food security, along with smart rules, are crucial. Educational initiatives are needed to promote waste recycling for agriculture, while strategies to minimize food waste stress improved storage and transportation practices. In conclusion, the chapter offers actionable recommendations, emphasizing the need for coordinated efforts across sectors.

The heavy rainfall in several Indian states has caused extensive damage to crops, threatening food security. Karnataka is the worst-hit, with 1.4 million hectares affected, followed by Bihar, West Bengal, Rajasthan, and Maharashtra, where hundreds of thousands of hectares of agricultural land have been damaged.

The impact of these losses underscores the vulnerability of India's agriculture to extreme weather events. Excessive rainfall, exceeding 100% in some regions, has led to flooding and waterlogging, disrupting farming activities and affecting livelihoods. Urgent measures are needed, including improved drainage systems to prevent waterlogging, the development of climate-resilient crop varieties, and better post-harvest loss management to minimize losses due to improper storage and transportation. These efforts are crucial for ensuring food security in the face of climate change. By addressing these challenges, India can build a more resilient agricultural sector capable of withstanding the impacts of extreme weather events, ultimately ensuring food security for its population.

CONCLUSION

The concluding section of the chapter underscores the significance of adopting a holistic approach to address the challenges of waste management and food security. It emphasizes the importance of integrated policies and practices that promote environmental sustainability, resource efficiency, and resilient food systems. By recognizing the interdependence between waste management and food security, stakeholders can develop strategies that maximize the value of waste as a resource rather than viewing it as a burden. This shift in perspective can lead to innovative solutions and contribute significantly to achieving sustainable development goals related to waste management and food security.

Further research could delve into specific regions or countries to provide a more nuanced understanding of waste management and food security challenges, considering local practices, policies, and cultural factors. In-depth case studies could be conducted to analyze the effectiveness of certain waste management practices, such as community-based initiatives or public-private partnerships, in enhancing food security. Additionally, exploring innovative technologies and approaches in waste management, such as bioremediation or decentralized composting, could offer valuable insights into sustainable waste management practices.

The chapter's analysis relies on existing literature and data, which may not fully capture the most recent developments or localized challenges in waste management and food security. The broad scope of the chapter covers various aspects of waste management and food security, which could limit the depth of analysis for each topic. The effectiveness of proposed solutions may vary depending on the context, requiring further feasibility studies and considerations for implementation.

REFERENCES

Abdel Bari, E. M. (2016). Qatar's vision of global problems, challenges and solutions. *Qscience Proceedings*, 2016(4), 9.

Adelodun, B., Kim, S. H., Odey, G., & Choi, K. S. (2021). Assessment of environmental and economic aspects of household food waste using a new Environmental-Economic Footprint (EN-EC) index: A case study of Daegu, South Korea. *The Science of the Total Environment*, 776, 145928. 10.1016/j.scitotenv.2021.14592833640543

Aldaco, R., Hoehn, D., Laso, J., Margallo, M., Ruiz-Salmón, J., Cristobal, J., Kahhat, R., Villanueva-Rey, P., Bala, A., Batlle-Bayer, L., Fullana-i-Palmer, P., Irabien, A., & Vazquez-Rowe, I. (2020). Food waste management during the COVID-19 outbreak: A holistic climate, economic and nutritional approach. *The Science of the Total Environment*, 742, 140524. 10.1016/j.scitotenv.2020.14052432619842

. Bakharev, V. V., Mityashin, G. Y., & Stepanova, T. V. (2023). Food security, food waste and food sharing: The conceptual analysis. *Food systems, 6*(3), 390-396.

Bilal, M., & Iqbal, H. M. N. (2019). Sustainable Bioconversion of Food Waste into High-Value Products by Immobilized Enzymes to Meet Bio-Economy Challenges and Opportunities—A Review. *Food Research International*, 123, 226–240. 10.1016/j.foodres.2019.04.06631284972

Blakeney, M. (2019). Food loss and waste and food security. In *Food loss and food waste* (pp. 1–26). Edward Elgar Publishing. 10.4337/9781788975391.00006

Cristofoli, N. L., Lima, A. R., Tchonkouang, R. D. N., Quintino, A. C., & Vieira, M. C. (2023). Advances in the Food Packaging Production from Agri-Food Waste and By-Products: Market Trends for a Sustainable Development. *Sustainability (Basel)*, 15(7), 6153. 10.3390/su15076153

Durán-Sandoval, D., Durán-Romero, G., & Uleri, F. (2023). How much food loss and waste do countries with problems with food security generate? *Agriculture*, 13(5), 966. 10.3390/agriculture13050966

Durán-Sandoval, D., Uleri, F., Durán-Romero, G., & López, A. M. (2023). Food, Climate Change, and the Challenge of Innovation. *Encyclopedia*, 3(3), 839–852. 10.3390/encyclopedia3030060

Ebikade, E. O., Sadula, S., Gupta, Y., & Vlachos, D. G. (2021). A Review of Thermal and Thermocatalytic Valorization of Food Waste. *Green Chemistry*, 23(8), 2806–2833. 10.1039/D1GC00536G

Eldridge, S. M., Yin Chan, K., Donovan, N. J., Saleh, F., Orr, L., & Barchia, I. (2018). Agronomic and economic benefits of green-waste compost for peri-urban vegetable production: Implications for food security. *Nutrient Cycling in Agroecosystems*, 111(2-3), 155–173. 10.1007/s10705-018-9931-9

Hidayat, W. (2021). Analisis Pengaruh Kualitas Pelayanan Go Food Terhadap Tingkat KepuasanKonsumen. *Frontiers in Neuroscience*, 14(1).

Hilborn, R. (2017). Food for Thought Are MPAs effective? –. *ICES Journal of Marine Science*, 75.

Irani, Z., Sharif, A. M., Lee, H., Aktas, E., Topaloğlu, Z., van't Wout, T., & Huda, S. (2018). Managing food security through food waste and loss: Small data to big data. *Computers & Operations Research*, 98, 367–383. 10.1016/j.cor.2017.10.007

Jamaludin, H., Elmaky, H. S. E., & Sulaiman, S. (2022). The future of food waste: Application of circular economy. *Energy Nexus*, 7, 100098. 10.1016/j.nexus.2022.100098

Kabir, A. (2018). Application Of G.I.S In Site Selection For Solid Waste Collection Points In Kofar Kaura New Layout Katsina. *World Development*, 1(1).

Lin, A. Y. C., Huang, S. T. Y., & Wahlqvist, M. L. (2009). Waste management to improve food safety and security for health advancement. *Asia Pacific Journal of Clinical Nutrition*, 18(4), 538–545.19965345

Lin, A. Y. C., Huang, S. T. Y., & Wahlqvist, M. L. (2009). Waste management to improve food safety and security for health advancement. *Asia Pacific Journal of Clinical Nutrition*, 18(4).19965345

Lo Turco, V., Potortì, A. G., Tropea, A., Dugo, G., & Di Bella, G. (2020). Element Analysis of Dried Figs (Ficus carica L.) from the Mediterranean Areas. *Journal of Food Composition and Analysis*, 90, 103503. 10.1016/j.jfca.2020.103503

Margiotta, M., & Baudoin, W. (1999). *Environmental issues and peri-urban agriculture in Mauritania*. Peri-Urban Agriculture in Sub-Saharan African.

Massari, S., Principato, L., Antonelli, M., & Pratesi, C. A. (2022). Learning from and designing after pandemics. CEASE: A design thinking approach to maintaining food consumer behaviour and achieving zero waste. *Socio-Economic Planning Sciences*, 82, 101143. 10.1016/j.seps.2021.101143

N., P. S., Chaudhari, S., Barde, S., & Devices. (2021). Hubungan Tingkat Pengetahuan dan Kebiasaan-Konsumsi Junk Food dengan Status Gizi. *Frontiers in Neuroscience, 14*(1).

Papargyropoulou, E., Lozano, R., Steinberger, J. K., & Wright, N. (2014). bin Ujang, Z. The Food Waste Hierarchy as a Framework for the Management of Food Surplus and Food Waste. *Journal of Cleaner Production*, 76, 106–115. 10.1016/j.jclepro.2014.04.020

Pnuma, W. (2021). *UNEP Food Waste Index Report 2021*. UN Environment Programme.

Santeramo, F. G., & Lamonaca, E. (2021). Food loss–food waste–food security: A new research agenda. *Sustainability (Basel)*, 13(9), 4642. 10.3390/su13094642

Scarpare, F. V. (2013). Bioenergy and water: Brazilian sugarcane ethanol. *Bioenergy and Water, 89.*

Siddiqui, S., Bajaj, S., & Mathew, C. (2021). Exploring Biodiversity: Flora and Fauna in Maitri Bagh, Bhilai, Chhattisgarh, India. *NeuroQuantology : An Interdisciplinary Journal of Neuroscience and Quantum Physics*, 19(12), 705.

Stefano, D. (2007). Psychology and economics: Evidence from the field. *Journal of Economic Literature*, 1(November).

Stunžėnas, E., & Kliopova, I. (2021). Industrial ecology for optimal food waste management in a region. *Environmental Research, Engineering and Management*, 77(1), 7–24. 10.5755/j01.erem.77.1.27605

Surbakti, E. P. C. B. (2021). HubunganPengetahuan dan SikapTerhadap Tindakan KonsumsiMakanan-Cepat Saji (fast Food) Pada Remaja Di SMA Negeri 1 Tigapanah. *Frontiers in Neuroscience*, 14(1).

Turner, K., Georgiou, S., Clark, R., Brouwer, R., & Burke, J. (2004). The role of water in agricultural development. In *Economic valuation of water resources in agriculture*. Research Gate.

Wani, N. R., Rather, R. A., Farooq, A., Padder, S. A., Baba, T. R., Sharma, S., Mubarak, N. M., Khan, A. H., Singh, P., & Ara, S. (2023). New insights in food security and environmental sustainability through waste food management. *Environmental Science and Pollution Research International*, 31(12), 1–23. 10.1007/s11356-023-26462-y36988800

Chapter 17
Adoption Challenges of Industry 4.0 in Agrisector and Designing a Framework to Reduce It

Meghana Mishra
http://orcid.org/0000-0002-6419-5420
KIIT University, India

Suchismita Satapathy
http://orcid.org/0000-0002-4805-1793
KIIT University, India

ABSTRACT

Agriculture 4.0 technology allows farmers to use trend analysis to predict future weather conditions and crop yields in the coming days. IoT in agriculture helps farmers maintain crop quality and soil fertility, thereby increasing yield and quality. The data collected is used to leverage technological advances to enable better decision-making. By recording data from sensors, IoT devices provide real-time information about plant health. Hence the barriers of Industry 4.0 must be mitigated to improve agrisector. Still people prefer traditional farming process, hence in this chapter, a study is carried out to find barriers of adoption of Industry 4.0 in agrisector and designed a framework to mitigate the challenges.

INTRODUCTION

Industry 4.0 or Fourth Industrial Revolution, has brought many changes to manufacturing sector as well as agrisector. The integration of advanced technologies like automation, artificial intelligence (AI), Internet of Things (IoT), big data analytics, and robotics changed the agrisector to agriculture 4.0. It has changed agrisector into smarter, more connected, and more efficient systems. In the agricultural sector, Industry 4.0 holds immense potential to revolutionize farming practices.It has groomed agrisector by developing modern equipment that were automated and self operated without human effort. Due to advanced and modern equipment productivity increased, resource usage is optimized with increasing sustainability and enhancing food security.Starting from soil quality checking to all kind of farming operations are conducted by modern IoT and AI. Many difficult tasks were conducted in many coun-

DOI: 10.4018/979-8-3693-3583-3.ch017

tries with the help of robots and robotic technology. Difficult farming tasks like seeding, harvesting, ploughing etc are done within a fraction of second with new devices.Protection of seeds, grains, vegetables, packaging, processing of foods and transportation of agri products are now very easy with the help of different kind of IoT based devices and easy of monitoring and tracking has also enhanced the agribusiness.Due to this farming and agro industries are gaining back their popularity and further young generations were attracted towards it.

Industry 4.0 is transforming the agrisector, By use of sensors and IoT devices soil conditions, crop health, weather patterns, and equipment status can be monitored and generated data can be stored.Variable Rate Technology (VRT) helps to apply seeds, fertilizers, and pesticides at variable rates across a field, It will help to utilize resouces optimally.

The labour oriented difficult tasks like planting, harvesting, spraying, and monitoring tasks can be done automatically with the help of Drones and tractors.Which has improved efficiency with reducing physcal labour and labour cost.Even for plucking use of robotic arms has reduced labour cost increasing harvesting efficiency.

Predictive analysis and farm management softwares are help full in managing and prediction crop health, detect diseases with the help of previous historical data.Remote sensing techniques help to know details about crop health, soil conditions etc.Use of drones for pesticides and fertilizer spraying has reduced risk for farmers.Smart irrigation systems use sensors and weather data to optimize water use, reduce waste and improve crop health. Similarly, nutrient management can also be easy by using sensors for detecting soil fertility.Blockchain technology and market-driven data not only protect crops till they reach the consumer, they also update farmers regarding market demand and prices.Specifically, precision agriculture saves the environment by reducing its carbon footprint.

So, with adoption of industry 4.0, increases the efficiency of crop production and decreases labour and other farming costs. It also provides /updates farmers with advanced technology and provides jobs/ employment opportunities for skilled workforce.

Still for adopting Industry4.0 there are many challenges,which farmers have to confront in future. Due to these challenges till agrisector remains undeveloped and farmers prefer old and conventional methods of farming.Hence, in this chapter a study is carried on to find these challenges and a framework is designed to mitigate these challenges Quality function deployment method.

BACKGROUND

Dwivedi et al(2019) have discussed that AI acts as an transformative technology across almost all industries.The transformative potential of artificial intelligence (AI) in agriculture is highlighted by Liakos et al. (2018), Elbasi et al. (2023), and Sharma et al. (2021). These studies demonstrate AI's capacity for automation, self-learning, human emulation, prediction, and augmentation. These qualities allow AI to reduce environmental impact, improve crop monitoring, and optimize resource allocation—all of which have the potential to transform agricultural output (Elbasi et al., 2023).

Chanchaichujit et al.(2024) have explored the barriers of agri supply chain adoption of Industry 4.0 by ISM method .As per Chanchaichujit Lack of information about technologies and lack of compatibility with traditional methods emerged as the two main barriers that influence each other.Zhai et al.(2020) and Erdoğan(2022) have written that industry 4.0 is the solution for every industry.Latino et al.(2022),

Łukowska et al.(2019) and Zambon(2019) have explained that Industry 4,0 notonly develops advance machines for agrisector it also develops the production line and efficiency/productivity.

RESEARCH METHODOLOGY

A study is carried on to design a standard questionnaire of 18 questions about the barriers of adoption of industry 4.0 in agrisector and a survey is conducted and expert analysis is done. Around 78 experts of agrisector and industry 4.0 answered in likert scale (1to 3) (I.e.,dis agree, no opinion and agree).In Table.1 shows question for barriers.Then factor analysis is carried on and quality function delpoy ment is implemented to design a framework to mitigate these barriers.

Table.1 Questionnaire for barriers

	Question	1	2	3
1.	Cost of drone, IoT, Robotics very high			
2.	Due to lack of standards, difficult to adopt			
3.	lack of IT infrastructure			
4.	data safety and security issue			
5	Establishing the facility in splitted farms very difficult			
6.	Farmers may be hesitant to invest without clear evidence of financial gains.			
7.	loans or financial support for purchasing advanced technology can be challenging			
8	making the transition to smart farming complex			
9	maintenance and support, which can be difficult to manage, especially in remote or rural areas			
10	skill levels of many farmers and farm workers			
11.	Lack of training and education			
12	Farmers who are accustomed to traditional methods may be reluctant to adopt new practices.			
13	poor internet infrastructure, hindering the effective use of smart farming technologies.			
14	Consistent and reliable power supply is necessary			
15	for deploying new technologies, especially those involving data privacy, drones, and genetically modified organisms (GMOs), can be complex			
16	absence of standardized protocols and framework			
17	the environmental impact of deploying new technologies, such as increased electronic waste and energy consumption,			
18	resistance from local communities due to concerns about job displacement			

RESULTS AND DISCUSSION

To find the adoption challenges, factor analysis is conducted and 17challenges are selected under 7 dimensions (I.e., Technological challenges, Financial challenges, Operational challenges, Social challenges, Infrastructure challenges, Policy challenges and Environmental challenges). The items selected having Chrobanch's alpha value more than 0.5, items having less value than 0.5 are discarded.

Table.2 Factor analysis

Dimension	Itmes	F1	F2	F3	F4	F5	
Technological challenges	1,2,3,4	0.671,0.523,0.543,0.511					
Financial challengs	6,7		0.563,0.531				
Operational challenge	8,9,10			0.501,0.570,0,525			
Social challenges	11,12				-0.526,0.518		
Infrastructure challenges	13,14					0.520,0.531	
Policy challenges	15,16					0.516,0.522	
Environmental challenges	17,18						0.547,0.619

QFD(Quality function deployment) is used to find the relation between WHat and HoW(I.e. customer needs and designer requirements).Hence for developing a framework to fulfill customer need is done by QFD. In this chapter, design requirements/Hows are selected from expert suggestion and the challenges found after factor analysis are the WHats /customer requirements. WHat vses How are plotted by house of quality in QFD by finding relation between whats vses whats at customer rating matrix(left matrix). Then What/How as central matrix later How vses How as roof matrix.The symbols like [0] [Θ][Δ], and [●] are used, whose values are 0.8, 0.6, 0.4 and 0.2 respectively.The symbols [0] [Θ][Δ], and [●] represent good, sensitive, bad and very bad relations in the HoQ matrix..Chan and Wu (2005) have discussed the terms like customer needs/whats, design requirements/how and also established the HoQ relation and named them as left, central and roof matrix.The design requirements are calculated by using formula $\sum_j^n AijXj$,(where Aij represent the relation between ith consumer's demand to j-th customers demand, and Xj represents consumer ratings and n is the numer of consumers).By using this formula, consumer score can be found.Similarly, $Z_j + \frac{1}{n-1}(\sum B_{ij}Z_j)$ is used to find central matrix . In this formula, Bij represents customers' rating and Zj shows the relation between customer needs and design requirements. From this initial rating is calculated . Similarly, further formula $\sum_j^n AijXj$ is used to find revised rating by using initial rating and I to j relation between design requirements(roof matrix).Lastly, normalized matrix is calculated and ranking of the matrix is done. Few researchers have used this method among them Yazdani et al., (2016) have used this QFD to study the supplier evaluation process by using customer attitude. Song et al.(2020) have used QFD for controlling the quality of few service and manufacturing industries. Memenza et al., (2020) have used the QFD tool for quality management process in strawberry farming process in the Huaura region. Lombardi and Fargnoli (2018) have implemented the Quality Function Deployment (QFD) tool for hazard prevention by using specific tasks. a

Table.3 shows customer requirements and table.4 states design requirements to mitigate challenges as per expert suggestion.Table.5 shows designing framework for industry 4.0 adoption challenges.Figure.1 Shows HOQ(house of Quality)

Table.3 Customer requirement

Sl.No	Customer requirement
1	Technological challenges
2	Financial challenges
3	Operational challenge

continued on following page

Table. Continued

Sl.No	Customer requirement
4	Social challenges
5	Infrastructure challenges
6	Policy challenges
7	Environmental challenges

Table.4 Design requirement

Sl.No	Design requirement
1	Govt. Policies for supporting new technologies
2	Govt. Funds to support infrastructure
3	Training programme for developing skill of farmers
4	Infrasture development in villages
5	Industrial collaborations to develop technologies
6	Public awareness programme
7	Framing rules for e-waste management
8	Development of data security system

Table.5 Quality function deployment for designing a framework

	requirement	1	2	3	4	5	6	7	8	Customer rating
1	1									4.886
2	4									5.839
3	2									9.954
4	3									5.123
5	2									6.788
6	1									11.012
7	5									5.835
Initial rating		**42.49**	**51.98**	**43.95**	**33.83**	**34.71**	**35.79**	**43.76**	**40.58**	
Revised Rating		**21.76**	**14.38**	**44.98**	**35.89**	**13.1**	**36.23**	**21.31**	**22.46**	
Normalized Rating		**0.483770565**		**3.433587786**	**0.990615512**	**0.614734866**	**1.613089938**	**0.473766118**	**0.625801059**	
		6	**8**	**1**	**3**	**5**	**2**	**7**	4	

Figure1. HoQ (house of quality) for adoption of industry 4.0 in farming sector

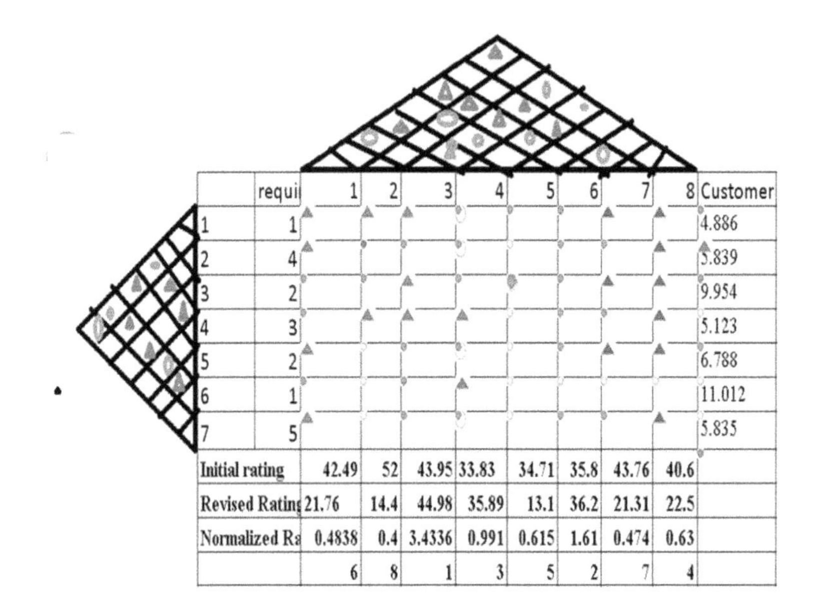

From Table.5 it is found that to mitigate barriers of industry 4.0, the most important steps are training and education programs to develop skills of farmers, public awareness program and proper infrastructure development in villages. These preliminary steps must be taken to mitigate barriers. Then ISM (interpretieve structural modelling) is implimented to find interrelation between the design framework.

The Interpretive Structural Modeling (ISM) approach used is an interactive learning process that organizes a collection of elements into a whole system. ISM helps you understand the order and function of complex interactions between different elements in a system. Interpretive structural equation modeling is used to find the relation between predictive behavior for successful business performance improvement (Anderson et al. (1994) . Kumar et al. (2009) have explained that with the ISM method the complex relationship between various elements can be established

For computer assisted learning practice. Satapathy (2014) has found the consumer satisfaction in electricity utility service by establishing the relation between service delivery by the ISM approach. ISM is used in many researches .The steps of ISM are written below.

ISM Steps

Step-1 Design a Self-Structured Interactive Matrix

If i is the basis for obtaining factor j, then the relation is V. If j is the basis of factor I, then the relationship is equal to A. In both cases, the relationship between I and factor j is X, and if there is no relationship, it is written as O. Table 6 shows the SSIM matrix.

Step-2 Structure of the Initial Accessibility Matrix

The reach ability matrix is created using1 or 0 for V, X, O, and A. The relationship from I to j is written as 1, and the value of the relationship from j to i is written as 0. It is set to 1 if both relationships exist, and 0 if there is no relationship. Table 7 shows the initial accessibility matrix and Table 8 shows the final accessibility matrix. Next, overall drive and dependency values are calculated. We will see the transitivity relationship later.

Step -3 Calculate Level Classification

After computing the reachability matrix, the antecedent matrix is also computed. The intersection set shows the commonalities between the reachability matrix and the antecedent matrix. Then, an iterative step is performed to select parameters at each level. Table.6 to 12 shows the steps of ISM.

Table. 6 SSIM matrix

	8	**7**	**6**	**5**	**4**	**3**	**2**	**1**
1	V	V	O	A	V	O	V	
2	O	V	V	O	V	A		
3	V	O	V	V	O			
4	A	O	V	A				
5	V	O	A					
6	O	A						
7	V							
8								

Table 7. Initial reach-ability matrix

	8	**7**	**6**	**5**	**4**	**3**	**2**	**1**
1	1	1	0	1	1	1	1	1
2	1	1	0	1	1	1	1	1
3	1	1	1	1	1	1	1	1
4	1	1	0	1	1	1	1	1
5	1	0	0	1	1	1	0	1
6	0	1	0	0	1	0	1	1
7	1	1	0	1	1	1	1	1
8	0	1	0	1	0	1	0	1

Table. 8 Final reach-ability matrix

	8	7	6	5	4	3	2	1	Drive power
1	1	1	0	1*	1	1	1	1	7
2	1	1	0	1*	1	1	1	1	7
3	1	1*	1	1	1	1	1	1	8
4	1	1	0	1	1	1	1	1	7
5	1	0	0	1	1	1	0	1	5
6	0	1	0	0	1*	0	1	1	4
7	1	1	0	1	1	1	1	1	7
8	0	1	0	1	0	1*	0	1	4
Dependency	6	7	1	7	7	7	6	8	

Table 9. Level partition

Challenges	Reach ability set	Antecedent Set	Intersection set	Level
1	1,2,4,5,6,7,8	1,2,3,4,5,7	1,2,5,7	
2	1,2,4,5,6,7,8	1,2,3,4,6,7,8,	1,2,4,6,7,8	
3	1,2,3,4,5,6,7, 8	3	3	
4	1,2,4,5,6,7,8	1,2,3,4,5,7,8	1,2,4,5,7,8	
5	1,4,5,6,8	1,2,3,4,5,6,7	1,4,5,6	
6	2,5,7,8	1,2,3,4,5,7,8	2,5,7,8	
7	1,2,4,5,6,7,8	1,2,3,4,6,7	1,2,4,6,7	
8	2,4,6,8	1,2,3,4,5,6,7,8	2,4,6,8	I

Table 10. Iteration-1

Challenges	Reach ability set	Antecedent Set	Intersection set	Level
1	1,2,4,5,6,7	1,2,3,4,5,7	1,2,5,7	
2	1,2,4,5,6,7	1,2,3,4,6,7	1,2,4,6,7	
3	1,2,3,4,5,6,7	3	3	
4	1,2,4,5,6,7	1,2,3,4,5,7	1,2,4,5,7	
5	1,4,5,6	1,2,3,4,5,6,7	1,4,5,6	II
6	2,5,7	1,2,3,4,5,7	2,5,7	II
7	1,2,4,5,6,7	1,2,3,4,6,7	1,2,4,6,7	

continued on following page

Table 11. Continued

Table 11. Iteration-1

Challenges	Reach ability set	Antecedent Set	Intersection set	Level
1	1,2,4,7	1,2,3,4,7	1,2,4,7	III
2	1,2,4,7	1,2,3,4,7	1,2,4,7	III
3	1,2,3,4,7	3	3	
4	1,2,4,7	1,2,3,4,7	1,2,4,7	III
7	1,2,4,7	1,2,3,4,7	1,2,4,7	III

Table 12. Iteration-1

Challenges	Reach ability set	Antecedent Set	Intersection set	Level
3	3	3	3	IV

Figure 2. ISM model

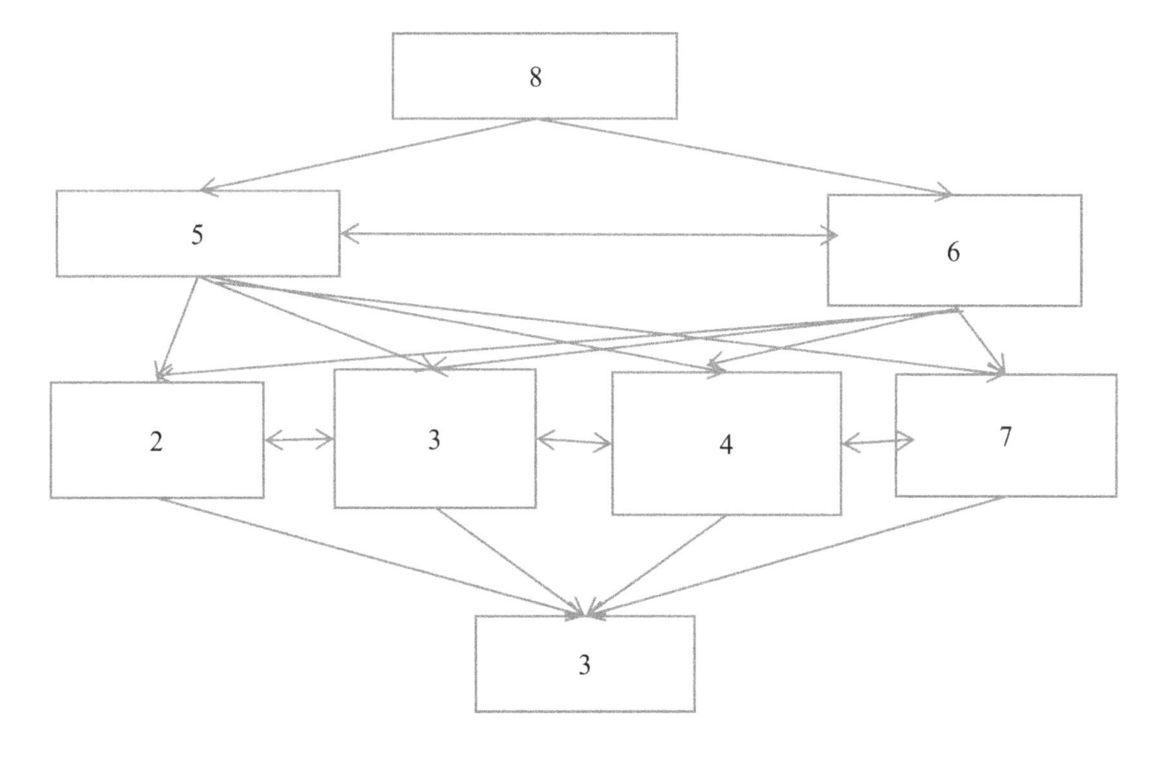

After implementing the ISM model(figure.2) the relation between the variables found as per the level. Element no 8 is in its first level, 56 in 2nd level, 2,3, 4, 7 in third level and 3 comes in last level.

CONCLUSION

Like manufacturing sector agrisector can also develop by industry 4.0. So, there is no turning back from the inevitable trend known as "agriculture 4.0" in the agricultural sector. These days, farmers can connect farming equipment via their personal mobile phones and operate all kinds of difficult farming tasks. Because of its affordability and ease of use, the integration of communication technology. The current state of agriculture is extremely precarious, mainly as a result of climate change, resource shortages, and population expansion on a worldwide scale. Agriculture 4.0 lowers the environmental effect of farming methods by using less chemicals, fertilizers, pesticides. Establishing proper waste management techniques, enhancing crop management, and lowering costs associated with fuel, water, and power with the help of Agriculture 4.0, agriculture industry is now more productive and profit-making compared to previous traditional farming. It is therefore valued in this particular circumstance. In order to ensure farmers' livelihoods and food security in a changing climate, smart agriculture seeks to be adapted.

To mitigate climate change by sequestering carbon in the soil, lowering greenhouse gas emissions, and enhancing the productivity and profitability of agricultural systems. IoT and precision farming techniques will combine to create a new agricultural equipment network that is networked. Precision agriculture may address many agricultural issues in a number of ways by utilizing IoT. In order to plan for their property and this year, farmers can gather information about their farms more quickly. In order to increase productivity and effectiveness, precision farming techniques including irrigation, livestock management, and vehicle tracking are essential. Utilizing Agriculture 4.0 technologies, you can evaluate soil

REFERENCES

Anderson, J. C., Rungtusanthanam, M., & Schroeder, R. G. (1994). A theory of quality management underlying the Deming management method. *Academy of Management Review*, 19(3), 472–509. 10.2307/258936

Chan, L. K., & Wu, M. L. (2002). Quality Function Deployment: A Literature Review. *European Journal of Operational Research*, 143(3), 463–497. 10.1016/S0377-2217(02)00178-9

Chanchaichujit, J., Balasubramanian, S., & Shukla, V. (2024). Barriers to Industry 4.0 technology adoption in agricultural supply chains: A Fuzzy Delphi-ISM approach. *International Journal of Quality & Reliability Management*. 10.1108/IJQRM-07-2023-0222

Dwivedi, Y., Hughes, L., Ismagilova, E., Aarts, G., Coombs, C., Crick, T., & Duan, Y. (2019). Artificial Intelligence (AI): Multidisciplinary perspectives on emerging challenges, opportunities, and agenda for research, practice and policy. *International Journal of Information Management*. 10.1016/j.ijinfomgt.2019.08.002

Elbasi, E., Mostafa, N., AlArnaout, Z., Zreikat, A., Cina, E., Varghese, G., Shdefat, A., Topcu, A., Abdelbaki, W., Mathew, S., & Zaki, C. (2023). Artificial Intelligence Technology in the Agricultural Sector: A Systematic Literature Review. *IEEE Access : Practical Innovations, Open Solutions*, 11, 171–202. 10.1109/ACCESS.2022.3232485

Erdoğan, M. (2022). Assessing farmers' perception to Agriculture 4.0 technologies: A new interval-valued spherical fuzzy sets based approach. *International Journal of Intelligent Systems*, 37(2), 1751–1801. 10.1002/int.22756

Kumar, A., Varghese, M., & Mohan, D. (2000). Equipment-related injuries in agriculture: An international perspective. *Injury Control and Safety Promotion*, 7(3), 1–12. 10.1076/1566-0974(200009)7:3;1-N;FT175

Latino, M. E., Menegoli, M., Lazoi, M., & Corallo, A. (2022). Voluntary traceability in food supply chain: A framework leading its implementation in Agriculture 4.0. *Technological Forecasting and Social Change*, 178, 121564. 10.1016/j.techfore.2022.121564

Liakos, K., Busato, P., Moshou, D., Pearson, S., & Bochtis, D. (2018). Machine Learning in Agriculture: A Review. *Sensors (Basel)*, 18(8), 2674. 10.3390/s1808267430110960

Lombardi, M., & Fargnoli, M. (2018). Prioritization of hazards by means of a QFD-based procedure. *Safety and Security Studies, 163.*

Łukowska, A., Tomaszuk, P., Dzierżek, K., & Magnuszewski, Ł. (2019, May). Soil sampling mobile platform for Agriculture 4.0. In *2019 20th International Carpathian Control Conference (ICCC)* (pp. 1-4). IEEE.

Satapathy, S. (2014). ANN, QFD and ISM approach for framing electricity utility service in India for consumer satisfaction. *International Journal of Services and Operations Management*, 18(4), 404–428. 10.1504/IJSOM.2014.063243

Sharma, R. (2021). Artificial Intelligence in Agriculture: A Review. *2021 5th International Conference on Intelligent Computing and Control Systems (ICICCS)*, (pp. 937-942). IEEE. 10.1109/ICICCS51141.2021.9432187

Song, C., Wang, J. Q., & Li, J. B. (2014). New framework for quality function deployment using linguistic Z-numbers. *Mathematics*, 8(2), 224. 10.3390/math8020224

Yazdani, M., Hashemkhani Zolfani, S., & Zavadskas, E. K. (2016). New integration of MCDM methods and QFD in the selection of green suppliers. *Journal of Business Economics and Management*, 17(6), 1097–1113. 10.3846/16111699.2016.1165282

Zambon, I., Cecchini, M., Egidi, G., Saporito, M. G., & Colantoni, A. (2019). Revolution 4.0: Industry vs. agriculture in a future development for SMEs. *Processes (Basel, Switzerland)*, 7(1), 36. 10.3390/pr7010036

Zhai, Z., Martínez, J. F., Beltran, V., & Martínez, N. L. (2020). Decision support systems for agriculture 4.0: Survey and challenges. *Computers and Electronics in Agriculture*, 170, 105256. 10.1016/j.compag.2020.105256

Compilation of References

Abdel Bari, E. M. (2016). Qatar's vision of global problems, challenges and solutions. *Qscience Proceedings*, 2016(4), 9.

Abdirad, M., & Krishnan, K. (2021). Industry 4.0 in logistics and supply chain management: A systematic literature review. *Engineering Management Journal*, 33(3), 187–201. 10.1080/10429247.2020.1783935

Abdul Aziz, M. F., Bukhari, W. M., Sukhaimie, M. N., Izzuddin, T. A., Norasikin, M. A., Rasid, A. F. A., & Bazilah, N. F. (2021). Development of smart sorting machine using artificial intelligence for chili fertigation industries. *Journal of Automation Mobile Robotics and Intelligent Systems*, 15(4), 44–52.

Abdulai, A. (2022). *Toward digitalization futures in smallholder farming systems in Sub-Sahara Africa: A social practice proposal., 6*. Frontiers. .10.3389/fsufs.2022.866331

Achour, B., Belkadi, M., Filali, I., Laghrouche, M., & Lahdir, M. (2020). Image analysis for individual identification and feeding behaviour monitoring of dairy cows based on Convolutional Neural Networks (CNN). *Biosystems Engineering*, 198, 31–49. 10.1016/j.biosystemseng.2020.07.019

Adam, C., Bevan, D., Gollin, D., & Mkenda, B. (2012). *Transportation costs, food markets and structural transformation: The case of Tanzania (Working Paper)*. International Growth Center.

Adelodun, B., Kim, S. H., Odey, G., & Choi, K. S. (2021). Assessment of environmental and economic aspects of household food waste using a new Environmental-Economic Footprint (EN-EC) index: A case study of Daegu, South Korea. *The Science of the Total Environment*, 776, 145928. 10.1016/j.scitotenv.2021.14592833640543

Aditya, K. S., Jha, G. K., Sonkar, V. K., Saroj, S., Singh, K. M., & Singh, R. K. P. (2019). Determinants of access to and intensity of formal credit: evidence from a survey of rural households in eastern India. *Agricultural economics research review, 32*(conf), 93-102.

Adli, H., Remli, M., Wong, K., Ismail, N., González-Briones, A., Corchado, J., & Mohamad, M. (2023). Recent Advancements and Challenges of AIoT Application in Smart Agriculture: A Review. *Sensors (Basel)*, 23(7), 3752. 10.3390/s2307375237050812

Aggarwal, M., Khullar, V., & Goyal, N. (2023). Exploring classification of rice leaf diseases using machine learning and deep learning. *3rd International Conference on Innovative Practices in Technology and Management (ICIPTM)*, (pp. 1-6). IEEE. 10.1109/ICIPTM57143.2023.10117854

Ahlqvist, V., Norrman, A., & Jahre, M. (2020). *Supply chain risk governance: towards a conceptual multi-level framework.*

Ahmad, M., Hameed, A., Ullah, F., Wahid, I., Rehman, S. U., & Khattak, H. A. (2020). A bio-inspired clustering in mobile adhoc networks for internet of things based on honey bee and genetic algorithm. *Journal of Ambient Intelligence and Humanized Computing*, 11(11), 4347–4361. 10.1007/s12652-018-1141-4

Ahmed, K., Shahidi, T. R., Alam, S. M. I., & Momen, S. (2019, December). Rice leaf disease detection using machine learning techniques. In *2019 International Conference on Sustainable Technologies for Industry 4.0 (STI)* (pp. 1-5). IEEE. 10.1109/STI47673.2019.9068096

Ahmed, J. U., Mpanme, D., Momin, C. C., Shamsan, A. H., & Singh, K. D. (2023). Factors Affecting Access to Agricultural Finance in India: An Empirical Validation from Farmersâ€™ Perspectives. *International Journal of Regional Development*, 10(1), 1–41.

Aiello, G., Giovino, I., Vallone, M., Catania, P., & Argento, A. (2018). A decision support system based on multisensor data fusion for sustainable greenhouse management. *Journal of Cleaner Production*, 172, 4057–4065. 10.1016/j.jclepro.2017.02.197

Ajaykumar, K., & Madhavi, S. (2022). Review on Crop Yield Prediction with Deep Learning and Machine Learning Algorithms. *2022 4th International Conference on Inventive Research in Computing Applications (ICIRCA)*, 903-909. 10.1109/ICIRCA54612.2022.9985016

Akansha, K. (2022). Email Security. *Journal of Image Processing and Intelligent remote sensing,2*(6).

Akkaş, M. A., & Sokullu, R. (2017). An IoT-based greenhouse monitoring system with Micaz motes. *Procedia Computer Science*, 113, 603–608. 10.1016/j.procs.2017.08.300

Alamiedy, T. A., Anbar, M., Alqattan, Z. N., & Alzubi, Q. M. (2020). Anomaly-based intrusion detection system using multi-objective grey wolf optimization algorithm. *Journal of Ambient Intelligence and Humanized Computing*, 11(9), 3735–3756. 10.1007/s12652-019-01569-8

Alazzai, W. K., Obaid, M. K., Abood, B. S. Z., & Jasim, L. (2024). Smart Agriculture Solutions: Harnessing AI and IoT for Crop Management. In *E3S Web of Conferences* (*Vol. 477*, p. 00057). EDP Sciences.

Aldaco, R., Hoehn, D., Laso, J., Margallo, M., Ruiz-Salmón, J., Cristobal, J., Kahhat, R., Villanueva-Rey, P., Bala, A., Batlle-Bayer, L., Fullana-i-Palmer, P., Irabien, A., & Vazquez-Rowe, I. (2020). Food waste management during the COVID-19 outbreak: A holistic climate, economic and nutritional approach. *The Science of the Total Environment*, 742, 140524. 10.1016/j.scitotenv.2020.14052432619842

Aleshkovski, I. (2022). *Social Risks and Negative Consequences of Diffusion of Artificial Intelligence Technologies.* ISTORIYA., 10.18254/S207987840019849-2

Ali, A., Qadri, S., Mashwani, W. K., Brahim Belhaouari, S., Naeem, S., Rafique, S., Jamal, F., Chesneau, C., & Anam, S. (2020). Machine learning approach for the classification of corn seed using hybrid features. *International Journal of Food Properties*, 23(1), 1110–1124. 10.1080/10942912.2020.1778724

Alora, A., & Barua, M. K. (2019). An integrated structural modelling and MICMAC analysis for supply chain disruption risk classification and prioritisation in India. *International Journal of Value Chain Management*, 10(1), 1–25. 10.1504/IJVCM.2019.096538

Al-Raqadi, A. M., Rahim, A. A., Masrom, M., & Al-Riyami, B. S. N. (2017). Cooperation and direction as potential components for controlling stress on the perceptions of improving organisation's performance. *International Journal of System Assurance Engineering and Management*, 8(1), 327–341. 10.1007/s13198-015-0337-7

Alshammari, B. M., & Guesmi, T. (2020). New chaotic sunflower optimization algorithm for optimal tuning of power system stabilizers. *Journal of Electrical Engineering & Technology*, 15(5), 1985–1997. 10.1007/s42835-020-00470-1

Altuntaş, Y., Cömert, Z., & Kocamaz, A. F. (2019). Identification of haploid and diploid maize seeds using convolutional neural networks and a transfer learning approach. *Computers and Electronics in Agriculture*, 163, 104874. 10.1016/j.compag.2019.104874

Amentae, T. K., & Gebresenbet, G. (2021). Digitalization and future agro-food supply chain management: A literature-based implications. *Sustainability (Basel)*, 13(21), 12181. 10.3390/su132112181

Amertet Finecomess, S., Gebresenbet, G., & Alwan, H. M. (2024). Utilizing an Internet of Things (IoT) Device, Intelligent Control Design, and Simulation for an Agricultural System. *IoT.*, 5(1), 58–78. 10.3390/iot5010004

Ammulu, D. (2020). The Impact of Artificial Intelligence in Agriculture. *International Journal of Advanced Research in Science. Tongxin Jishu.* 10.48175/IJARSCT-739

Anderson, J. C., Rungtusanthanam, M., & Schroeder, R. G. (1994). A theory of quality management underlying the Deming management method. *Academy of Management Review*, 19(3), 472–509. 10.2307/258936

Anjikar, A. D. & Jha, V. C. (2022). *Operational Improvement with Advanced Design of Agri Robot in The Era of Agriculture*. Academic Press.

Anoop, V., & Bipin, P. R. (2021). Exploitation whale optimization based optimal offloading approach and topology optimization in a mobile ad hoc cloud environment. *Journal of Ambient Intelligence and Humanized Computing*, 1–20.

Anthimopoulos, M., Christodoulidis, S., Ebner, L., Christe, A., & Mougiakakou, S. (2016). Lung pattern classification for interstitial lung diseases using a deep convolutional neural network. *IEEE Transactions on Medical Imaging*, 35(5), 1207–1216. 10.1109/TMI.2016.253586526955021

Anwer, M., Farooqi, S., & Qureshi, Y. (2015). Agriculture sector performance: An analysis through the role of agriculture sector share in GDP. *Journal of Agricultural Economics. Extension and Rural Development*, 3(3), 270–275.

Araújo, S. O., Peres, R. S., Barata, J., Lidon, F., & Ramalho, J. C. (2021). Characterising the agriculture 4.0 landscape—Emerging trends, challenges and opportunities. *Agronomy (Basel)*, 11(4), 667. 10.3390/agronomy11040667

Arboleda, E. R., Fajardo, A. C., & Medina, R. P. (2018, May). An image processing technique for coffee black beans identification. *In2018 IEEE International Conference on Innovative Research and Development (ICIRD)*, (pp. 1-5). IEEE. 10.1109/ICIRD.2018.8376325

Archana, K. S., & Sahayadhas, A. (2018). Automatic rice leaf disease segmentation using image processing techniques. *Int. J. Eng. Technol, 7*(3.27), 182-185.

Archana, S., & Kumar, P. S. (2023). A Survey on Deep Learning Based Crop Yield Prediction. *Nature Environment & Pollution Technology*, 22(2), 579–592. 10.46488/NEPT.2023.v22i02.004

Arefi, A., Motlagh, A. M., & Teimourlou, R. F. (2011). Wheat class identification using computer vision system and artificial neural networks. *International Agrophysics*, 25(4).

Ariyani, M., Pitoi, M. M., Koesmawati, T. A., Maulana, H., & Endah, E. S. (2020). *Pyrethroid residues on tropical soil of an Indonesian tea plantation: analytical method development, monitoring, and risk assessment* (Vol. 7).

Arora, B., Bhagat, N., Saritha, L. R., & Arcot, S. (2020, February). Rice grain classification using image processing & machine learning techniques. *In 2020 International Conference on Inventive Computation Technologies (ICICT).* IEEE.

Arun, B. (2017). Indian agriculture-status, importance and role in Indian economy. *Journal for Studies in Management and Planning*, 3(12), 212–213.

Ashoka, P., Avinash, G., Apoorva, M., Raj, P., Sekhar, M., Singh, S., Kumar, R., & Singh, B. (2023). *Efficient Detection of Soil Nutrient Deficiencies through Intelligent Approaches.* BIONATURE., 10.56557/bn/2023/v43i21877

Assadzadeh, S., Walker, C. K., & Panozzo, J. F. (2022). Deep learning segmentation in bulk grain images for prediction of grain market quality. *Food and Bioprocess Technology*, 15(7), 1615–1628. 10.1007/s11947-022-02840-1

Atila, Ü., Uçar, M., Akyol, K., & Uçar, E. (2021). Plant leaf disease classification using EfficientNet deep learning model. *Ecological Informatics*, 61, 101182. 10.1016/j.ecoinf.2020.101182

Aukkapinyo, K., Sawangwong, S., Pooyoi, P., & Kusakunniran, W. (2020). Localization and classification of rice-grain images using region proposals-based convolutional neural network. *International Journal of Automation and Computing*, 17(2), 233–246. 10.1007/s11633-019-1207-6

Aydemir, O. (2020). Detection of highly motivated time segments in brain computer interface signals. *Journal of the Institution of Electronics and Telecommunication Engineers*, 66(1), 3–13. 10.1080/03772063.2018.1476190

Ayouba, K., & Vigeant, S. (2020). Can We Really Use Prices to Control Pesticide Use? Results from a Nonparametric Model. Springer. 10.1007/s10666-020-09714-w

Babalık, A., Baykan, Ö. K., İşcan, H., Babaoğlu, İ., & Fındık, O. (2010). Effects of feature selection using binary particle swarm optimization on wheat variety classification. *Advances in Information Technology: 4th International Conference, IAIT 2010*, (pp. 11-17). Springer Berlin Heidelberg. 10.1007/978-3-642-16699-0_2

Bader, F., & Jagtap, S. (2020). Internet of things-linked wearable devices for managing food safety in the healthcare sector. In *Wearable and Implantable Medical Devices* (pp. 229–253). Academic Press. 10.1016/B978-0-12-815369-7.00010-0

Bader, F., & Rahimifard, S. (2020). A methodology for the selection of industrial robots in food handling. *Innovative Food Science & Emerging Technologies*, 64, 102379. 10.1016/j.ifset.2020.102379

Balafas, V., Karantoumanis, E., Louta, M., & Ploskas, N. (2023). Machine learning and deep learning for plant disease classification and detection. *IEEE Access : Practical Innovations, Open Solutions*, 11, 114352–114377. 10.1109/ACCESS.2023.3324722

Balasubramanian, S. (2025). A comprehensive assessment of agricultural sustainability through robotic farming. *Journal ID*, 4625, 6352.

Balasubranian, R. (2020). *Callister's Materials Science and Engineering.* Wiley.

Bali, N., & Singla, A. (2021). Deep learning based wheat crop yield prediction model in Punjab region of North India. *Applied Artificial Intelligence*, 35(15), 1304–1328. 10.1080/08839514.2021.1976091

Bangar, S., Shelar, P., Alhat, P., & Budgujar, R. (2019). *Multipurpose Agri Robot.* Semantic Scholar.

Bayan, B. (2018). Factors influencing extent of adoption of artificial insemination (AI) technology among cattle farmers in Assam. *Indian Journal of Economics and Development*, 14(3), 528–534. 10.5958/2322-0430.2018.00166.X

Behnam, M., & Pourghassem, H. (2017). Spectral correlation power-based seizure detection using statistical multi-level dimensionality reduction and PSO-PNN optimization algorithm. *Journal of the Institution of Electronics and Telecommunication Engineers*, 63(5), 736–753. 10.1080/03772063.2017.1308845

Belan, P. A., de Macedo, R. A. G., Alves, W. A. L., Santana, J. C. C., & Araújo, S. A. (2020). Machine vision system for quality inspection of beans. *International Journal of Advanced Manufacturing Technology*, 111(11-12), 3421–3435. 10.1007/s00170-020-06226-5

Belhadi, A., Kamble, S., Subramanian, N., Singh, R. K., & Venkatesh, M. (2024). Digital capabilities to manage agri-food supply chain uncertainties and build supply chain resilience during compounding geopolitical disruptions. *International Journal of Operations & Production Management.* Advance online publication. 10.1108/IJOPM-11-2022-0737

Benaissa, S., Tuyttens, F. A. M., Plets, D., Trogh, J., Martens, L., Vandaele, L., Joseph, W., & Sonck, B. (2020). Calving and estrus detection in dairy cattle using a combination of indoor localization and accelerometer sensors. *Computers and Electronics in Agriculture*, 168, 105153. 10.1016/j.compag.2019.105153

Bengal, S. W., Dutta, T., & Nayak, C. (2019). S-Transferase Enzyme Activities and Their Correlation with Genotypic Variations Based on GST M1 and GST T1 Loci in Long Term-Pesticide-Exposed Tea Garden. *Toxicology and Environmental Health Sciences*, 11(1), 63–72. 10.1007/s13530-019-0389-1

Benos, L., Tagarakis, A., Dolias, G., Berruto, R., Kateris, D., & Bochtis, D. (2021). Machine Learning in Agriculture: A Comprehensive Updated Review. *Sensors (Basel)*, 21(11), 3758. 10.3390/s2111375834071553

Benyezza, H., Bouhedda, M., Kara, R., & Rebouh, S. (2023). Smart platform based on IoT and WSN for monitoring and control of a greenhouse in the context of precision agriculture. *Internet of Things : Engineering Cyber Physical Human Systems*, 23, 100830. 10.1016/j.iot.2023.100830

Bera, S. (2021). An application of operational analytics: for predicting sales revenue of restaurant. *Machine learning algorithms for industrial applications*, 209-235.

Bestelmeyer, B., Marcillo, G., McCord, S., Mirsky, S., Moglen, G., Neven, L., Peters, D., Sohoulande, C., & Wakie, T. (2020). Scaling Up Agricultural Research With Artificial Intelligence. *IT Professional*, 22(3), 33–38. 10.1109/MITP.2020.2986062

Bhargav, S. (2017). Agricultural Marketing in Growth of Rural India. International Journal of Management. *IT and Engineering*, 7(7), 306–319.

Bharti, V., Bhan, S., Meetali, , & Deepshikha, . (2018). Impact of artificial intelligence for agricultural sustainability. *Journal of Soil and Water Conservation*, 17(4), 393–399. 10.5958/2455-7145.2018.00060.7

Bhat, S. A., & Huang, N. F. (2021). Big data and ai revolution in precision agriculture: Survey and challenges. *IEEE Access : Practical Innovations, Open Solutions*, 9, 110209–110222. 10.1109/ACCESS.2021.3102227

Bhimavarapu, U., Battineni, G., & Chintalapudi, N. (2023). Improved optimization algorithm in LSTM to predict crop yield. *Computers*, 12(1), 10. 10.3390/computers12010010

Bhos, C. D., Deshmukh, S. M., Bhise, P. A., & Avhad, S. B. (2020). Solar Powered Multi-Function Agri-Robot. *International Research Journal of Engineering and Technology (IRJET), 7*(6).

Bi, L., & Hu, G. (2021). A genetic algorithm-assisted deep learning approach for crop yield prediction. *Soft Computing*, 25(16), 10617–10628. 10.1007/s00500-021-05995-9

Bilal, M., & Iqbal, H. M. N. (2019). Sustainable Bioconversion of Food Waste into High-Value Products by Immobilized Enzymes to Meet Bio-Economy Challenges and Opportunities—A Review. *Food Research International*, 123, 226–240. 10.1016/j.foodres.2019.04.06631284972

Birla, R., & Chauhan, A. P. S. (2015). An efficient method for quality analysis of rice using machine vision system. *Journal of advances in Information Technology, 6*(3).

Blakeney, M. (2019). Food loss and waste and food security. In *Food loss and food waste* (pp. 1–26). Edward Elgar Publishing. 10.4337/9781788975391.00006

Bondori, A. (2021). *Modeling farmers ' intention for safe pesticide use: the role of risk perception and use of information sources.* Research gate.

Bonvoisin, T., Utyasheva, L., Knipe, D., Gunnell, D., & Eddleston, M. (2020). *Suicide by pesticide poisoning in India: a review of pesticide regulations and their impact on suicide trends.*

Bottani, E., Murino, T., Schiavo, M., & Akkerman, R. (2019). Resilient food supply chain design: Modelling framework and metaheuristic solution approach. *Computers & Industrial Engineering*, 135, 177–198. 10.1016/j.cie.2019.05.011

Bruzzone, L., & Quaglia, G. (2012). Review article: Locomotion systems for ground mobile robots in unstructured environments. *Mech. Sci.*, 3(2), 49–62. 10.5194/ms-3-49-2012

Bujang, A. S., & Bakar, B. H. A. (2019). *Agriculture 4.0: Data-driven approach to galvanize Malaysia's agro-food sector development. FFTC Agriculture Policy Platform*. FFTCAP.

Bünger, L. (2021). *Robotic waste sorting*. Worcester Polytechnic Institute.

Bute, P. V., Deshmukh, S., Rai, G., Patil, C., & Deshmukh, V. (2018). Design and Fabrication of Multipurpose Agro System. *International Journal of Emerging Trends in Engineering Research*.

Butuner, R., Cinar, I., Taspinar, Y. S., Kursun, R., Calp, M. H., & Koklu, M. (2023). Classification of deep image features of lentil varieties with machine learning techniques. *European Food Research and Technology*, 249(5), 1303–1316. 10.1007/s00217-023-04214-z

Căescu, Ş. C., & Dumitru, I. (2011). Particularities Of The Competitive Environment In The Business To Business Field. *Management & Marketing, 6*(2).

Çakmak, Y. S., & Boyacı, İ. H. (2011). Quality evaluation of chickpeas using an artificial neural network integrated computer vision system. *International Journal of Food Science & Technology*, 46(1), 194–200. 10.1111/j.1365-2621.2010.02482.x

Carayannis, E. G., Rozakis, S., & Grigoroudis, E. (2018). Agri-science to agri-business: The technology transfer dimension. *The Journal of Technology Transfer*, 43(4), 837–843. 10.1007/s10961-016-9527-y

Castillo, O., & Meliif, P. (1970). Automated quality control in the food industry combining artificial intelligence techniques with fractal theory. *WIT Transactions on Information and Communication Technologies, 10.*

Cavazza, A., Mas, F., Campra, M., & Brescia, V. (2023). Artificial intelligence and new business models in agriculture: The "ZERO" case study. *Management Decision*. 10.1108/MD-06-2023-0980

Cedric, L. S., Adoni, W. Y. H., Aworka, R., Zoueu, J. T., Mutombo, F. K., Krichen, M., & Kimpolo, C. L. M. (2022). Crops yield prediction based on machine learning models: Case of West African countries. *Smart Agricultural Technology*, 2, 100049. 10.1016/j.atech.2022.100049

Chakrabarty, K. 2. (2013). Financial Inclusion in India: Journey so far and way forward. *Key note address at Finance Inclusion Conclave Organised by CNBC TV, 18.*

Chanchaichujit, J., Balasubramanian, S., & Shukla, V. (2024). Barriers to Industry 4.0 technology adoption in agricultural supply chains: A Fuzzy Delphi-ISM approach. *International Journal of Quality & Reliability Management*. 10.1108/IJQRM-07-2023-0222

Chan, L. K., & Wu, M. L. (2002). Quality Function Deployment: A Literature Review. *European Journal of Operational Research*, 143(3), 463–497. 10.1016/S0377-2217(02)00178-9

Chauhan, H., Satapathy, S. & Sahoo, A.K. (2021). Mental stress minimization in farmers: an approach using REBA, PSO, and SA. *Int J Syst Assur EngManag*. .10.1007/s13198-021-01167-y

Chen, X., Xun, Y., Li, W., & Zhang, J. (2010). Combining discriminant analysis and neural networks for corn variety identification. *Computers and Electronics in Agriculture*, 71, S48–S53. 10.1016/j.compag.2009.09.003

Cherkaoui, B., Beni-hssane, A., & Erritali, M. (2020). Variable control chart for detecting black hole attack in vehicular ad-hoc networks. *Journal of Ambient Intelligence and Humanized Computing*, 11(11), 5129–5138. 10.1007/s12652-020-01825-2

Chiasson, M. È., Imbeau, D., Aubry, K., & Delisle, A. (2012). Comparing the results of eight methods used to evaluate risk factors associated with musculoskeletal disorders. *International Journal of Industrial Ergonomics*, 42(5), 478–488. 10.1016/j.ergon.2012.07.003

Chowdary, A. V., Saini, N., Kumar, A., Kumar, S., Ballabh, J., Bhatt, S. S., Bhatt, A., Prakash, S., & Patel, A. (2023). Influence of integrated nutrient management on physiological parameters of lentil (Lens culinaris Medik.). *Plant Science Today*, 10(3), 94–97. 10.14719/pst.2124

Cinar, I., & Koklu, M. (2019). Classification of rice varieties using artificial intelligence methods. *International Journal of Intelligent Systems and Applications in Engineering*, 7(3), 188–194. 10.18201/ijisae.2019355381

Cinar, I., & Koklu, M. (2022). Identification of rice varieties using machine learning algorithms. *Journal of Agricultural Sciences*, 9–9.

Civele, C. (2019). Development of an IOT based tractor tracking device to be used as a precision agriculture tool for Turkey's agricultural tractors. *Sch. J. Agric. Vet. Sci*, 6, 199–203.

Cockburn, I., Henderson, R., & Stern, S. (2018). *The Impact of Artificial Intelligence on Innovation. IRPN: Innovation & Cyberlaw & Policy*. Topic. 10.3386/w24449

Cole, S., & Xiong, W. (2017). Agricultural Insurance and Economic Development. *Annual Review of Economics*, 9(1), 235–262. 10.1146/annurev-economics-080315-015225

Coskun, M., Yildirim, O., & Demir, Y. (2021). *Efficient deep neural network model for classification of grasp types using sEMG signals*. J Ambient Intell Human Comput., 10.1007/s12652-021-03284-9

Cristofoli, N. L., Lima, A. R., Tchonkouang, R. D. N., Quintino, A. C., & Vieira, M. C. (2023). Advances in the Food Packaging Production from Agri-Food Waste and By-Products: Market Trends for a Sustainable Development. *Sustainability (Basel)*, 15(7), 6153. 10.3390/su15076153

Cubric, M. (2020). Drivers, barriers and social considerations for AI adoption in business and management: A tertiary study. *Technology in Society*, 62, 101257. 10.1016/j.techsoc.2020.101257

Dal Mas, F., Massaro, M., Ndou, V., & Raguseo, E. (2023). Blockchain technologies for sustainability in the agrifood sector: A literature review of academic research and business perspectives. *Technological Forecasting and Social Change*, 187, 122155. 10.1016/j.techfore.2022.122155

Damalas, C. A. (2021). Farmers ' intention to reduce pesticide use: the role of perceived risk of loss in the model of the planned behavior theory. Research Gate.

Dara, R., Fard, S., & Kaur, J. (2022). Recommendations for ethical and responsible use of artificial intelligence in digital agriculture. *Frontiers in Artificial Intelligence*, 5, 884192. 10.3389/frai.2022.88419235968036

Das, A. (2020). An appraisal of agribusiness industries in india and their market growth scenario. *Indian Journal of Economics and Development*, 16(2s), 496–499.

David, D. (2023). Weather Based Prediction Models for Disease and Pest Using Machine Learning: A Review. *Asian Journal of Agricultural Extension. Economia e Sociologia*, 41(11), 334–345. 10.9734/ajaees/2023/v41i112290

Deininger, K. (2003). *Land Policies for Growth and Poverty Reduction*. World Bank Publications.

Dekhne, A., Hastings, G., Murnane, J., & Neuhaus, F. (2019). Automation in logistics: Big opportunity, bigger uncertainty. *The McKinsey Quarterly*, 24.

Delgado, J. A., Short, N. M.Jr, Roberts, D. P., & Vandenberg, B. (2019). Big data analysis for sustainable agriculture on a geospatial cloud framework. *Frontiers in Sustainable Food Systems*, 3, 54. 10.3389/fsufs.2019.00054

Dellinger, G., Terfous, A., Garambois, P. A., & Ghenaim, A. (2016). Experimental investigation and performance analysis of Archimedes screw generator. *Journal of Hydraulic Research*, 54(2), 197–209. 10.1080/00221686.2015.1136706

Devaraja, T. S. (2011). *Rural credit in India-An overview of history and perspectives.*

Devi, T. G., Neelamegam, P., & Sudha, S. (2017, September). *Machine vision-based quality analysis of rice grains. In 2017 IEEE international conference on power, control, signals and instrumentation engineering (ICPCSI).* IEEE.

Dewi, T., Risma, P., & Oktarina, Y. (2020). Fruit sorting robot based on color and size for an agricultural product packaging system. *Bulletin of Electrical Engineering and Informatics*, 9(4), 1438–1445. 10.11591/eei.v9i4.2353

Dhaka, V. S., Meena, S. V., Rani, G., Sinwar, D., Ijaz, M. F., & Woźniak, M. (2021). A survey of deep convolutional neural networks applied for prediction of plant leaf diseases. *Sensors (Basel)*, 21(14), 4749. 10.3390/s2114474934300489

Dharani, M. K., Thamilselvan, R., Natesan, P., Kalaivaani, P. C. D., & Santhoshkumar, S. (2021). Review on Crop Prediction Using Deep Learning Techniques. *Journal of Physics Conference Series, 1767*, 012-026. 10.1088/1742-6596/1767/1/012026

Dhawale, K. N. (2019, May 31). Review on IoT Based Smart Agriculture System. *International Journal for Research in Applied Science and Engineering Technology*, 7(5), 4063–4066. 10.22214/ijraset.2019.5684

Dhillon, R., & Moncur, Q. (2023). Small-Scale Farming: A Review of Challenges and Potential Opportunities Offered by Technological Advancements. *Sustainability (Basel)*, 15(21), 15478. 10.3390/su152115478

Dixit, J. (2015). Iris Recognition by Daugman's Algorithm – an Efficient Approach. *Journal of applied Research and Social Sciences.*

Dixit, J. (2015). Iris Recognition by Daugman's Method. *International Journal of Latest Technology in Engineering, Management &. Applied Sciences (Basel, Switzerland)*, 4(6), 90–93.

Drixit, A. (2014). A Review paper on Iris Recognition. *Journal GSD International society for green. Sustainable Engineering and Management*, 1(14), 71–81.

Duong, L. N., Al-Fadhli, M., Jagtap, S., Bader, F., Martindale, W., Swainson, M., & Paoli, A. (2020). A review of robotics and autonomous systems in the food industry: From the supply chains perspective. *Trends in Food Science & Technology*, 106, 355–364. 10.1016/j.tifs.2020.10.028

Durán-Sandoval, D., Durán-Romero, G., & Uleri, F. (2023). How much food loss and waste do countries with problems with food security generate? *Agriculture*, 13(5), 966. 10.3390/agriculture13050966

Durán-Sandoval, D., Uleri, F., Durán-Romero, G., & López, A. M. (2023). Food, Climate Change, and the Challenge of Innovation. *Encyclopedia*, 3(3), 839–852. 10.3390/encyclopedia3030060

Dwijendra, N. (2022). Machine Learning Time Series Models For Tea Pest Helopeltis Infestation In India. *Webology, 19*(2). https://www.webology.org/abstract.php?id=1625

Dwivedi, D. N., & Anand, A. (2021). The Text Mining of Public Policy Documents in Response to COVID-19: A Comparison of the United Arab Emirates and the Kingdom of Saudi Arabia. *Public Governance / Zarządzanie Publiczne, 55*(1), 8-22. 10.15678/ZP.2021.55.1.02

Dwivedi, D. N., Mahanty, G., & Vemareddy, A. (2023). Sentiment Analysis and Topic Modeling for Identifying Key Public Concerns of Water Quality/Issues. In: Harun, S., Othman, I.K., Jamal, M.H. (eds) *Proceedings of the 5th International Conference on Water Resources (ICWR). Lecture Notes in Civil Engineering*. Springer, Singapore. 10.1007/978-981-19-5947-9_28

Dwivedi, D. N., Pandey, A. K., & Dwivedi, A. D. (2023). Examining the emotional tone in politically polarized Speeches in India: An In-Depth analysis of two contrasting perspectives. *SOUTH INDIA JOURNAL OF SOCIAL SCIENCES, 21*(2), 125-136. https://journal.sijss.com/index.php/home/article/view/65

Dwivedi, D. N., & Anand, A. (2021). Trade Heterogeneity in the EU: Insights from the Emergence of COVID-19 Using Time Series Clustering. *Zeszyty Naukowe Uniwersytetu Ekonomicznego w Krakowie*, 3(993), 9–26. 10.15678/ZNUEK.2021.0993.0301

Dwivedi, D. N., & Anand, A. (2022). A Comparative Study of Key Themes of Scientific Research Post COVID-19 in the United Arab Emirates and WHO Using Text Mining Approach. In Tiwari, S., Trivedi, M. C., Kolhe, M. L., Mishra, K., & Singh, B. K. (Eds.), *Advances in Data and Information Sciences. Lecture Notes in Networks and Systems* (Vol. 318). Springer. 10.1007/978-981-16-5689-7_30

Dwivedi, D. N., & Batra, S. (2024). Case Studies in Big Data Analysis: A Novel Computer Vision Application to Detect Insurance Fraud. In Darwish, D. (Ed.), *Big Data Analytics Techniques for Market Intelligence* (pp. 441–450). IGI Global. 10.4018/979-8-3693-0413-6.ch018

Dwivedi, D. N., Batra, S., & Pathak, Y. K. (2024). Enhancing Customer Experience: Exploring Deep Learning Models for Banking Customer Journey Analysis. In Sharma, H., Chakravorty, A., Hussain, S., & Kumari, R. (Eds.), *Artificial Intelligence: Theory and Applications. AITA 2023. Lecture Notes in Networks and Systems* (Vol. 843). Springer. 10.1007/978-981-99-8476-3_39

Dwivedi, D. N., & Gupta, A. (2022). Artificial intelligence-driven power demand estimation and short-, medium-, and long-term forecasting. In *Artificial Intelligence for Renewable Energy Systems* (pp. 231–242). Woodhead Publishing. 10.1016/B978-0-323-90396-7.00013-4

Dwivedi, D. N., & Mahanty, G. (2024). Mental Health in Messages: Unravelling Emotional Patterns Through Advanced Text Analysis. In Rai, M., & Pandey, J. (Eds.), *Using Machine Learning to Detect Emotions and Predict Human Psychology* (pp. 187–208). IGI Global. 10.4018/979-8-3693-1910-9.ch009

Dwivedi, D. N., Mahanty, G., & Vemareddy, A. (2022). How Responsible Is AI?: Identification of Key Public Concerns Using Sentiment Analysis and Topic Modeling. [IJIRR]. *International Journal of Information Retrieval Research*, 12(1), 1–14. 10.4018/IJIRR.298646

Dwivedi, D. N., & Pathak, S. (2022). Sentiment Analysis for COVID Vaccinations Using Twitter: Text Clustering of Positive and Negative Sentiments. In Hassan, S. A., Mohamed, A. W., & Alnowibet, K. A. (Eds.), *Decision Sciences for COVID-19. International Series in Operations Research & Management Science* (Vol. 320). Springer. 10.1007/978-3-030-87019-5_12

Dwivedi, D. N., Wójcik, K., & Vemareddyb, A. (2022). Identification of Key Concerns and Sentiments Towards Data Quality and Data Strategy Challenges Using Sentiment Analysis and Topic Modeling. In Jajuga, K., Dehnel, G., & Walesiak, M. (Eds.), *Modern Classification and Data Analysis. SKAD 2021. Studies in Classification, Data Analysis, and Knowledge Organization*. Springer. 10.1007/978-3-031-10190-8_2

Dwivedi, D., Kapur, P. N., & Kapur, N. N. (2023). Machine Learning Time Series Models for Tea Pest Looper Infestation in Assam, India. In Sharma, A., Chanderwal, N., & Khan, R. (Eds.), *Convergence of Cloud Computing, AI, and Agricultural Science* (pp. 280–289). IGI Global. 10.4018/979-8-3693-0200-2.ch014

Dwivedi, D., & Vemareddy, A. (2023). Sentiment Analytics for Crypto Pre and Post Covid: Topic Modeling. In Molla, A. R., Sharma, G., Kumar, P., & Rawat, S. (Eds.), Lecture Notes in Computer Science: Vol. 13776. *Distributed Computing and Intelligent Technology. ICDCIT 2023*. Springer. 10.1007/978-3-031-24848-1_21

Dwivedi, Y., Hughes, L., Ismagilova, E., Aarts, G., Coombs, C., Crick, T., & Duan, Y. (2019). Artificial Intelligence (AI): Multidisciplinary perspectives on emerging challenges, opportunities, and agenda for research, practice and policy. *International Journal of Information Management*. 10.1016/j.ijinfomgt.2019.08.002

Eastwood, C. R. (2008). *Innovative precision dairy systems: A case study of farmer learning and technology co-development*. University of Melbourne, Melbourne School of Land and Environment.

Ebikade, E. O., Sadula, S., Gupta, Y., & Vlachos, D. G. (2021). A Review of Thermal and Thermocatalytic Valorization of Food Waste. *Green Chemistry*, 23(8), 2806–2833. 10.1039/D1GC00536G

Echelmeyer, W., Kirchheim, A., & Wellbrock, E. (2008, September). Robotics-logistics: Challenges for automation of logistic processes. In *2008 IEEE International Conference on Automation and Logistics* (pp. 2099-2103). IEEE. 10.1109/ICAL.2008.4636510

Ekanayake, & R. D. Nawarathna. (2021). Novel deep learning approaches for crop leaf disease classification: A review, *International Research Conference on Smart Computing and Systems Engineering (SCSE)*, (pp. 49-52). IEEE. .10.1109/SCSE53661.2021.9568324

El Hoummaidi, L., Larabi, A., & Alam, K. (2021). Using unmanned aerial systems and deep learning for agriculture mapping in Dubai. *Heliyon*, 7(10), e08154. 10.1016/j.heliyon.2021.e0815434703924

Elavarasan, D., & Durai Raj Vincent, P. M. (2021). Fuzzy deep learning-based crop yield prediction model for sustainable agronomical frameworks. *Neural Computing & Applications*, 33(20), 1–20. 10.1007/s00521-021-05950-7

Elbasi, E., Mostafa, N., AlArnaout, Z., Zreikat, A., Cina, E., Varghese, G., Shdefat, A., Topcu, A., Abdelbaki, W., Mathew, S., & Zaki, C. (2023). Artificial Intelligence Technology in the Agricultural Sector: A Systematic Literature Review. *IEEE Access : Practical Innovations, Open Solutions*, 11, 171–202. 10.1109/ACCESS.2022.3232485

Eldridge, S. M., Yin Chan, K., Donovan, N. J., Saleh, F., Orr, L., & Barchia, I. (2018). Agronomic and economic benefits of green-waste compost for peri-urban vegetable production: Implications for food security. *Nutrient Cycling in Agroecosystems*, 111(2-3), 155–173. 10.1007/s10705-018-9931-9

Eli-Chukwu, N. (2019). Applications of Artificial Intelligence in Agriculture: A Review. *Engineering, Technology &. Applied Scientific Research*, 9(4), 4377–4383. 10.48084/etasr.2756

Elyasi, A., & Teimoury, E. (2022). Applying Critical Systems Practice meta-methodology to improve sustainability in the rice supply chain of Iran. *Sustainable Production and Consumption*, 35, 453–468. 10.1016/j.spc.2022.11.024

Emmi, L., Fernández, R., & Guerrero, J. (2023). Editorial: Robotics for smart farms. *Frontiers in Robotics and AI*, 9, 1113440. 10.3389/frobt.2022.111344036686213

Emmi, L., Gonzalez-de-Soto, M., Pajares, G., & Gonzalez-de-Santos, P. (2014). New trends in robotics for agriculture: Integration and assessment of a real fleet of robots. *TheScientificWorldJournal*, 2014, 2014. 10.1155/2014/40405925143976

Erdoğan, M. (2022). Assessing farmers' perception to Agriculture 4.0 technologies: A new interval-valued spherical fuzzy sets based approach. *International Journal of Intelligent Systems*, 37(2), 1751–1801. 10.1002/int.22756

FAO. (1999). *Better Practices in Agricultural Lending*. UN Food and Agricultural Organization.

FAO. (2017). *Addressing Land Tenure Issues in Development Cooperation*. Food and Agriculture Organization.

Fazel-Niari, Z., Afkari-Sayyah, A. H., Abbaspour-Gilandeh, Y., Herrera-Miranda, I., Hernández-Hernández, J. L., & Hernández-Hernández, M. (2022). Quality assessment of components of wheat seed using different classifications models. *Applied Sciences (Basel, Switzerland)*, 12(9), 4133. 10.3390/app12094133

Ferentinos, K. P. (2018). Deep learning models for plant disease detection and diagnosis. *Computers and Electronics in Agriculture*, 145, 311–318. 10.1016/j.compag.2018.01.009

Filho, F., Heldens, W., Kong, Z., & Lange, E. (2019). Drones: Innovative Technology for Use in Precision Pest Management. *Journal of Economic Entomology*, 113(1), 1–25. 10.1093/jee/toz26831811713

Finagina, O., Prodanova, L., Zinchenko, O., Buriak, I., Gavrylovskyi, O., & Khoroshun, Y. (2021). Improving investment management in agribusiness. *Estudios de Economía Aplicada*, 39(5). 10.25115/eea.v39i5.4981

Fountas, S., Espejo-García, B., Kasimati, A., Mylonas, N., & Darra, N. (2020). The Future of Digital Agriculture: Technologies and Opportunities. *IT Professional*, 22(1), 24–28. 10.1109/MITP.2019.2963412

Galati, F., & Bigliardi, B. (2019). Industry 4.0: Emerging themes and future research avenues using a text mining approach. *Computers in Industry*, 109, 100–113. 10.1016/j.compind.2019.04.018

Gallia, G. K. N., & Kephaliacos, C. (2021). Ecological - economic modeling of pollination complexity and pesticide use in agricultural crops. *Journal of Bioeconomics*, 23(3), 297–323. 10.1007/s10818-021-09317-9

Ganeshkumar, C., Jena, S., Sivakumar, A., & Nambirajan, T. (2021). Artificial intelligence in agricultural value chain: Review and future directions. *Journal of Agribusiness in Developing and Emerging Economies*. 10.1108/JADEE-07-2020-0140

Gao, D., Sun, Q., Hu, B., & Zhang, S. (2020). A Framework for Agricultural Pest and Disease Monitoring Based on Internet-of-Things and Unmanned Aerial Vehicles. *Sensors (Basel)*, 20(5), 1487. 10.3390/s2005148732182732

García, M., Candelo-Becerra, J. E., & Hoyos, F. E. (2019). Quality and defect inspection of green coffee beans using a computer vision system. *applied sciences, 9*(19), 4195.

Gardeazabal, A., Lunt, T., Jahn, M. M., Verhulst, N., Hellin, J., & Govaerts, B. (2023). Knowledge management for innovation in agri-food systems: A conceptual framework. *Knowledge Management Research and Practice*, 21(2), 303–315. 10.1080/14778238.2021.1884010

Ghadge, A., Er Kara, M., Moradlou, H., & Goswami, M. (2020). The impact of Industry 4.0 implementation on supply chains. *Journal of Manufacturing Technology Management*, 31(4), 669–686. 10.1108/JMTM-10-2019-0368

Gill, A., Kaur, T., & Devi, Y. K. (2023). 38 A review for identification and detection of plant disease using machine learning. *Recent Advances in Computing Sciences:Proceedings of RACS 2022*, 216. Research Gate.

Giunta, F., Pruneddu, G., & Motzo, R. (2019). Grain yield and grain protein of old and modern durum wheat cultivars grown under different cropping systems. *Field Crops Research*, 230, 107–120. 10.1016/j.fcr.2018.10.012

Gomez, N. A., Sanchez, K., & Arguello, H. (2019, April). Non-destructive method for classification of cocoa beans from spectral information. *In 2019 XXII Symposium on Image, Signal Processing and Artificial Vision (STSIVA)*. IEEE.

Gong, L., Yu, M., Jiang, S., Cutsuridis, V., & Pearson, S. (2021). Deep learning based prediction on greenhouse crop yield combined TCN and RNN. *Sensors (Basel)*, 21(13), 4537. 10.3390/s2113453734283083

González-Nucamendi, A., Noguez, J., Neri, L., Robledo-Rella, V., & García-Castelán, R. M. (2023). Predictive analytics study to determine undergraduate students at risk of dropout. *Frontiers in Education*, 8, 1244686. 10.3389/feduc.2023.1244686

Government of India. (2023). *Annual Report Ministry of Food Processing Industries*. https://www.mofpi.gov.in/documents/reports/annual-report.

Gowda, V., Prabhu, M., Ramesha, M., Kudari, J. M., & Samal, A. (2021, November 1). Smart Agriculture and Smart Farming using IoT Technology. *Journal of Physics: Conference Series*, 2089(1), 012038–012038. 10.1088/1742-6596/2089/1/012038

Guan, T. (2021). Research on the application of robot welding technology in modern architecture. *International Journal of System Assurance Engineering and Management*, 1-10.

Guevara-Hernandez, F., & Gil, J. G. (2011). A machine vision system for classification of wheat and barley grain kernels. *Spanish Journal of Agricultural Research*, 9(3), 672–680. 10.5424/sjar/20110903-140-10

Gujjar, H. S., & Siddappa, D. M. (2013). A method for identification of basmati rice grain of india and its quality using pattern classification. *International Journal of Engineering Research and Applications*, 3(1), 268–273.

Gul, F., Rahiman, W., Alhady, S. S., Ali, A., Mir, I., & Jalil, A. (2021). Meta-heuristic approach for solving multi-objective path planning for autonomous guided robot using PSO–GWO optimization algorithm with evolutionary programming. *Journal of Ambient Intelligence and Humanized Computing*, 12(7), 7873–7890. 10.1007/s12652-020-02514-w

Gund, V. D. (2023). PIR Sensor-Based Arduino Home Security System. *Journal of Instrumentation and Innovation Sciences*, 8(3), 33–37.

Gunes, E., & Movassaghi, H. (2017). Agricultural Credit Market and Farmers 'Response: A Case Study of Turkey. *Turkish Journal of Agriculture-Food Science and Technology*, 5(1), 84–92. 10.24925/turjaf.v5i1.84-92.951

Guo, X., Ge, Y., Liu, F., & Yang, J. (2023). Identification of maize and wheat seedlings and weeds based on deep learning. *Frontiers in Earth Science (Lausanne)*, 11, 1146558. 10.3389/feart.2023.1146558

Guo, Y., Zhang, J., Yin, C., Hu, X., Zou, Y., Xue, Z., & Wang, W. (2020). Plant disease identification based on deep learning algorithm in smart farming. *Discrete Dynamics in Nature and Society*, 2020, 1–11. 10.1155/2020/2479172

Gupta, A. (2021). Understanding Consumer Product Sentiments through Supervised Models on Cloud: Pre and Post COVID. *Webology, 18*(1). .10.14704/WEB/V18I1/WEB18097

Gupta, A., Dwivedi, D. N., & Shah, J. (2023). Applying Artificial Intelligence on Investigation. In: *Future of Business and Finance.* Springer, Singapore. 10.1007/978-981-99-2571-1_9

Gupta, A., Dwivedi, D. N., & Shah, J. (2023). Applying Machine Learning for Effective Customer Risk Assessment. In: *Future of Business and Finance.* Springer, Singapore. 10.1007/978-981-99-2571-1_6

Gupta, A., Dwivedi, D. N., & Shah, J. (2023). Artificial Intelligence-Driven Effective Financial Transaction Monitoring. In: *Future of Business and Finance*. Springer, Singapore. 10.1007/978-981-99-2571-1_7

Gupta, A., Dwivedi, D. N., & Shah, J. (2023). Data Organization for an FCC Unit. In *Future of Business and Finance*. Springer, Singapore. 10.1007/978-981-99-2571-1_4

Gupta, A., Dwivedi, D. N., & Shah, J. (2023). Ethical Challenges for AI-Based Applications. In: *Future of Business and Finance.* Springer, Singapore. 10.1007/978-981-99-2571-1_10

Gupta, A., Dwivedi, D. N., & Shah, J. (2023). Financial Crimes Management and Control in Financial Institutions. In: *Artificial Intelligence Applications in Banking and Financial Services*. pringer, Singapore. 10.1007/978-981-99-2571-1_2

Gupta, A., Dwivedi, D. N., & Shah, J. (2023). Machine Learning-Driven Alert Optimization. In: *Future of Business and Finance.* Springer, Singapore. 10.1007/978-981-99-2571-1_8

Gupta, A., Dwivedi, D. N., & Shah, J. (2023). Overview of Money Laundering. In: Artificial Intelligence Applications in Banking and Financial Services. *Future of Business and Finance*. Springer, Singapore. 10.1007/978-981-99-2571-1_1

Gupta, A., Dwivedi, D. N., & Shah, J. (2023). Overview of Technology Solutions. In: *Future of Business and Finance*. Springer, Singapore. 10.1007/978-981-99-2571-1_3

Gupta, A., Dwivedi, D. N., & Shah, J. (2023). Planning for AI in Financial Crimes. In: *Future of Business and Finance*. Springer, Singapore. 10.1007/978-981-99-2571-1_5

Gupta, A., Dwivedi, D. N., & Shah, J. (2023). Setting up a Best-In-Class AI-Driven Financial Crime Control Unit (FCCU). In: *Future of Business and Finance*. Springer, Singapore. 10.1007/978-981-99-2571-1_11

Gupta, A., Dwivedi, D.N. & Jain, A. (2021). Threshold fine-tuning of money laundering scenarios through multi-dimensional optimization techniques. *Journal of Money Laundering Control*. 10.1108/JMLC-12-2020-0138

Gupta, V., & Mittal, M. (2019). QRS complex detection using STFT, chaos analysis, and PCA in standard and real-time ECG databases. *Journal of The Institution of Engineers (India): Series B, 100*(5), 489-497.

Gupta, S. (2023). *Artificial Intelligence in Smart Agriculture: Applications and Challenges*. CURRENT APPLIED SCIENCE AND TECHNOLOGY. 10.55003/cast.2023.254427

Gupta, S., Singh, N., & Kashyap, S. (2023). Management of agriculture through artificial intelligence in adverse climatic conditions. *Environment Conservation Journal*, 24(2), 408–412. Advance online publication. 10.36953/ECJ.23602638

Gupta, V., Mittal, M., & Mittal, V. (2020). R-peak detection-based chaos analysis of ECG signal. *Analog Integrated Circuits and Signal Processing*, 102(3), 479–490. 10.1007/s10470-019-01556-1

Gutierrez-Galan, D., Dominguez-Morales, J. P., Cerezuela-Escudero, E., Rios-Navarro, A., Tapiador-Morales, R., Rivas-Perez, M., Dominguez-Morales, M., Jimenez-Fernandez, A., & Linares-Barranco, A. (2018). Embedded neural network for real-time animal behavior classification. *Neurocomputing*, 272, 17–26. 10.1016/j.neucom.2017.03.090

Hajjaj, S. S. H., & Sahari, K. S. M. (2016, December). Review of agriculture robotics: Practicality and feasibility. In *2016 IEEE International Symposium on Robotics and Intelligent Sensors (IRIS)* (pp. 194-198). IEEE. 10.1109/IRIS.2016.8066090

Halli Umar, M. (2022). Nanotechnology in IoT Security. *Journal of Nanoscience. Nanoengineering & Applications*, 12(3), 11–16.

Han, X., Liu, F., He, X., & Ling, F. (2022). Research on rice yield prediction model based on deep learning. *Computational Intelligence and Neuroscience*, 2022, 1–9.. 10.1155/2022/192256135515497

Han, Y. Y., Wang, K. Y., Liu, Z. Q., Zhang, Q., Pan, S. H., Zhao, X. Y., & Wang, S. F. (2017). A crop trait information acquisition system with multitag-based identification technologies for breeding precision management. *Computers and Electronics in Agriculture*, 135, 71–80. 10.1016/j.compag.2017.01.004

Harika, S., Sandhyarani, G., Sagar, D., & Reddy, G. V. S. (2023). *Image-based black gram crop disease detection. International Conference on Inventive Computation Technologies (ICICT)*, Lalitpur, Nepal. 10.1109/ICICT57646.2023.10134027

Harisha, B. N. (2017). Agricultural credit in India: Issues and challenges, 2(6).

Hassan, M., Iqbal, S., Garg, H., Hassan, S. G., & Yan, Y. (2023). An Integrated FCEM-AHP Approach for Borrower's Satisfaction and Perception Analysis of Microfinance Institution. *CMES-Computer Modeling in Engineering & Sciences, 134*(1).

Hatefi, S. M., Moshashaee, S. M., & Mahdavi, I. (2019). A bi-objective programming model for reliable supply chain network design under facility disruption. *International Journal of Integrated Engineering*, 11(6), 80–92. 10.30880/ijie.2019.11.06.009

He, J., Wang, J., He, D., Dong, J., & Wang, Y. (2011). The design and implementation of an integrated optimal fertilization decision support system. *Mathematical and Computer Modelling*, 54(3-4), 1167–1174. 10.1016/j.mcm.2010.11.050

Hellas, A., Ihantola, P., Petersen, A., Ajanovski, V., Gutica, M., Hynninen, T., Knutas, A., Leinonen, J., Messom, C., & Liao, S. Predicting academic performance: a systematic literature review. *Proceedings Companion of the 23rd Annual ACM Conference on Innovation and Technology in Computer Science Education*. (175-199). ACM. 10.1145/3293881.3295783

Heshmati, F. (2021). *Simultaneous multi- determination of pesticide residues in black tea leaves and infusion: a risk assessment study.*

Hidayat, W. (2021). Analisis Pengaruh Kualitas Pelayanan Go Food Terhadap Tingkat KepuasanKonsumen. *Frontiers in Neuroscience*, 14(1).

Hilborn, R. (2017). Food for Thought Are MPAs effective? –. *ICES Journal of Marine Science*, 75.

Hohma, E., & Lütge, C. (2023). From Trustworthy Principles to a Trustworthy Development Process: The Need and Elements of Trusted Development of AI Systems. *AI*, 4(4), 904–926. Advance online publication. 10.3390/ai4040046

Holzinger, A., Saranti, A., Angerschmid, A., Retzlaff, C., Gronauer, A., Pejaković, V., Medel-Jiménez, F., Krexner, T., Gollob, C., & Stampfer, K. (2022). Digital Transformation in Smart Farm and Forest Operations Needs Human-Centered AI: Challenges and Future Directions. *Sensors (Basel)*, 22(8), 3043. Advance online publication. 10.3390/s2208304335459028

Hossain, M. D., Aftab, A., Al Imam, M. H., Mahmud, I., Chowdhury, I. A., Kabir, R. I., & Sarker, M. (2018). Prevalence of work related musculoskeletal disorders (WMSDs) and ergonomic risk assessment among readymade garment workers of Bangladesh: A cross sectional study. *PLoS One*, 13(7), e0200122. 10.1371/journal.pone.020012229979734

Hossen, M. A., Talukder, M. R. A., Al Mamun, M. R., Rahaman, H., Paul, S., Rahman, M. M., Miaruddin, M., Ali, M. A., & Islam, M. N. (2020). Mechanization status, promotional activities and government strategies of Thailand and Vietnam in comparison to Bangladesh. *AgriEngineering*, 2(4), 489–510. 10.3390/agriengineering2040033

Hota, J., & Verma, V. (2022). Challenges to Adoption of Digital Agriculture in India. *2022 International Conference on Maintenance and Intelligent Asset Management (ICMIAM)*, (pp. 1-6). IEEE. 10.1109/ICMIAM56779.2022.10147002

Huang, N. F., Chou, D. L., & Lee, C. A. (2019, July). Real-time classification of green coffee beans by using a convolutional neural network. In *2019 3rd International Conference on Imaging, Signal Processing and Communication (ICISPC).* IEEE. 10.1109/ICISPC.2019.8935644

Huang, M., He, C., Zhu, Q., & Qin, J. (2016). Maize seed variety classification using the integration of spectral and image features combined with feature transformation based on hyperspectral imaging. *Applied Sciences (Basel, Switzerland)*, 6(6), 183. 10.3390/app6060183

Hussain, E., Mahanta, L. B., Das, C. R., & Talukdar, R. K. (2020). A comprehensive study on the multi-class cervical cancer diagnostic prediction on Pap smear images using a fusion-based decision from ensemble deep convolutional neural network. *Tissue and Cell, 65*, 101347.10.1016/j.tice.2020.101347

Iannone, L., Palmisano, I., & Fanizzi, N. (2007). An algorithm based on counterfactuals for concept learning in the semantic web. *Applied Intelligence*, 26(2), 139–159. 10.1007/s10489-006-0011-5

Ingram, J., & Maye, D. (2020). What are the implications of digitalisation for agricultural knowledge? *Frontiers in Sustainable Food Systems*, 4, 66. 10.3389/fsufs.2020.00066

Irani, Z., Sharif, A. M., Lee, H., Aktas, E., Topaloğlu, Z., van't Wout, T., & Huda, S. (2018). Managing food security through food waste and loss: Small data to big data. *Computers & Operations Research*, 98, 367–383. 10.1016/j.cor.2017.10.007

Irfan, M., Wang, M., & Akhtar, N. (2019). Impact of IT capabilities on supply chain capabilities and organizational agility: A dynamic capability view. *Operations Management Research : Advancing Practice Through Research*, 12(3-4), 113–128. 10.1007/s12063-019-00142-y

Islam, T., Sah, M., Baral, S., & Choudhury, R. R. (2018, April). A faster technique on rice disease detectionusing image processing of affected area in agro-field. In *2018 Second International Conference on Inventive Communication and Computational Technologies (ICICCT)* (pp. 62-66). IEEE. 10.1109/ICICCT.2018.8473322

Ivanov, D. (2020). Predicting the impacts of epidemic outbreaks on global supply chains: A simulation-based analysis on the coronavirus outbreak (COVID-19/SARS-CoV-2) case. *Transportation Research Part E, Logistics and Transportation Review*, 136, 101922. 10.1016/j.tre.2020.10192232288597

Jagtap, S., & Duong, L. N. K. (2019). Improving the new product development using big data: A case study of a food company. *British Food Journal*, 121(11), 2835–2848. 10.1108/BFJ-02-2019-0097

Jagtap, S., Rahimifard, S., & Duong, L. N. (2022). Real-time data collection to improve energy efficiency: A case study of food manufacturer. *Journal of Food Processing and Preservation*, 46(8), e14338. 10.1111/jfpp.14338

Jain, R., Malangmeih, L., Raju, S. S., Srivastava, S. K., Immaneulraj, K., & Kaur, A. P. (2018). Optimization techniques for crop planning: A review. *Indian Journal of Agricultural Sciences*, 88(12), 1826–1835. 10.56093/ijas.v88i12.85423

Jamaludin, H., Elmaky, H. S. E., & Sulaiman, S. (2022). The future of food waste: Application of circular economy. *Energy Nexus*, 7, 100098. 10.1016/j.nexus.2022.100098

Janandi, R., & Cenggoro, T. W. (2020, August). An Implementation of Convolutional Neural Network for Coffee Beans Quality Classification in a Mobile Information System. In *2020 International Conference on Information Management and Technology (ICIMTech) IEEE*. IEEE. 10.1109/ICIMTech50083.2020.9211257

Javaid, M., Haleem, A., Khan, I. H., & Suman, R. (2023). Understanding the potential applications of Artificial Intelligence in Agriculture Sector. *Advanced Agrochem*, 2(1), 15–30. 10.1016/j.aac.2022.10.001

Javaid, M., Haleem, A., Singh, R. P., & Suman, R. (2022). Enhancing smart farming through the applications of Agriculture 4.0 technolgies. *Int. J. Intell. Netw.*, 3, 150–164.

Jena, L., Behera, S. K., & Sethy, P. K. (2021, May). Identification of wheat grain using geometrical feature and machine learning. *In 2021 2nd International conference for emerging technology (INCET)*. IEEE.

Jeong, S., Ko, J., & Yeom, J. M. (2022). Predicting rice yield at pixel scale through synthetic use of crop and deep learning models with satellite data in South and North Korea. *The Science of the Total Environment*, 802, 149726. 10.1016/j.scitotenv.2021.14972634464811

Jha, P. (2023). Comparative analysis of crop diseases detection using machine learning algorithm. *Third International Conference on Artificial Intelligence and Smart Energy (ICAIS)*. IEEE. 10.1109/ICAIS56108.2023.10073831

Jha, K., Doshi, A., Patel, P., & Shah, M. (2019). A comprehensive review on automation in agriculture using artificial intelligence. *Artificial Intelligence in Agriculture*, 2, 1–12. 10.1016/j.aiia.2019.05.004

Jia, C., Hew, K., Bai, S. & Huang, W. (2021). Adaptation of a conventional flipped course to an online flipped format during the Covid-19 pandemic: Student learning performance and engagement. *Journal of Research on Technology in Education, 54*(2). 10.1080/15391523.2020.1847220

Jiang, B., He, J., Yang, S., Fu, H., Li, T., Song, H., & He, D. (2019). Fusion of machine vision technology and AlexNet-CNNs deep learning network for the detection of postharvest apple pesticide residues. *Artificial Intelligence in Agriculture*, 1, 1–8. 10.1016/j.aiia.2019.02.001

Jiang, Y., Hao, K., Cai, X., & Ding, Y. (2018). An improved reinforcement-immune algorithm for agricultural resource allocation optimization. *Journal of Computational Science*, 27, 320–328. 10.1016/j.jocs.2018.06.011

Jin, X., Zhang, J., Kong, J., Su, T., & Bai, Y. (2022). A Reversible Automatic Selection Normalization (RASN) Deep Network for Predicting in the Smart Agriculture System. *Agronomy (Basel)*, 12(3), 591. 10.3390/agronomy12030591

Jogekar, R. N., & Tiwari, N. (2021). A review of deep learning techniques for identification and diagnosis of plant leaf disease. *Smart Trends in Computing and Communications:Proceedings of SmartCom 2020*, (pp. 435-441). Springer. 10.1007/978-981-15-5224-3_43

Jordan, M., & Mitchell, T. (2015). Machine learning: Trends, perspectives, and prospects. *Science*, 349(6245), 255–260. 10.1126/science.aaa841526185243

Joshi, A., Guevara, D., & Earles, M. (2022). Standardizing and Centralizing Datasets for Efficient Training of Agricultural Deep Learning Models. *Plant Phenomics (Washington, D.C.)*, 5, 0084. 10.34133/plantphenomics.008437680999

Joshi, S., Singh, R. K., & Sharma, M. (2023). Sustainable agri-food supply chain practices: Few empirical evidences from a developing economy. *Global Business Review*, 24(3), 451–474. 10.1177/0972150920907014

Joshua, S. V., Priyadharson, A. S. M., Kannadasan, R., Khan, A. A., Lawanont, W., Khan, F. A., & Ali, M. J. (2022). Crop yield prediction using machine learning approaches on a wide spectrum. *Computers, Materials & Continua*, 72(3), 5663–5679. 10.32604/cmc.2022.027178

Kabir, A. (2018). Application Of G.I.S In Site Selection For Solid Waste Collection Points In Kofar Kaura New Layout Katsina. *World Development*, 1(1).

Kagalkar, S., Agashe, A., Paralkar, T. A., & Deogaonkar, A. (2023). Narrative Synthesis of the Economic Impact of Agricultural Supply Chain and Distribution Networks on Output: Economic Impact of Agricultural Supply Chains. International Journal of Professional Business Review: Int. *J. Prof. Bus. Rev.*, 8(6), 22.

Kambo, R., & Yerpude, A. (2014). *Classification of basmati rice grain variety using image processing and principal component analysis*. arXiv preprint arXiv,1405,7626.

Kamilaris, A., & Prenafeta-Boldú, F. X. (2018). Deep learning in agriculture: A survey. *Computers and Electronics in Agriculture*, 147, 70–90. 10.1016/j.compag.2018.02.016

Kang, K., & Wang, S. (2018). Analyze and predict student dropout from online programs. In *Proceedings of the 2nd International Conference on Compute and Data Analysis* (pp. 6-12). ACM. 10.1145/3193077.3193090

Karabegović, I., Karabegović, E., Mahmić, M., & Husak, E. J. A. I. P. E. (2015). The application of service robots for logistics in manufacturing processes. *Advances in Production Engineering & Management*, 10(4), 185–194. 10.14743/apem2015.4.201

Karale Aishwarya A. (2023). Smart Billing Cart Using RFID, YOLO and Deep Learning for Mall Administration. *International Journal of Instrumentation and Innovation Sciences,8*(2).

Kaur, H., & Singh, B. (2013). Classification and grading rice using multi-class SVM. *International Journal of Scientific and Research Publications*, 3(4), 1–5.

Kaya, E., & Saritas, İ. (2019). Towards a real-time sorting system: Identification of vitreous durum wheat kernels using ANN based on their morphological, colour, wavelet and gaborlet features. *Computers and Electronics in Agriculture*, 166, 105016. 10.1016/j.compag.2019.105016

Kazi K S L. (2022). IoT-based weather Prototype using WeMos. *Journal of Control and Instrumentation Engineering*, 9(1), 10 - 22.

Kazi K. (2022a). Model for Agricultural Information system to improve crop yield using IoT. *Journal of open Source development*, *9*(2), 16 – 24.

Kazi Kutubuddin S. L. (2022). Business Mode and Product Life Cycle to Improve Marketing in Healthcare Units. *E-Commerce for future & Trends*, *9*(3), 1-9.

Kazi Kutubuddin, S. L. (2022). A novel Design of IoT based 'Love Representation and Remembrance' System to Loved One's. *Gradiva Review Journal*, 8(12), 377–383.

Kazi Kutubuddin, S. L. (2022). Predict the Severity of Diabetes cases, using K-Means and Decision Tree Approach. *Journal of Advances in Shell Programming*, 9(2), 24–31.

Kazi, K. (2017). Lassar Methodology for Network Intrusion Detection. *Scholarly Research Journal for Humanity science and English Language*, *4*(24), 6853 - 6861.

Kazi, K. (2022). Hybrid optimum model development to determine the Break. *Journal of Multimedia Technology & Recent Advancements*, 9(2), 24–32.

Kazi, K. (2024). AI-Driven IoT (AIIoT) in Healthcare Monitoring. In Nguyen, T., & Vo, N. (Eds.), *Using Traditional Design Methods to Enhance AI-Driven Decision Making* (pp. 77–101). IGI Global. 10.4018/979-8-3693-0639-0.ch003

Kazi, K. S. (2017). Significance And Usage Of Face Recognition System. *Scholarly Journal For Humanity Science and English Language*, 4(20), 4764–4772.

Kazi, K. S. (2022). IoT-Based Healthcare Monitoring for COVID-19 Home Quarantined Patients. *Recent Trends in Sensor Research & Technology*, 9(3), 26–32.

Kazi, K. S. (2023). Detection of Malicious Nodes in IoT Networks based on Throughput and ML. *Journal of Electrical and Power System Engineering*, 9(1), 22–29.

Kazi, K. S. (2024b). Computer-Aided Diagnosis in Ophthalmology: A Technical Review of Deep Learning Applications. In Garcia, M., & de Almeida, R. (Eds.), *Transformative Approaches to Patient Literacy and Healthcare Innovation* (pp. 112–135). IGI Global.10.4018/979-8-3693-3661-8.ch006

Kazi, K. S. L. (2018). Significance of Projection and Rotation of Image in Color Matching for High-Quality Panoramic Images used for Aquatic study. *International Journal of Aquatic Science*, 09(02), 130–145.

Kazi, S. (2023). Fruit Grading, Disease Detection, and an Image Processing Strategy. *Journal of Image Processing and Artificial Intelligence*, 9(2), 17–34.

Kazi, S. S. L. (2023). Electronics with Artificial Intelligence Creating a Smarter Future: A Review. *Journal of Communication Engineering and Its Innovations*, 9(3), 38–42.

Kazi, S. S. L. (2023). IoT Changing the Electronics Manufacturing Industry. *Journal of Analog and Digital Communications*, 8(3), 13–17.

Kazi, S. S. L. (2023). IoT in Electrical Vehicle: A Study. *Journal of Control and Instrumentation Engineering*, 9(3), 15–21.

Kazi, V. (2023). Deep Learning, YOLO and RFID based smart Billing Handcart. *Journal of Communication Engineering & Systems*, 13(1), 1–8.

Khan, N. M., Ray, R. L., Sargani, G. R., Ihtisham, M., Khayyam, M., & Ismail, S. (2021, April 27). Current Progress and Future Prospects of Agriculture Technology: Gateway to Sustainable Agriculture. *Sustainability (Basel)*, 13(9), 4883–4883. 10.3390/su13094883

Kim, S., Yoo, E., & Kim, S. (2023). *A Study on the Prediction of University Dropout Using Machine Learning*. arXiv preprint arXiv:2310.10987. DOI:/arXiv.2310.1098710.48550

Kireyenka, N. (2021). *Models of agrarian business development in international practice., 59*, 22-40. .10.29235/1817-7204-2021-59-1-22-40

Koklu, M., & Ozkan, I. A. (2020). Multiclass classification of dry beans using computer vision and machine learning techniques. *Computers and Electronics in Agriculture*, 174, 105507. 10.1016/j.compag.2020.105507

Kosgiker, G. M. (2018). Machine Learning- Based System, Food Quality Inspection and Grading in Food industry. *International Journal of Food and Nutritional Sciences*, 11(10), 723–730.

Kottalil, A. M., Krishnan, B. B., Anto, A., & Alex, B. (2016). Automatic sorting machine. *Journal for Research*, 2(04).

Kozłowski, M., Górecki, P., & Szczypiński, P. M. (2019). Varietal classification of barley by convolutional neural networks. *Biosystems Engineering*, 184, 155–165. 10.1016/j.biosystemseng.2019.06.012

Kramer, K. (2020). Seasonal abundance of tea mosquito bug, Looper Antoni Signoret infesting name. *Journal of Entomology and Zoology Studies*, 8(6), 2006–2009. christianaid.org.uk. 10.22271/j.ento.2020.v8.i6aa.8118

Krizhevsky, A., Sutskever, I., & Hinton, G. E. (2012). Imagenet classification with deep convolutional neural networks. *Advances in Neural Information Processing Systems*, 25.

Krizhevsky, A., Sutskever, I., & Hinton, G. E. (2017). ImageNet classification with deep convolutional neural networks. *Communications of the ACM*, 60(6), 84–90. 10.1145/3065386

KS. (2023). IoT based Healthcare system for Home Quarantine People. *Journal of Instrumentation and Innovation sciences, 8*.

KSSL. (2023). IoT in the Electric Power Industry. *Journal of Controller and Converters*, 8(3), 1–7.

Kuchekar, N. A., & Yerigeri, V. V. (2018). Rice grain quality grading using digital image processing techniques. *IOSR J Electronics Communication Eng, 13*(3).

Kumar, G., Ramachandran, K., Sharma, S., Ramesh, R., Qureshi, K., & Ganesh, K. (2023). AI-Assisted Resource Allocation for Improved Business Efficiency and Profitability. *2023 3rd International Conference on Advance Computing and Innovative Technologies in Engineering (ICACITE)*, 54-58. IEEE. 10.1109/ICACITE57410.2023.10182679

Kumar, J., Chawla, R., Katiyar, D., Chouriya, A., Nath, D., Sahoo, S., Ali, A., & Singh, B. (2023). Optimizing Irrigation and Nutrient Management in Agriculture through Artificial Intelligence Implementation. *International Journal of Environment and Climate Change*. .10.9734/ijecc/2023/v13i103077

Kumar, K., & Kayalvizhi, S. (2015). Real Time Industrial Colour Shape and Size Detection System Using Single Board. *International Journal of Science, Engineering and Technology Research (IJSETR), 4*(3).

Kumar, T., & Prakash, N. (2020). ADOPTION OF AI IN AGRICULTURE: THE GAME-CHANGER FOR INDIAN FARMERS. *Proceedings of the International Conferences on ICT, Society and Human Beings (ICT 2020), Connected Smart Cities (CSC 2020) and Web Based Communities and Social Media (WBC 2020)*. ICT. 10.33965/ict_csc_wbc_2020_202008C025

Kumar, A., Das, N., Sharma, H., & Singh, R. (2024). Impact of Agrochemicals on the Environment and Ecological Alternatives. In Karmaoui, A. (Ed.), *Water-Soil-Plant-Animal Nexus in the Era of Climate Change* (pp. 317–328). IGI Global. 10.4018/978-1-6684-9838-5.ch015

Kumar, A., Kumar, S., Juyal, R., Sharma, H., & Bisht, M. (2023). Nanoscience in Agricultural Steadiness. In Dar, G. H., Bhat, R. A., & Mehmood, M. A. (Eds.), *Microbiomes for the Management of Agricultural Sustainability* (pp. 285–296). Springer. 10.1007/978-3-031-32967-8_17

Kumar, A., Mathur, N. N., Varghese, M., Mohan, D., Singh, J. K., & Mahajan, P. (2005). Effect of tractor driving on hearing loss in farmers in India. *American Journal of Industrial Medicine*, 47(4), 341–348. 10.1002/ajim.2014315776468

Kumar, A., Sharma, H., Shakeb, A., & Joshi, M. (2022). Environmental Sustainability: An Emerging Concept. *International Year of Millets*, 2023, 29.

Kumar, A., Varghese, M., & Mohan, D. (2000). Equipment-related injuries in agriculture: An international perspective. *Injury Control and Safety Promotion*, 7(3), 1–12. 10.1076/1566-0974(200009)7:3;1-N;FT175

Kumar, A., Varghese, M., Mohan, D., Mahajan, P., Gulati, P., & Kale, S. (1999). Effect of whole-body vibration on the low back: A study of tractor-driving farmers in north India. *Spine*, 24(23), 2506. 10.1097/00007632-19991201 0-0001310626314

Kumar, M., Raut, R. D., Jagtap, S., & Choubey, V. K. (2023). Circular economy adoption challenges in the food supply chain for sustainable development. *Business Strategy and the Environment*, 32(4), 1334–1356. 10.1002/bse.3191

Kumar, N. R., & Nagabhooshanam, E. (2021). EKF with Artificial Bee Colony for Precise Positioning of UAV Using Global Positioning System. *Journal of the Institution of Electronics and Telecommunication Engineers*, 67(1), 60–73. 10.1080/03772063.2018.1528186

Kumar, R., Sachan, S., Kumar, R., Parmar, K., Mishra, A. K., Kumar, A., & Mishra, A. K. (2023). Unveiling Use of Information and Communication Technology (ICT) in Western Uttar Pradesh, India: The Farmers' Perspective. *Journal of Global Innovations in Agricultural Sciences*, 11(4), 521–531. 10.22194/JGIAS/23.1197

Kumar, R., Singh, A., Yadav, R. B., Kumar, A., Kumar, S., Shahi, U. P., & Singh, A. P. (2017). Growth, development and yield response of rice (Oryza sativa L.) as influenced by efficient nitrogen management under subtropical climatic condition. *Journal of Pharmacognosy and Phytochemistry*, 6(6S), 791–797.

Kumar, S., Kumar, V., & Sharma, R. K. (2015). Sugarcane yield forecasting using artificial neural network models. *International Journal of Artificial Intelligence & Applications*, 6(5), 51–68. 10.5121/ijaia.2015.6504

Kumar, S., Raut, R., Nayal, K., Kraus, S., Yadav, V., & Narkhede, B. (2021). To identify industry 4.0 and circular economy adoption barriers in the agriculture supply chain by using ISM-ANP. *Journal of Cleaner Production*, 293, 126023. 10.1016/j.jclepro.2021.126023

Kumar, T. A., Lalswamy, B., Raghavendra, Y., Usharani, S. G., & Usharani, S. (2018). Intelligent food and grain storage management system for the warehouse and cold storage. *Int. J. Res. Eng. Sci. Manag*, 1(4), 130–132.

Kumawat, L., Jat, L., Kumar, A., Yadav, M., Ram, B., & Dudwal, B. L. (2022). Effect of organic nutrient sources on growth, yield attributes and yield of wheat under rice (Oryza sativa L.) wheat (Triticum aestivum L.) cropping system. *The Pharma Innovation Journal*, 11(2), 1618–1623.

Kunhimohammed, C. K., Saifudeen, K. M., Sahna, S., Gokul, M. S., & Abdulla, S. U. (2015). *Automatic color sorting machine using TCS230 color sensor and PIC microcontroller*. Research Gate.

Kuo, S., Merkley, G., & Liu, C. (2000). Decision support for irrigation project planning using a genetic algorithm. *Agricultural Water Management*, 45(3), 243–266. 10.1016/S0378-3774(00)00081-0

Kutter, T., Tiemann, S., Siebert, R., & Fountas, S. (2011). The role of communication and co-operation in the adoption of precision farming. *Precision Agriculture*, 12(1), 2–17. 10.1007/s11119-009-9150-0

Kutub, K. (2022). Reverse Engineering's Neural Network Approach to human brain. *Journal of Communication Engineering & Systems*, 12(2), 17–24.

Kutubuddin, K. (2022). Big data and HR Analytics in Talent Management: A Study. *Recent Trends in Parallel Computing*, 9(3), 16–26.

Kutubu, K. (2022). Detection of Malicious Nodes in IoT Networks based on packet loss using ML. *Journal of Mobile Computing, Communication & mobile. Networks*, 9(3), 9–16.

Kwakye, M. O., Mengistie, B., & Ofosu, J. (2019). Pesticide registration, distribution and use practices. *Environment, Development and Sustainability*, 21(6), 2647–2671. 10.1007/s10668-018-0154-7

Laabassi, K., Belarbi, M. A., Mahmoudi, S., Mahmoudi, S. A., & Ferhat, K. (2021). Wheat varieties identification based on a deep learning approach. *Journal of the Saudi Society of Agricultural Sciences*, 20(5), 281–289. 10.1016/j.jssas.2021.02.008

Lainjo, B. (2023). Mitigating Academic Institution Dropout Rates with Predictive Analytics Algorithms. *International Journal of Education, Teaching, and Social Sciences.*

Lainjo, B., & Tsmouche, H. (2023). Impact of Artificial Intelligence On Higher Learning Institutions. *International Journal of Education, Teaching, and Social Sciences.*

Lamsal, R., Karthikeyan, P., Otero, P., & Ariza, A. (2023). Design and Implementation of Internet of Things (IoT) Platform Targeted for Smallholder Farmers: From Nepal Perspective. *Agriculture*, 13(10), 1900. Advance online publication. 10.3390/agriculture13101900

Latino, M. E., Menegoli, M., Lazoi, M., & Corallo, A. (2022). Voluntary traceability in food supply chain: A framework leading its implementation in Agriculture 4.0. *Technological Forecasting and Social Change*, 178, 121564. 10.1016/j.techfore.2022.121564

Latu, . (2021). Sustainable Development: The Role of Gis and Visualisation. *The Electronic Journal on Information Systems in Developing Countries. EJISDC*, 38(5), 1–17.

Layaq, M. W., Goudz, A., Noche, B., & Atif, M. (2019). Blockchain technology as a risk mitigation tool in supply chain. *Int. J. Transp. Eng. Technol*, 5(3), 50–59. 10.11648/j.ijtet.20190503.12

Lee, H., Moon, A., Moon, K., & Lee, Y. (2017, July). Disease and pest prediction IoT system in orchard: A preliminary study. In *2017 Ninth International Conference on Ubiquitous and Future Networks (ICUFN)* (pp. 525-527). IEEE. 10.1109/ICUFN.2017.7993840

Lee, T. Y., Reza, M. N., Chung, S. O., Kim, D. U., Lee, S. Y., & Choi, D. H. (2023). Application of fuzzy logics for smart agriculture: A review. *Precision Agriculture*, 5, 1.

Leong, Y., Lim, E., Subri, N., & Jalil, N. (2023). Transforming Agriculture: Navigating the Challenges and Embracing the Opportunities of Artificial Intelligence of Things. *2023 IEEE International Conference on Agrosystem Engineering, Technology & Applications (AGRETA)*, (pp. 142-147). IEEE. 10.1109/AGRETA57740.2023.10262747

Le, V. S., Lesueur, D., Herrmann, L., Hudek, L., Ngoc, L., & Lambert, Q. (2021). Sus- tainable tea production through agroecological management practices in Vietnam: A review. *Environmental Sustainability*, 4(0123456789), 589–604. 10.1007/s42398-021-00182-w

Liakos, K., Busato, P., Moshou, D., Pearson, S., & Bochtis, D. (2018). Machine Learning in Agriculture: A Review. *Sensors (Basel)*, 18(8), 2674. 10.3390/s1808267430110960

Li, G., & Buckle, P. (1998, October). A practical method for the assessment of work-related musculoskeletal risks-Quick Exposure Check (QEC). *Proceedings of the Human Factors and Ergonomics Society Annual Meeting*, 42(19), 1351–1355. 10.1177/154193129804201905

Lin, A. Y. C., Huang, S. T. Y., & Wahlqvist, M. L. (2009). Waste management to improve food safety and security for health advancement. *Asia Pacific Journal of Clinical Nutrition*, 18(4), 538–545.19965345

Linaza, M., Posada, J., Bund, J., Eisert, P., Quartulli, M., Döllner, J., Pagani, A., Olaizola, I., Barriguinha, A., Moysiadis, T., & Lucat, L. (2021). Data-Driven Artificial Intelligence Applications for Sustainable Precision Agriculture. *Agronomy (Basel)*. 10.3390/agronomy11061227

Lingwal, S., Bhatia, K. K., & Tomer, M. S. (2021). Image-based wheat grain classification using convolutional neural network. *Multimedia Tools and Applications*, 80(28-29), 1–25. 10.1007/s11042-020-10174-3

Linh, T. N., Long, H. T., Chi, L. V., Tam, L. T., & Lebailly, P. (2019). Access to rural credit markets in developing countries, the case of Vietnam: A literature review. *Sustainability (Basel)*, 11(5), 1468. 10.3390/su11051468

Lin, J. (1991). Education and Innovation Adoption in Agriculture: Evidence from Hybrid Rice in China. *American Journal of Agricultural Economics*, 73(3), 713–723. 10.2307/1242823

Lin, Q., Li, T., Shakeel, P. M., & Samuel, R. D. J. (2021). Advanced artificial intelligence in heart rate and blood pressure monitoring for stress management. *Journal of Ambient Intelligence and Humanized Computing*, 12(3), 3329–3340. 10.1007/s12652-020-02650-3

Lin, W., Liu, C., & Li, M. (2022). Research On Factors Influencing the Efficiency of Agricultural Science and Technology Innovation. *Proceedings of the International Conference on Information Economy, Data Modeling and Cloud Computing, ICIDC 2022*, Qingdao, China. 10.4108/eai.17-6-2022.2322808

Li, S., Xu, L. D., & Zhao, S. (2015). The internet of things: A survey. *Information Systems Frontiers*, 17(2), 243–259. 10.1007/s10796-014-9492-7

Liu, D., Mishra, A. K., & Yu, Z. (2016). Evaluating uncertainties in multi-layer soil moisture estimation with support vector machines and ensemble Kalman filtering. *Journal of Hydrology (Amsterdam)*, 538, 243–255. 10.1016/j.jhydrol.2016.04.021

Liu, Y., Ji, D., Zhang, L., An, J., & Sun, W. (2021). Rural Financial Development Impacts on Agricultural Technology Innovation: Evidence from China. *International Journal of Environmental Research and Public Health*, 18(3), 1110. 10.3390/ijerph1803111033513778

Li, X. (2023). Application of Intelligent Algorithm in the Research of Logistics Distribution Positioning System. *2023 Asia-Europe Conference on Electronics, Data Processing and Informatics (ACEDPI)*, (pp. 460-464). IEEE. 10.1109/ACEDPI58926.2023.00094

Liyakat, K. K. S. (2023). *Detecting Malicious Nodes in IoT Networks Using Machine Learning and Artificial Neural Networks. 2023 International Conference on Emerging Smart Computing and Informatics (ESCI)*, Pune, India.10.1109/ESCI56872.2023.10099544

Liyakat, K. K. S. (2023). Machine Learning Approach Using Artificial Neural Networks to Detect Malicious Nodes in IoT Networks. In Shukla, P. K., Mittal, H., & Engelbrecht, A. (Eds.), *Computer Vision and Robotics. CVR 2023. Algorithms for Intelligent Systems*. Springer. 10.1007/978-981-99-4577-1_3

Liyakat, K. K. S. (2024). Machine Learning Approach Using Artificial Neural Networks to Detect Malicious Nodes in IoT Networks. In Udgata, S. K., Sethi, S., & Gao, X. Z. (Eds.), *Intelligent Systems. ICMIB 2023. Lecture Notes in Networks and Systems* (Vol. 728). Springer. https://link.springer.com/chapter/10.1007/978-981-99-3932-9_12, 10.1007/978-981-99-3932-9_12

Liyakat, K. S. (2023). Integrating IoT and Mechanical Systems in Mechanical Engineering Applications. *Journal of Mechanisms and Robotics*, 8(3), 1–6.

Liyakat, S. S. (2023). IoT Based Arduino-Powered Weather Monitoring System. *Journal of Telecommunication Study*, 8(3), 25–31. 10.46610/JTC.2023.v08i03.005

Lo Turco, V., Potortì, A. G., Tropea, A., Dugo, G., & Di Bella, G. (2020). Element Analysis of Dried Figs (Ficus carica L.) from the Mediterranean Areas. *Journal of Food Composition and Analysis*, 90, 103503. 10.1016/j.jfca.2020.103503

Lombardi, M., & Fargnoli, M. (2018). Prioritization of hazards by means of a QFD-based procedure. *Safety and Security Studies, 163.*

Lopes de Sousa Jabbour, A. B., Jabbour, C. J. C., Godinho Filho, M., & Roubaud, D. (2018). Industry 4.0 and the circular economy: A proposed research agenda and original roadmap for sustainable operations. *Annals of Operations Research*, 270(1-2), 273–286. 10.1007/s10479-018-2772-8

Lourens, A., & Bleazard, D. (2016). Applying predictive analytics in identifying students at risk: A case study. *South African Journal of Higher Education*, 30(2), 129–142. 10.20853/30-2-583

Lu, J., Tan, L., & Jiang, H. (2021). Review on convolutional neural network (CNN) applied to plant leaf disease classification. *Agriculture*, 11(8), 707. 10.3390/agriculture11080707

Łukowska, A., Tomaszuk, P., Dzierżek, K., & Magnuszewski, Ł. (2019, May). Soil sampling mobile platform for Agriculture 4.0. In *2019 20th International Carpathian Control Conference (ICCC)* (pp. 1-4). IEEE.

Luo, L., Shen, G. Q., Xu, G., Liu, Y., & Wang, Y. (2019, March). Stakeholder-associated supply chain risks and their interactions in a prefabricated building project in Hong Kong. *Journal of Management Engineering*, 35(2), 05018015. 10.1061/(ASCE)ME.1943-5479.0000675

Ma, C., Li, Y., Yin, G., & Ji, J. (2012, July). The monitoring and information management system of pig breeding process based on internet of things. In *2012 Fifth International conference on information and computing science* (pp. 103-106). IEEE. 10.1109/ICIC.2012.61

Machado, C. G., Winroth, M. P., & Ribeiro da Silva, E. H. D. (2020). Sustainable manufacturing in Industry 4.0: An emerging research agenda. *International Journal of Production Research*, 58(5), 1462–1484. 10.1080/00207543.2019.1652777

Mahajan, S., & Kaur, S. (2014). Quality analysis of Indian basmati rice grains using top-hat transformation. *International Journal of Computer Applications*, 94(15), 42–48. 10.5120/16423-6085

Mahale, B., & Korde, S. (2014, April). Rice quality analysis using image processing techniques. In *International Conference for Convergence for Technology.* IEEE. 10.1109/I2CT.2014.7092300

Mahanty, G., Dwivedi, D. N., & Gopalakrishnan, B. N. (2021). The Efficacy of Fiscal Vs Monetary Policies in the Asia-Pacific Region: The St. Louis Equation Revisited. *Vision (Basel)*, (November). 10.1177/09722629211054148

Mahlein, A. K., Oerke, E. C., Steiner, U., & Dehne, H. W. (2012). Recent advances in sensing plant diseases for precision crop protection. *European Journal of Plant Pathology*, 133(1), 197–209. 10.1007/s10658-011-9878-z

Mahmoodabadi, M. J., & Shahangian, M. M. (2019). A new multi-objective artificial bee colony algorithm for optimal adaptive robust controller design. *Journal of the Institution of Electronics and Telecommunication Engineers*, 1–14.

Majumdar, J., Naraseeyappa, S., & Ankalaki, S. (2017). Analysis of agriculture data using data mining techniques: Application of big data. *Journal of Big Data*, 4(1), 20. 10.1186/s40537-017-0077-4

Mallik, P., & Ghosh, T. (2008). Impact of climate on tea production: A study of the Dooars region in India. *Theor. Apple. Climatol.* 10.1007/s00704-021-03848-x

Manavalan, E., & Jayakrishna, K. (2019). A review of Internet of Things (IoT) embedded sustainable supply chain for industry 4.0 requirements. *Computers & Industrial Engineering*, 127, 925–953. 10.1016/j.cie.2018.11.030

Manderson, A., & Hunt, C. (2013). Introducing the Agri-Rover: An Autonomous on-the-go sensing rover for science and farming. In *Proceedings of the 26th Annual FLRC Workshop Held at Massey University.* Massey University.

Mao, J., Zhu, Q., Wachenheim, C. J., & Hanson, E. D. (2020). A Credit Scoring Model for Farmer Lending Decisions in Rural China. *International Journal of Agricultural Management*, 8(4), 134–141.

Maraveas, C. (2022). Incorporating Artificial Intelligence Technology in Smart Greenhouses: Current State of the Art. *Applied Sciences (Basel, Switzerland)*, 13(1), 14. 10.3390/app13010014

Margiotta, M., & Baudoin, W. (1999). *Environmental issues and peri-urban agriculture in Mauritania.* Peri-Urban Agriculture in Sub-Saharan African.

Marinoudi, V., Sørensen, C., Pearson, S., & Bochtis, D. (2019). Robotics and labour in agriculture. A context consideration. *Biosystems Engineering*, 184, 111–121. 10.1016/j.biosystemseng.2019.06.013

Marra, M., Pannell, D., & Ghadim, A. (2003). The economics of risk, uncertainty and learning in the adoption of new agricultural technologies: Where are we on the learning curve? *Agricultural Systems*, 75(2-3), 215–234. 10.1016/S0308-521X(02)00066-5

Massari, S., Principato, L., Antonelli, M., & Pratesi, C. A. (2022). Learning from and designing after pandemics. CEASE: A design thinking approach to maintaining food consumer behaviour and achieving zero waste. *Socio-Economic Planning Sciences*, 82, 101143. 10.1016/j.seps.2021.101143

Maulik, S., De, A., & Iqbal, R. (2012, July). Work related musculoskeletal disorders among medical laboratory technicians. In *2012 Southeast Asian Network of Ergonomics Societies Conference (SEANES)* (pp. 1-6). IEEE. 10.1109/SEANES.2012.6299585

Medar, et.al., (2020). A Survey on Data Mining Techniques for Crop Yield Prediction. *International Journal of Advance Research in Computer Science and Management Studies*, *2*(9).

Mehrabi, Z., McDowell, M., Ricciardi, V., Levers, C., Martinez, J., Mehrabi, N., Wittman, H., Ramankutty, N., & Jarvis, A. (2020). The global divide in data-driven farming. *Nature Sustainability*, 4(2), 154–160. 10.1038/s41893-020-00631-0

Mekonnen, N., & Burton, S. (2020). Machine learning techniques in wireless sensor network based precision agriculture. *Journal of the Electrochemical Society*, 167(3), 037522. 10.1149/2.0222003JES

Mentsiev, A., & Gatina, F. (2021). Data analysis and digitalisation in the agricultural industry. *IOP Conference Series. Earth and Environmental Science*, 677(3), 032101. 10.1088/1755-1315/677/3/032101

Mentzer, J. T., DeWitt, W., Keebler, J. S., Min, S., Nix, N. W., Smith, C. D., & Zacharia, Z. G. (2001). Defining supply chain management. *Journal of Business Logistics*, 22(2), 1–25. 10.1002/j.2158-1592.2001.tb00001.x

Menychtas, D., Glushkova, A., & Manitsaris, S. (2020). Analyzing the kinematic and kinetic contributions of the human upper body's joints for ergonomics assessment. *Journal of Ambient Intelligence and Humanized Computing*, 11(12), 6093–6105. 10.1007/s12652-020-01926-y

Miao, A., Zhuang, J., Tang, Y., He, Y., Chu, X., & Luo, S. (2018). Hyperspectral image-based variety classification of waxy maize seeds by the t-SNE model and Procrustes analysis. *Sensors (Basel)*, 18(12), 4391. 10.3390/s1812439130545028

Migunov, R., Syutkina, A., Zaruk, N., Kolomeeva, E., & Arzamastseva, N. (2023). *Global Challenges and Barriers to Sustainable Economic Growth in the Agribusiness Sector*. WSEAS TRANSACTIONS ON BUSINESS AND ECONOMICS., 10.37394/23207.2023.20.85

Milesi, C., & Kukunuri, M. (2022). Crop yield estimation at gram panchayat scale by integrating field, weather and satellite data with crop simulation models. *Photonirvachak (Dehra Dun)*, 50(2), 239–255. 10.1007/s12524-021-01372-z

Mishra, D., Muduli, K., Raut, R., Narkhede, B., Shee, H., & Jana, S. (2023). Challenges Facing Artificial Intelligence Adoption during COVID-19 Pandemic: An Investigation into the Agriculture and Agri-Food Supply Chain in India. *Sustainability (Basel)*, 15(8), 6377. 10.3390/su15086377

Misra, N., Dixit, Y., Al-Mallahi, A., Bhullar, M., Upadhyay, R., & Martynenko, A. (2020). IoT, Big Data, and Artificial Intelligence in Agriculture and Food Industry. *IEEE Internet of Things Journal*, 9(9), 6305–6324. 10.1109/JIOT.2020.2998584

Mohan, D., & Raj, M. G. (2020). Quality analysis of rice grains using ANN and SVM. *Journal of Critical Reviews, 7*(1), 395-402.

Mondal, B., Bhushan, M., Dawar, I., Rana, M., Negi, A., & Layek, S. (2023). Crop disease prediction using machine learning and deep learning: an exploratory study. *International Conference on Sustainable Computing and Smart Systems (ICSCSS)*, (pp. 278-283). IEEE. 10.1109/ICSCSS57650.2023.10169612

Moon, M.-H., & Kim, G. (2023). Predicting University Dropout Rates Using Machine Learning Algorithms. *Journal of Economics and Finance Education*, 32(2), 57–68. 10.46967/jefe.2023.32.2.57

Mulani, A. (2019). Effect of Rotation and Projection on Real time Hand Gesture Recognition system for Human Computer Interaction. *Journal of The Gujrat Research Society*, 21(16), 3710–3718.

Mulani, A. O., & Patil, R. M. (2023). Discriminative Appearance Model For Robust Online Multiple Target Tracking. *Telematique*, 22(1), 24–43.

Müller, J. M., & Voigt, K. I. (2018). The impact of industry 4.0 on supply chains in engineer-to-order industries-an exploratory case study. *IFAC-PapersOnLine*, 51(11), 122–127. 10.1016/j.ifacol.2018.08.245

N., P. S., Chaudhari, S., Barde, S., & Devices. (2021). Hubungan Tingkat Pengetahuan dan KebiasaanKonsumsi Junk Food dengan Status Gizi. *Frontiers in Neuroscience, 14*(1).

NABARD. (2020). *National Bank for Agriculture and Rural Development Report*. National Bank for Agriculture and Rural Development.

Nabavi-Pelesaraei, A., Rafiee, S., Mohtasebi, S., Hosseinzadeh-Bandbafha, H., & Chau, K. (2018). Integration of artificial intelligence methods and life cycle assessment to predict energy output and environmental impacts of paddy production. *The Science of the Total Environment*, 631-632, 1279–1294. 10.1016/j.scitotenv.2018.03.08829727952

Narendra, V. G., Krishanamoorthi, M., Shivaprasad, G., Amitkumar, V. G., & Kamath, P. (2021). Almond kernel variety identification and classification using decision tree. *Journal of Engineering Science and Technology*, 16(5), 3923–3942.

Nasirahmadi, A., & Behroozi-Khazaei, N. (2013). Identification of bean varieties according to color features using artificial neural network. *Spanish Journal of Agricultural Research*, 11(3), 670–677. 10.5424/sjar/2013113-3942

Navarro, E. D. M., Costa, N., & Pereira, A. (2020, July 29). A Systematic Review of IoT Solutions for Smart Farming. *Sensors (Basel)*, 20(15), 4231–4231. 10.3390/s2015423132751366

Nedumaran, G., & Manida, M. (2019). Impact of FDI in agriculture sector in India: Opportunities and challenges. *International Journal of Recent Technology and Engineering*, 8(3), 380–383.

Neeraja, P., Kumar, R. G., Kumar, M. S., Liyakat, K. K. S., & Vani, M. S. (2024). *DL-Based Somnolence Detection for Improved Driver Safety and Alertness Monitoring. 2024 IEEE International Conference on Computing, Power and Communication Technologies (IC2PCT)*. Greater Noida. https://ieeexplore.ieee.org/document/10486714, 10.1109/IC2PCT60090.2024.10486714

Neethirajan, S. (2023). Harnessing the Metaverse for Livestock Welfare: Unleashing Sensor Data and Navigating Ethical Frontiers. *Preprints*, 2023040409.

Negia, C. S., & Kumarb, S. (2020). *Promoting agro-based industry in India (issues and challenges)*. INTERNATIONAL JOURNALOF TRADE & COMMERCE-IIARTC. 10.46333/ijtc/9/1/27

Nerkar, P., & Shinde, S. (2023). Monitoring Fresh Fruit and Food Using Iot and Machine Learning to Improve Food Safety and Quality. *Tuijin Jishu/Journal of Propulsion Technology*, *44*(3), 2927 – 2931.

Nerkar, P. M., & Dhaware, B. U. (2023). Predictive Data Analytics Framework Based on Heart Healthcare System (HHS) Using Machine Learning. *Journal of Advanced Zoology,44*(2).

Nguyen, T. T. (2021). Enhanced sunflower optimization for placement distributed generation in distribution system. *Iranian Journal of Electrical and Computer Engineering*, 11(1), 107. 10.11591/ijece.v11i1.pp107-113

Nikita, K., & Supriya, J. (2020). Design of Vehicle system using CAN Protocol. *International Journal for Research in Applied Science and Engineering Technology*, 8(V), 1978–1983. 10.22214/ijraset.2020.5321

Nikolaidis, P., Ismail, M., Shuib, L., Khan, S., & Dhiman, G. (2022). *Predicting Student Attrition in Higher Education through the Determinants of Learning Progress: A Structural Equation Modelling Approach. Sustainability*. MDPI. https://www.mdpi.com/2071-1050/14/20/1358410.3390/su142013584

Nikolidakis, S. A., Kandris, D., Vergados, D. D., & Douligeris, C. (2015). Energy efficient automated control of irrigation in agriculture by using wireless sensor networks. *Computers and Electronics in Agriculture*, 113, 154–163. 10.1016/j.compag.2015.02.004

Niwa, H. (2021). Detection of organic tea farms based on the density of spider webs using aerial photoFigurey with an unmanned aerial vehicle (UAV). *Landscape and Ecological Engineering*, 17(4), 541–546. 10.1007/s11355-021-00454-x

Nosirov, K., Begmatov, S., Arabboev, M., Kuchkorov, T., Chedjou, J. C., Kyamakya, K., & Abhiram, K. (2020). The greenhouse control based-vision and sensors. In *Developments of Artificial Intelligence Technologies in Computation and Robotics: Proceedings of the 14th International FLINS Conference (FLINS 2020)* (pp. 1514-1523). World Scientific. 10.1142/9789811223334_0181

Nugegoda, N. (2016). *Rice grains classification using image processing technics*. Academic Press.

Nurmalitasari, A. (2023). *The Predictive Learning Analytics for Student Dropout Using Data Mining Technique: A Systematic Literature Review. Advances in Technology Transfer Through IoT and IT Solutions*. Springer. . (9-17). 10.1007/978-3-031-25178-8_2

Nyéki, A., & Neményi, M. (2022). Crop Yield Prediction in Precision Agriculture. *Agronomy (Basel)*, 12(10), 2460. 10.3390/agronomy12102460

Nyoni, J. (2021). Achieving Economic Growth and Economic Development through Agriculture Transformation: a Framework for Enhancing Agriculture Production and Agriculture Producvity. Academia Letters, 2-5.

Oikonomidis, A., Catal, C., & Kassahun, A. (2022). Hybrid deep learning-based models for crop yield prediction. *Applied Artificial Intelligence*, 36(1), 2031822. 10.1080/08839514.2022.2031823

Oikonomidis, A., Catal, C., & Kassahun, A. (2023). Deep learning for crop yield prediction: A systematic literature review. *New Zealand Journal of Crop and Horticultural Science*, 51(1), 1–26. 10.1080/01140671.2022.2032213

Okengwu, U., Onyejegbu, L., Oghenekaro, L., Musa, M., & Ugbari, A. (2023). *Environmental and ethical negative implications of AI in agriculture and proposed mitigation measures*. Scientia Africana. 10.4314/sa.v22i1.13

Olofintuyi, S. S., Olajubu, E. A., & Olanike, D. (2023). An ensemble deep learning approach for predicting cocoa yield. *Heliyon*, 9(4), e15245. 10.1016/j.heliyon.2023.e1524537089327

Onishchuk, M. O. (2020). *Opto-mechanical sorting of municipal solid waste* [Doctoral dissertation, BHTY].

Oqaidi, K., Aouhassi, S., & Mansouri, K. (2022). A Comparison between Using Fuzzy Cognitive Mapping and Machine Learning to Predict Students' Performance in Higher Education *2022 IEEE 3rd International Conference on Electronics, Control, Optimization and Computer Science (ICECOCS)*. IEEE. https://ieeexplore.ieee.org/document/9983470/10.1109/ICECOCS55148.2022.9983470

Osumba, J. J., Recha, J. W., & Oroma, G. W. (2021). Transforming agricultural extension service delivery through innovative bottom–up climate-resilient agribusiness farmer field schools. *Sustainability (Basel)*, 13(7), 3938. 10.3390/su13073938

OuYang, A. G., Gao, R. J., Sun, X. D., Pan, Y. Y., & Dong, X. L. (2010, August). An automatic method for identifying different variety of rice seeds using machine vision technology. In *2010 Sixth International Conference on Natural Computation,* (pp. 84-88). IEEE. 10.1109/ICNC.2010.5583370

Palepu. (2021). An Analysis of Agricultural Soils by using Data Mining Techniques. *International Journal of Engineering Science and Computing,7*(10).

Pal, K. (2019). Bio-control of Pests in Tea: Effect of Environmental. *International Journal of Applied and Computational Mathematics*, 5(3), 1–9. 10.1007/s40819-019-0666-3

Pandey, P. C., & Pandey, M. (2023). Highlighting the role of agriculture and geospatial technology in food security and sustainable development goals. *Sustainable Development (Bradford)*, 31(5), 3175–3195. 10.1002/sd.2600

Panetto, H., Lezoche, M., Hormazabal, J. E. H., Diaz, M. D. M. E. A., & Kacprzyk, J. (2020). Special issue on Agri-Food 4.0 and digitalization in agriculture supply chains-New directions, challenges and applications. *Computers in Industry*, 116, 103188. 10.1016/j.compind.2020.103188

Pant, J., Pant, R. P., Singh, M. K., Singh, D. P., & Pant, H. (2021). Analysis of agricultural crop yield prediction using statistical techniques of machine learning. *Materials Today: Proceedings*, 46, 10922–10926. 10.1016/j.matpr.2021.01.948

Papargyropoulou, E., Lozano, R., Steinberger, J. K., & Wright, N. (2014). bin Ujang, Z. The Food Waste Hierarchy as a Framework for the Management of Food Surplus and Food Waste. *Journal of Cleaner Production*, 76, 106–115. 10.1016/j.jclepro.2014.04.020

Pardo, L. A. (2020). Pesticide exposure and risk of aggressive prostate cancer among private pesticide applicators. *Reading the tea leaves Climate change and the British cuppa*. 10.1186/s12940-020-00583-0

Patel, K. K., Kar, A., Jha, S. N., & Khan, M. A. (2012). Machine vision system: A tool for quality inspection of food and agricultural products. *Journal of Food Science and Technology*, 49(2), 123–141. 10.1007/s13197-011-0321-423572836

Patil, R. R., & Kumar, S. (2020). A Bibliometric Survey on the Diagnosis of Plant Leaf Diseases using Artificial Intelligence. *Library Philosophy and Practice*, 1-26.

Patil, N. K., Malemath, V. S., & Yadahalli, R. M. (2011). Color and texture based identification and classification of food grains using different color models and haralick features. *International Journal on Computer Science and Engineering*, 3(12), 3669.

Pazoki, A. R., Farokhi, F., & Pazoki, Z. (2014). Classification of rice grain varieties using two Artificial Neural Networks (MLP and Neuro-Fuzzy). *The Journal of Animal & Plant Sciences*, 24(1), 336–343.

Pereira, A. C., & Romero, F. (2017). A review of the meanings and the implications of the Industry 4.0 concept. *Procedia Manufacturing*, 13, 1206–1214. 10.1016/j.promfg.2017.09.032

Perich, G., Turkoglu, M. O., Graf, L. V., Wegner, J. D., Aasen, H., Walter, A., & Liebisch, F. (2023). Pixel-based yield mapping and prediction from Sentinel-2 using spectral indices and neural networks. *Field Crops Research*, 292, 108824. 10.1016/j.fcr.2023.108824

Peters, D., Rivers, A., Hatfield, J., Lemay, D., Liu, S., & Basso, B. (2020). Harnessing AI to Transform Agriculture and Inform Agricultural Research. *IT Professional*, 22(3), 16–21. 10.1109/MITP.2020.2986124

Phadnis, A. (2023). Implementation of Prediction of Crop Using SVM Algorithm. *International Journal for Research in Applied Science and Engineering Technology*, 11(5), 3812–3816. 10.22214/ijraset.2023.52265

Phillips, J. (1994). Farmer Education and Farmer Efficiency: A Meta-Analysis. *Economic Development and Cultural Change*, 43(1), 149–165. 10.1086/452139

Picheny, V., Trépos, R., Poublan, B., & Casadebaig, P. (2015). Sunflower phenotype optimization under climatic uncertainties using crop models. *arXiv preprint arXiv:1509.05697.*

Pinki, F. T., Khatun, N., & Islam, S. M. (2017, December). Content based paddy leaf disease recognition and remedy prediction using support vector machine. In *2017 20th international conference of computer and information technology (ICCIT)* (pp. 1-5). IEEE. 10.1109/ICCITECHN.2017.8281764

Pinto, C., Furukawa, J., Fukai, H., & Tamura, S. (2017, August). Classification of Green coffee bean images base on defect types using convolutional neural network (CNN). In *2017 International Conference on Advanced Informatics, Concepts, Theory, and Applications (ICAICTA)*. IEEE.

Pnuma, W. (2021). *UNEP Food Waste Index Report 2021.* UN Environment Programme.

Podvezko, V. (2009). Application of AHP technique. *Journal of Business Economics and Management*, 10(2), 181–189. 10.3846/1611-1699.2009.10.181-189

Prabakaran, N., Naresh, K., Kannadasan, R., & Sainikhil, K. (2021). LiFi-based smart systems for industrial monitoring. *International Journal of Intelligent Enterprise*, 8(2-3), 177–184. 10.1504/IJIE.2021.114501

Prabakaran, N., Palaniappan, R., Kannadasan, R., Dudi, S. V., & Sasidhar, V. (2021). Forecasting the momentum using customised loss function for financial series. *International Journal of Intelligent Computing and Cybernetics*, 14(4), 702–713. 10.1108/IJICC-05-2021-0098

Prabha, R. K., Rai, B. N. J. P., & Singh, S. R. (2016). Role of government schemes in Indian agriculture and rural development. Indian Agriculture and Farmers, 92-102.

Pradana-López, S., Pérez-Calabuig, A. M., Cancilla, J. C., Lozano, M. Á., Rodrigo, C., Mena, M. L., & Torrecilla, J. S. (2021). Deep transfer learning to verify quality and safety of ground coffee. *Food Control*, 122, 107801. 10.1016/j.foodcont.2020.107801

Prajapati, H. B., Shah, J. P., & Dabhi, V. K. (2017). Detection and classification of rice plant diseases. *Intelligent Decision Technologies*, 11(3), 357–373. 10.3233/IDT-170301

Prasad, K., Roy, S., Sen, S., Neave, S., Nagpal, A., & Pandit, V. (2020). Impact of different pest management practices on natural enemy population in tea plantations of Assam special emphasis on spider fauna. Springer. 10.1007/s42690-020-00111-0

Prasanth, A., & Alqahtani, H. (2023). Predictive Models for Early Dropout Indicators in University Settings Using Machine Learning Techniques. In *2023 IEEE International Conference on Emerging Technologies and Applications in Sensors (ICETAS)*. IEEE. 10.1109/ICETAS59148.2023.10346531

Prashant, K. Magadum (2024). Machine Learning for Predicting Wind Turbine Output Power in Wind Energy Conversion Systems, *Grenze International Journal of Engineering and Technology, 10.*https://thegrenze.com/index.php?display=page&view=journalabstract&absid=2514&id=8

Praveena, M., Dubisetty, V. B., Varaprasad, K. V., Rama, M., Vadana, P. S., & Sai, T. S. R. (2023). An in-depth analysis of deep learning and machine learning methods for identifying rice leaf diseases. *4th International Conference on Smart Electronics and Communication (ICOSEC)*. IEEE. 10.1109/ICOSEC58147.2023.10276335

Preindl, R., Nikolopoulos, K., & Litsiou, K. (2020, January). Transformation strategies for the supply chain: The impact of industry 4.0 and digital transformation. In Supply Chain Forum [). Taylor & Francis.]. *International Journal (Toronto, Ont.)*, 21(1), 26–34.

Pretty, J. (2008). Agricultural sustainability: Concepts, principles and evidence. *Philosophical Transactions of the Royal Society of London. Series B, Biological Sciences*, 363(1491), 447–465. 10.1098/rstb.2007.216317652074

Pritam, B., & Nikola K. K., (2020). Spiking Neural Networks for Crop Yield Estimation Based on Spatiotemporal Analysis of Image Time Series. *IEEE Transactions On Geoscience And Remote Sensing*. IEEE.

Prodhan, F. A., Zhang, J., Sharma, T. P. P., Nanzad, L., Zhang, D., Seka, A. M., & Mohana, H. P. (2022). Projection of future drought and its impact on simulated crop yield over South Asia using ensemble machine learning approach. *The Science of the Total Environment*, 807, 151029. 10.1016/j.scitotenv.2021.15102934673078

Puhazhendhi, V., & Jayaraman, B. (1999). Rural credit delivery: Performance and challenges before banks. *Economic and Political Weekly*, 175–182.

Qadri, S., Aslam, T., Nawaz, S. A., Saher, N., Razzaq, A., Ur Rehman, M., Ahmad, N., Shahzad, F., & Furqan Qadri, S. (2021). Machine vision approach for classification of rice varieties using texture features. *International Journal of Food Properties*, 24(1), 1615–1630. 10.1080/10942912.2021.1986523

Qazi, S., Khawaja, B., & Farooq, Q. (2022). IoT-Equipped and AI-Enabled Next Generation Smart Agriculture: A Critical Review, Current Challenges and Future Trends. *IEEE Access : Practical Innovations, Open Solutions*, 10, 21219–21235. 10.1109/ACCESS.2022.3152544

Qian, C., Murphy, S., Orsi, R., & Wiedmann, M. (2022). How Can AI Help Improve Food Safety? *Annual Review of Food Science and Technology*. 10.1146/annurev-food-060721-01381536542755

Qiu, Z., Chen, J., Zhao, Y., Zhu, S., He, Y., & Zhang, C. (2018). Variety identification of single rice seed using hyperspectral imaging combined with convolutional neural network. *Applied Sciences (Basel, Switzerland)*, 8(2), 212. 10.3390/app8020212

Rad, C. R., Hancu, O., Takacs, I. A., & Olteanu, G. (2015). Smart monitoring of potato crop: A cyber-physical system architecture model in the field of precision agriculture. *Agriculture and Agricultural Science Procedia*, 6, 73–79. 10.1016/j.aaspro.2015.08.041

Rahman, C. R., Arko, P. S., Ali, M. E., Khan, M. A. I., Apon, S. H., Nowrin, F., & Wasif, A. (2020). Identification and recognition of rice diseases and pests using convolutional neural networks. *Biosystems Engineering*, 194, 112–120. 10.1016/j.biosystemseng.2020.03.020

Ramamoorthy, R., & Thangavelu, M. (2021). An enhanced hybrid ant colony optimization routing protocol for vehicular ad-hoc networks. *Journal of Ambient Intelligence and Humanized Computing*, 1–32.

Ramesh, M. V., & Das, R. N. (2012). *A public transport system based sensor network for fake alcohol detection.* In Wireless Communications and Applications: First International Conference, ICWCA 2011, Sanya, China. 10.1007/978-3-642-29157-9_13

Ramesh, S., Hebbar, R., Niveditha, M., Pooja, R., Shashank, N., & Vinod, P. V. (2018, April). Plant disease detection using machine learning. In *2018 International conference on design innovations for 3Cs compute communicate control (ICDI3C)* (pp. 41-45). IEEE. 10.1109/ICDI3C.2018.00017

Ramirez-Serrano, A., & Kuzyk, R. (2010, March). Modified mecanum wheels for traversing rough terrains. In *2010 Sixth International Conference on Autonomic and Autonomous Systems* (pp. 97-103). IEEE. 10.1109/ICAS.2010.35

Rani. (2020). The Impact of Data Analytics in Crop Management based on Weather Conditions. *International Journal of Engineering Technology Science and Research*, *4*(5).

Ratnaparkhi, S. T., Khan, S., Arya, C., Khapre, S., Singh, P., Diwakar, M., & Shankar, A. (2020, December 1). WITHDRAWN: Smart agriculture sensors in IOT: A review. *Materials Today: Proceedings*. 10.1016/j.matpr.2020.11.138

Ray, P. P. (2017, June 19). Internet of things for smart agriculture: Technologies, practices and future direction. *Journal of Ambient Intelligence and Smart Environments*, 9(4), 395–420. 10.3233/AIS-170440

Ray, P., Duraipandian, R., Kiranmai, G., Rao, R., & Jose, M. (2021). An Exploratory Study of Risks and Food Insecurity in the Agri Supply Chain. *Management Science*, 8(S1-Feb), 1–12. 10.34293/management.v8iS1-Feb.3752

Reddy, G. N., Reddy, G. M., Balaji, G., Harish, C. M., & Kethan, C. H. (2022). *IOT Based Seed Planting and Watering Rover*. Academic Press.

Ren, S., He, K., Girshick, R., & Sun, J. (2015). Faster r-cnn: Towards real-time object detection with region proposal networks. *Advances in Neural Information Processing Systems*, 28.

Reshma, R., Sathiyavathi, V., Sindhu, T., Selvakumar, K., & Sairamesh, L. (2020). IoT based Classification Techniques for Soil Content Analysis and Crop Yield Prediction. *2020 Fourth International Conference on I-SMAC (IoT in Social, Mobile, Analytics and Cloud) (I-SMAC)*, (pp. 156-160). IEEE. 10.1109/I-SMAC49090.2020.9243600

Riahi, J., Vergura, S., Mezghani, D., & Mami, A. (2020). Intelligent Control of the Microclimate of an Agricultural Greenhouse Powered by a Supporting PV System. *Applied Sciences (Basel, Switzerland)*, 10(4), 1350. 10.3390/app10041350

Riihijärvi, J., & Mähönen, P. (2018). Machine Learning for Performance Prediction in Mobile Cellular Networks. *IEEE Computational Intelligence Magazine*, 13(1), 51–60. 10.1109/MCI.2017.2773824

Ringler, C., Agbonlahor, M., Baye, K., Barron, J., Hafeez, M., Lundqvist, J., & Uhlenbrook, S. (2023). Water for food systems and nutrition. *Science and Innovations for Food Systems Transformation, 497.*

Robert Singh, K., & Chaudhury, S. (2020). A cascade network for the classification of rice grain based on single rice kernel. *Complex & Intelligent Systems*, 6(2), 321–334. 10.1007/s40747-020-00132-9

Rogers, H., & Fox, C. (2020). An open source seeding agri-robot. *Proceedings of the 3rd UK-RAS Conference.*

Rohilla, N., & Rai, M. (2021). Advanced machine learning techniques used for detecting and classification of disease in plants: a review. *3rd International Conference on Advances in Computing, Communication Control and Networking (ICAC3N).* IEEE. 10.1109/ICAC3N53548.2021.9725616

Roy, S., Ghosh, S., Beck, C. D., & Sinha, A. P. (2023). Supply chain traceability: A case of blockchain modelling application to agro-business product in India. *International Journal of Sustainable Agricultural Management and Informatics*, 9(4), 295–319. 10.1504/IJSAMI.2023.134067

Ryan, M., Isakhanyan, G., & Tekinerdogan, B. (2023). An interdisciplinary approach to artificial intelligence in agriculture. *NJAS: Impact in Agricultural and Life Sciences*, 95(1), 2168568. 10.1080/27685241.2023.2168568

Saaty, T. L., & Özdemir, M. S. (2014). How many judges should there be in a group? *Annals of Data Science*, 1(3-4), 359–368. 10.1007/s40745-014-0026-4

Sabanci, K., Kayabasi, A., & Toktas, A. (2017). Computer vision-based method for classification of wheat grains using artificial neural network. *Journal of the Science of Food and Agriculture*, 97(8), 2588–2593. 10.1002/jsfa.808027718230

Saba, T., & Rehman, A. (2013). Effects of artificially intelligent tools on pattern recognition. *International Journal of Machine Learning and Cybernetics*, 4(2), 155–162. 10.1007/s13042-012-0082-z

Saberi, S., Kouhizadeh, M., Sarkis, J., & Shen, L. (2019). Blockchain technology and its relationships to sustainable supply chain management. *International Journal of Production Research*, 57(7), 2117–2135. 10.1080/00207543.2018.1533261

Sahni, M. (2020). Challenges of agriculture credit in India. [IERJ]. *International Education and Research Journal*, 6(9), 29–31.

Saikenova, A. Z., Nurgassenov, T. N., Saikenov, B. R., Kudaibergenov, M. S., & Didorenko, S. V. (2021). Crop yield and quality of lentil varieties in the conditions of the southeast of kazakhstan. *Online Journal of Biological Sciences*, 21(1), 33–40. 10.3844/ojbsci.2021.33.40

Sairoel, A., Gebresenbet, G., Alwan, H. M., & Vladmirovna, K. O. (2023). Assessment of Smart Mechatronics Applications in Agriculture: A Review. *Applied Sciences (Basel, Switzerland)*, 13(12), 7315. 10.3390/app13127315

Sáiz-Rubio, V., & Rovira-Más, F. (2020). From Smart Farming towards Agriculture 5.0: A Review on Crop Data Management. *Agronomy (Basel)*, 10(2), 207. 10.3390/agronomy10020207

Salemink, K., Strijker, D., & Bosworth, G. (2017). Rural development in the digital age: A systematic literature review on unequal ICT availability, adoption, and use in rural areas. *Journal of Rural Studies*, 54, 360–371. 10.1016/j.jrurstud.2015.09.001

Salman, A. D., & Abdelaziz, M. A. (2020). Mobile robot monitoring system based on IoT. *Journal of Xi'An University of Architecture & Technology*, 12(3), 5438–5447.

Samad, A., Murdeshwar, P., & Hameed, Z. (2010). High-credibility RFID-based animal data recording system suitable for small-holding rural dairy farmers. *Computers and Electronics in Agriculture*, 73(2), 213–218. 10.1016/j.compag.2010.05.001

Sampaio, P. S., Almeida, A. S., & Brites, C. M. (2021). Use of artificial neural network model for rice quality prediction based on grain physical parameters. *Foods*, 10(12), 3016. 10.3390/foods1012301634945567

Sana, S. S., Ospina-Mateus, H., Arrieta, F. G., & Chedid, J. A. (2019). Application of genetic algorithm to job scheduling under ergonomic constraints in manufacturing industry. *Journal of Ambient Intelligence and Humanized Computing*, 10(5), 2063–2090. 10.1007/s12652-018-0814-3

Sanchez, W., Martinez, A., & Hernandez. (2018). A predictive model for stress recognition in desk jobs. *J Ambient Intell Human Comput*. .10.1007/s12652-018-1149-9

Sangeetha, R., Logeshwaran, J., Rocher, J., & Lloret, J. (2023). An improved agro deep learning model for detection of Panama wilts disease in banana leaves. *AgriEngineering*, 5(2), 660–679. 10.3390/agriengineering5020042

Sangeevan. (2021). Deep learning-based pesticides prescription system for leaf diseases of home garden crops in Sri Lanka. *International Research Conference on Smart Computing and Systems Engineering (SCSE)*. IEEE. .10.1109/SCSE53661.2021.9568308

Sanginga, P., Best, R., Chitsike, C., Delve, R., Kaaria, S., & Kirkby, R. (2004). Enabling rural innovation in Africa: An approach for integrating farmer participatory research and market orientation for building the assets of rural poor. *Uganda Journal of Agricultural Sciences*, 9, 934–949. 10.4314/UJAS.V9I1

Santeramo, F. G., & Lamonaca, E. (2021). Food loss–food waste–food security: A new research agenda. *Sustainability (Basel)*, 13(9), 4642. 10.3390/su13094642

Sarkar, M., Masud, S., Hossen, M., & Goh, M. (2022). A Comprehensive Study on the Emerging Effect of Artificial Intelligence in Agriculture Automation. *2022 IEEE 18th International Colloquium on Signal Processing & Applications (CSPA)*, (pp. 419-424). IEEE. 10.1109/CSPA55076.2022.9781883

Sarkar, N. C., Mondal, K., Das, A., Mukherjee, A., Mandal, S., Ghosh, S., & Huda, S. (2023). Enhancing livelihoods in farming communities through super-resolution agromet advisories using advanced digital agriculture technologies. *Journal of Agrometeorology*, 25(1), 68–78.

Satapathy, S. (2021). Work place discomfort and risk factors for construction site workers. *International Journal of System Assurance Engineering and Management*, 1-13.

Satapathy, S. (2014). ANN, QFD and ISM approach for framing electricity utility service in India for consumer satisfaction. *International Journal of Services and Operations Management*, 18(4), 404–428. 10.1504/IJSOM.2014.063243

Satheshkumar, K., & Mangai, S. (2021). EE-FMDRP: Energy efficient-fast message distribution routing protocol for vehicular ad-hoc networks. *Journal of Ambient Intelligence and Humanized Computing*, 12(3), 3877–3888. 10.1007/s12652-020-01730-8

Savary, S., Nelson, A., Djurle, A., Esker, P., Sparks, A., Amorim, L., Filho, A., Caffi, T., Castilla, N., Garrett, K., McRoberts, N., Rossi, V., Yuen, J., & Willocquet, L. (2017). Concepts, approaches, and avenues for modelling crop health and crop losses. *European Journal of Agronomy*. 10.1016/j.eja.2018.04.003

Sawik, T. (2019, July). Two-period vs. multi-period model for supply chain disruption management. *International Journal of Production Research*, 57(14), 4502–4518. 10.1080/00207543.2018.1504246

Sayyad, L. (2023), System for Love Healthcare for Loved Ones based on IoT. *Research Exploration: Transcendence of Research Methods and Methodology, 2.*

Scarpare, F. V. (2013). Bioenergy and water: Brazilian sugarcane ethanol. *Bioenergy and Water, 89.*

Schmidt, P., Biessmann, F., & Teubner, T. (2020). Transparency and trust in artificial intelligence systems. *Journal of Decision Systems*, 29(4), 260–278. 10.1080/12460125.2020.1819094

Scuotto, V., Caputo, F., Villasalero, M., & Del Giudice, M. (2017). A multiple buyer–supplier relationship in the context of SMEs' digital supply chain management. *Production Planning and Control*, 28(16), 1378–1388. 10.1080/09537287.2017.1375149

Seidel, E., & Kutieleh, S. (2017). Using predictive analytics to target and improve first year student attrition. *Australian Journal of Education*, 61(2), 200–218. 10.1177/0004944117712310

Sendin, K., Manley, M., & Williams, P. J. (2018). Classification of white maize defects with multispectral imaging. *Food Chemistry*, 243, 311–318. 10.1016/j.foodchem.2017.09.13329146343

Sethy, P. K. (2022). Identification of wheat tiller based on AlexNet-feature fusion. *Multimedia Tools and Applications*, 81(6), 8309–8316. 10.1007/s11042-022-12286-4

Sethy, P. K., & Chatterjee, A. (2018). Rice variety identification of western Odisha based on geometrical and texture feature. *International Journal of Applied Engineering Research: IJAER*, 13(4), 35–39.

Seyedghorban, Z., Tahernejad, H., Meriton, R., & Graham, G. (2020). Supply chain digitalization: Past, present and future. *Production Planning and Control*, 31(2-3), 96–114. 10.1080/09537287.2019.1631461

Shadrin, D., Menshchikov, A., Somov, A., Bornemann, G., Hauslage, J., & Fedorov, M. (2020). Enabling Precision Agriculture Through Embedded Sensing With Artificial Intelligence. *IEEE Transactions on Instrumentation and Measurement*, 69(7), 4103–4113. 10.1109/TIM.2019.2947125

Shafiq, D. A., Marjani, M., Habeeb, R. A. A., & Asirvatham, D. (2022). Predictive Analytics in Education: A Machine Learning Approach. In *2022 3rd International Multidisciplinary Conference on Computer and Energy Science (SpliTech)* (pp. 1-6). IEEE. [DOI:10.1109/MACS56771.2022

Shafi, U., Mumtaz, R., García-Nieto, J., Hassan, S., Zaidi, S., & Iqbal, N. (2019). Precision Agriculture Techniques and Practices: From Considerations to Applications. *Sensors (Basel)*, 19(17), 3796. 10.3390/s1917379631480709

Shah, J. P., Prajapati, H. B., & Dabhi, V. K. (2016, March). A survey on detection and classification of rice plant diseases. In *2016 IEEE International Conference on Current Trends in Advanced Computing (ICCTAC)* (pp. 1-8). IEEE. 10.1109/ICCTAC.2016.7567333

Shamshiri, R. R., Bojic, I., van Henten, E., Balasundram, S. K., Dworak, V., Sultan, M., & Weltzien, C. (2020). Model-based evaluation of greenhouse microclimate using IoT-Sensor data fusion for energy efficient crop production. *Journal of Cleaner Production*, 263, 121303. 10.1016/j.jclepro.2020.121303

Shantaiya, S., & Ansari, U. (2010, November). *Identification of food grains and its quality using pattern classification.* In IEEE international conference on communication technology, Raipur, India.

Sharma, R. (2021). Artificial Intelligence in Agriculture: A Review. *2021 5th International Conference on Intelligent Computing and Control Systems (ICICCS)*, (pp. 937-942). IEEE. 10.1109/ICICCS51141.2021.9432187

Sharma, A., Jain, A., Gupta, P., & Chowdary, V. (2021). Machine Learning Applications for Precision Agriculture: A Comprehensive Review. *IEEE Access : Practical Innovations, Open Solutions*, 9, 4843–4873. 10.1109/ACCESS.2020.3048415

Sharma, H., Kumar, A., Singh, R., & Kumar, S. (2023). Impact of Nanopriming and Omics-Based Applications for Sustainable Agriculture. In Singh, A., Rajput, V., Ghazaryan, K., Gupta, S., & Minkina, T. (Eds.), *Nanopriming Approach to Sustainable Agriculture* (pp. 175–191). IGI Global., 10.4018/978-1-6684-7232-3.ch008

Sharma, K., Sethi, G., & Bawa, R. (2020, March). State-of-the-Art in Automatic Rice Quality Grading System. In *Proceedings of the International Conference on Innovative Computing & Communications (ICICC)*. SSRN. 10.2139/ssrn.3564372

Sharma, L., & Garg, P. K. (Eds.). (2021). *Artificial intelligence: technologies, applications, and challenges*. 10.1201/9781003140351

Sharma, T. K., & Abraham, A. (2020). Artificial bee colony with enhanced food locations for solving mechanical engineering design problems. *Journal of Ambient Intelligence and Humanized Computing*, 11(1), 267–290. 10.1007/s12652-019-01265-7

Sheth, S., Kher, R., Shah, R., Dudhat, P., & Jani, P. (2010). Automatic sorting system using machine vision. In *Multi-Disciplinary International Symposium on Control, Automation & Robotics*.

Shetty, H. (2018). Food security through agricultural sustainability: In Indian context. *Int. J. Res. Eng. Sci. Manage*, 1, 685–691.

Shinde. (2020). Web Based Recommendation System for farmers. *International Journal on Recent and Innovation Trends in Computing and Communication*, *3*(3).

Shin, J., Mahmud, M. S., Rehman, T. U., Ravichandran, P., Heung, B., & Chang, Y. K. (2022). Trends and prospect of machine vision technology for stresses and diseases detection in precision agriculture. *AgriEngineering*, 5(1), 20–39. 10.3390/agriengineering5010003

Shobha, K., & Siji, K. (2018). Problems Faced by Farmers in Accessing Agricultural Credit with Special Reference to Malur in Kolar District of Karnataka. *IJCRT,6*(1).

Shook, J., Gangopadhyay, T., Wu, L., Ganapathysubramanian, B., Sarkar, S., & Singh, A. K. (2021). Crop yield prediction integrating genotype and weather variables using deep learning. *PLoS One*, 16(6), e0252402. 10.1371/journal.pone.025240234138872

Shorten, C., & Khoshgoftaar, T. M. (2019). A survey on image data augmentation for deep learning. *Journal of Big Data*, 6(1), 1–48. 10.1186/s40537-019-0197-0

Shrestha, A., & Mahmood, A. (2019). Review of Deep Learning Algorithms and Architectures. *IEEE Access : Practical Innovations, Open Solutions*, 7, 53040–53065. 10.1109/ACCESS.2019.2912200

Shreya, B., &Kalyani, B., (2020). Crop And Yield Prediction Model. *International Journal of Advance Scientific Research and Engineering Trends*, *1*.

Shrivastava, V. K., Pradhan, M. K., Minz, S., & Thakur, M. P. (2019). Rice plant disease classification using transfer learning of deep convolution neural network. *The International Archives of the Photogrammetry, Remote Sensing and Spatial Information Sciences*, 42(W6), 631–635. 10.5194/isprs-archives-XLII-3-W6-631-2019

Shuqfa, Z., & Harous, S. (2019). Data Mining Techniques Used in Predicting Student Retention in Higher Education: A Survey. *2019 International Conference on Electrical and Computing Technologies and Applications (ICECTA)*. IEEE. . https://ieeexplore.ieee.org/document/8959789/10.1109/ICECTA48151.2019.8959789

Siddh, M. M., Soni, G., Jain, R., Sharma, M. K., & Yadav, V. (2021). A framework for managing the agri-fresh food supply chain quality in Indian industry. *Management of Environmental Quality*, 32(2), 436–451. 10.1108/MEQ-05-2020-0085

Siddiqui, S., Bajaj, S., & Mathew, C. (2021). Exploring Biodiversity: Flora and Fauna in Maitri Bagh, Bhilai, Chhattisgarh, India. *NeuroQuantology : An Interdisciplinary Journal of Neuroscience and Quantum Physics*, 19(12), 705.

Sieling, C., & Westenberg, H. E. (1963). *The Agri-Robot A Revolution in Plowing (No. 630306).* SAE Technical Paper.

Silva, C. S., & Sonnadara, D. U. J. (2013). *Classification of rice grains using neural networks*. Academic Press.

Silva, S., Duarte, D., Barradas, R., Soares, S., Valente, A., & Reis, M. J. C. S. (2017). Arduino recursive backtracking implementation, for a robotic contest. In: *Human-Centric Robotics*, (pp. 169–178). World Scientific. 10.1142/9789813231047_0023

Singh, A., Mishra, R., Sarkar, A., Singh, A. K., Toppo, D., & Kumar, A. (2021). Design and Analysis of Agri-rover for Farming. In *Current Advances in Mechanical Engineering: Select Proceedings of ICRAMERD 2020* (pp. 657-663). Springer Singapore.

Singh, H., Halder, N., Singh, B., Singh, J., Sharma, S., & Shacham-Diamand, Y. (2023). Smart Farming Revolution: Portable and Real-Time Soil Nitrogen and Phosphorus Monitoring for Sustainable Agriculture. *Sensors (Basel)*, 23(13), 5914. 10.3390/s2313591437447764

Singh, K. R., & Chaudhury, S. (2016). Efficient technique for rice grain classification using back-propagation neural network and wavelet decomposition. *IET Computer Vision*, 10(8), 780–787. 10.1049/iet-cvi.2015.0486

Singh, S. K., & Kumar, P. (2020). A comprehensive survey on trajectory schemes for data collection using mobile elements in WSNs. *Journal of Ambient Intelligence and Humanized Computing*, 11(1), 291–312. 10.1007/s12652-019-01268-4

Singh, S. K., Parida, J. K., & Pattnaik, P. K. (2023). Ranking of Attributes for Commercial Banks Using Multi Criteria Decision Making. [EEL]. *European Economic Letters*, 13(5), 142–155.

Skitsko, O., Skladannyi, P., Shyrshov, R., Humeniuk, M., & Vorokhob, M. (2023). THREATS AND RISKS OF THE USE OF ARTIFICIAL INTELLIGENCE. *Cybersecurity: Education, Science, Technique*. .10.28925/2663-4023.2023.22.618

Sladojevic, S., Arsenovic, M., Anderla, A., Culibrk, D., & Stefanovic, D. (2016). Deep neural networks based recognition of plant diseases by leaf image classification. *Computational Intelligence and Neuroscience*, 2016, 2016. 10.1155/2016/328980127418923

Śmietanka, M., Koshiyama, A., & Treleaven, P. (2020). Algorithms in future insurance markets. *International Journal of Data Science and Big Data Analytics*. .10.2139/ssrn.3641518

Smith, M. (2020). Getting value from artificial intelligence in agriculture. *Animal Production Science*, 60(1), 46–54. 10.1071/AN18522

Sobota, J. PiŜl, R., Balda, P., & Schlegel, M. (2013). Raspberry Pi and Arduino boards in control education. *IFAC Proceedings Volumes, 46*(17), 7-12.

Song, C., Wang, J. Q., & Li, J. B. (2014). New framework for quality function deployment using linguistic Z-numbers. *Mathematics*, 8(2), 224. 10.3390/math8020224

Songol, M., Awuor, F., & Maake, B. (2021). Adoption of artificial intelligence in agriculture in the developing nations: A review. *Journal of Language. Technology & Entrepreneurship in Africa*, 12(2), 208–229.

Son, N. H., & Thai-Nghe, N. (2019, November). *Deep learning for rice quality classification. In 2019 international conference on advanced computing and applications (ACOMP).* IEEE.

Sowjanya, K. D., Sindhu, R., Parijatham, M., Srikanth, K., & Bhargav, P. (2017, April). Multipurpose autonomous agricultural robot. In *2017 International conference of Electronics, Communication and Aerospace Technology (ICECA)* (Vol. 2, pp. 696-699). IEEE. 10.1109/ICECA.2017.8212756

Soydan, D. K., Turgut, N., Yalç, M., & Turgut, C. (2021). *Evaluation of pesticide residues in fruits and vegetables from the Aegean region of Turkey and assessment of risk to consumers.*

Spanaki, K., Sivarajah, U., Fakhimi, M., Despoudi, S., & Irani, Z. (2021). Disruptive technologies in agricultural operations: A systematic review of AI-driven AgriTech research. *Annals of Operations Research*, 308(1-2), 491–524. 10.1007/s10479-020-03922-z

Sparrow, R., & Howard, M. (2020). Robots in agriculture: Prospects, impacts, ethics, and policy. *Precision Agriculture*, 22(3), 818–833. 10.1007/s11119-020-09757-9

Sridevi, G., & Chakkravarthy, M. (2021). A meta-heuristic multiple ensemble load balancing framework for real-time multi-task cloud scheduling process. *International Journal of System Assurance Engineering and Management*, 12(6), 1459–1476. 10.1007/s13198-021-01244-2

SSLK. (2023). IoT in Electrical Vehicle: A Study. *Journal of Control and Instrumentation Engineering*, 9(3), 15–21.

Stefano, D. (2007). Psychology and economics: Evidence from the field. *Journal of Economic Literature*, 1(November).

Stunžėnas, E., & Kliopova, I. (2021). Industrial ecology for optimal food waste management in a region. *Environmental Research, Engineering and Management*, 77(1), 7–24. 10.5755/j01.erem.77.1.27605

Subeesh, A., & Mehta, C. R. (2021). Automation and digitization of agriculture using artificial intelligence and internet of things. *Artificial Intelligence in Agriculture*, 5, 278–291. 10.1016/j.aiia.2021.11.004

Suganthi, M., Event, S., & Senthilkumar, P. (2020). Comparative bioefficacy of Bacil- lus and Pseudomonas continues against Looper severe in tea (Camellia sinensis (L.) O. Kuntze. *Physiology and Molecular Biology of Plants*, 26(10), 2053–2060. 10.1007/s12298-020-00875-233088049

Sujatha, R., Chatterjee, J. M., Jhanjhi, N. Z., & Brohi, S. N. (2021). Performance of deep learning vs machine learning in plant leaf disease detection. *Microprocessors and Microsystems*, 80, 103615. 10.1016/j.micpro.2020.103615

Sul, K. (2023b). ArduinoBased Weather Monitoring System. *Journal of Switching Hub*, 8(3), 24–29.

Sultanabanu, K. (2023). Arduino Based Weather Monitoring System. *Journal of Switching Hub*, 8(3), 24–29.

Sumaryanti, L., Musdholifah, A., & Hartati, S. (2015). Digital image-based identification of rice variety using image processing and neural network. *TELKOMNIKA Indonesian Journal of Electrical Engineering*, 16(1), 182–190. 10.11591/tijee.v16i1.1602

Sundararajan, N., Habeebsheriff, H. S., Dhanabalan, K., Cong, V. H., Wong, L. S., Rajamani, R., & Dhar, B. K. (2024). Mitigating global challenges: Harnessing green synthesized nanomaterials for sustainable crop production systems. *Global Challenges (Hoboken, NJ)*, 8(1), 2300187. 10.1002/gch2.20230018738223890

Sundari, T. (2018). Digital Transformation of Indian Agriculture. *Contemporary Social Science*, 27(4), 65–71. 10.29070/27/58309

Sung, P. C., Hsu, C. C., Lee, C. L., Chiu, Y. S. P., & Chen, H. L. (2015). Formulating grip strength and key pinch strength prediction models for Taiwanese: A comparison between stepwise regression and artificial neural networks. *Journal of Ambient Intelligence and Humanized Computing*, 6(1), 37–46. 10.1007/s12652-014-0245-8

Sun, R., Yang, W., Li, Y., & Sun, C. (2021). Multi - residue analytical methods for pesticides in teas: A review. *European Food Research and Technology*, 247(8), 1839–1858. 10.1007/s00217-021-03765-3

Surbakti, E. P. C. B. (2021). HubunganPengetahuan dan SikapTerhadap Tindakan KonsumsiMakananCepat Saji (fast Food) Pada Remaja Di SMA Negeri 1 Tigapanah. *Frontiers in Neuroscience*, 14(1).

Suresh Babu, C. V. (2023). *Artificial Intelligence and Expert Systems*. Anniyappa Publications.

Suresh Babu, C. V., Mahalashmi, J., Vidhya, A., Nila Devagi, S., & Bowshith, G. (2023). Save soil through machine learning. In Habib, M. (Ed.), *Global Perspectives on Robotics and Autonomous Systems: Development and Applications* (pp. 345–362). IGI Global. 10.4018/978-1-6684-7791-5.ch016

Suresh Babu, C. V., & Rahul, A. (2024). Securing the Future: Unveiling Risks and Safeguarding Strategies in Machine Learning-Powered Cybersecurity. In Almaiah, M., Maleh, Y., & Alkhassawneh, A. (Eds.), *Risk Assessment and Countermeasures for Cybersecurity* (pp. 80–95). IGI Global. 10.4018/979-8-3693-2691-6.ch005

Suresh Babu, C. V., Swapna, A., Chowdary, D. S., Vardhan, B. S., & Imran, M. (2023). Leaf disease detection using machine learning (ML). In Khang, A. (Ed.), *Handbook of Research on AI-Equipped IoT Applications in High-Tech Agriculture* (pp. 188–199). IGI Global. 10.4018/978-1-6684-9231-4.ch010

Suresh, A., & Sarath, T. (2019). An IoT Solution for Cattle Health Monitoring. *IOP Conference Series. Materials Science and Engineering*, 561(1), 012106. Advance online publication. 10.1088/1757-899X/561/1/012106

Swain, S., Nayak, S. K., & Barik, S. S. (2020). A review on plant leaf diseases detection and classification based on machine learning models. *Mukt shabd, 9*(6), 5195-5205.

Szczypiński, P. M., Klepaczko, A., & Zapotoczny, P. (2015). Identifying barley varieties by computer vision. *Computers and Electronics in Agriculture*, 110, 1–8. 10.1016/j.compag.2014.09.016

Szegedy, C., Liu, W., Jia, Y., Sermanet, P., Reed, S., Anguelov, D., & Rabinovich, A. (2015). Going deeper with convolutions. In *Proceedings of the IEEE conference on computer vision and pattern recognition* (pp. 1-9). IEEE.

Tajvar, A., Daneshmandi, H., Dortaj, E., Seif, M., Parsaei, H., Shakerian, M., & Choobineh, A. (2021). Common errors in selecting and implementing pen–paper observational methods by Iranian practitioners for assessing work-related musculoskeletal disorders risk: A systematic review. *International Journal of Occupational Safety and Ergonomics*, 1–7.33736566

Tambe, A. B., Mbanga, B. M. R., Nzefa, D. L., & Name, M. G. (2019). Pesticide usage and occupational hazards among farmers working in small-scale tomato farms in Cameroon. Research Gate. 10.1186/s42506-019-0021-x

Tan, W., Sun, L., Zhang, D., Ye, D., & Che, W. (2016, October). Classification of wheat grains in different quality categories by near infrared spectroscopy and support vector machine. *2016 2nd International Conference on Cloud Computing and Internet of Things (CCIOT)*. IEEE.

Taneja, A., Nair, G., Joshi, M., Sharma, S., Sharma, S., Jambrak, A., Roselló-Soto, E., Barba, F., Castagnini, J., Leksawasdi, N., & Phimolsiripol, Y. (2023). Artificial Intelligence: Implications for the Agri-Food Sector. *Agronomy (Basel)*, 13(5), 1397. 10.3390/agronomy13051397

Taspinar, Y. S., Dogan, M., Cinar, I., Kursun, R., Ozkan, I. A., & Koklu, M. (2022). Computer vision classification of dry beans (Phaseolus vulgaris L.) based on deep transfer learning techniques. *European Food Research and Technology*, 248(11), 2707–2725. 10.1007/s00217-022-04080-1

Teimouri, N., Omid, M., Mollazade, K., & Rajabipour, A. (2015). An artificial neural network-based method to identify five classes of almond according to visual features. *Journal of Food Process Engineering*, 39(6), 625–635. 10.1111/jfpe.12255

Tervonen, J. (2018). Experiment of the quality control of vegetable storage based on the Internet-of-Things. *Procedia Computer Science*, 130, 440–447. 10.1016/j.procs.2018.04.065

Tian, Y., Zhang, S., Liu, J., Chen, F., Li, L., & Xia, B. (2017). Research on a new omnidirectional mobile platform with heavy loading and flexible motion. *Advances in Mechanical Engineering*, 9(9), 1687814017726683. 10.1177/1687814017726683

Tin, M. M., Mon, K. L., Win, E. P., & Hlaing, S. S. (2019). Myanmar rice grain classification using image processing techniques. *In Big Data Analysis and Deep Learning Applications:Proceedings of the First International Conference on Big Data Analysis and Deep Learning 1st* (pp. 324-332). Springer Singapore.

Tinto, V. (1975). Dropout from Higher Education: A Theoretical Synthesis of Recent Research. *Review of Educational Research*, 45(1), 89–125. 10.3102/00346543045001089

Torén, A., Öberg, K., Lembke, B., Enlund, K., & Rask-Andersen, A. (2002). Tractor-driving hours and their relation to self-reported low-back and hip symptoms. *Applied Ergonomics*, 33(2), 139–146. 10.1016/S0003-6870(01)00061-812009120

Tripathi, S., Shukla, S., Attrey, S., Agrawal, A., & Bhadoria, V. S. (2020). Smart industrial packaging and sorting system. *Strategic system assurance and business analytics*, 245-254.

Tripathy, A. K. (2021). Data mining and wireless sensor network for agriculture pest/disease predictions. *Information and Communication Technologies (WICT), 2011 World Congress*. IEEE.

Trojnacki, M., & Dąbek, P. (2019). Mechanical properties of modern wheeled mobile robots. *Journal of Automation Mobile Robotics and Intelligent Systems*, 13(3), 3–13. 10.14313/JAMRIS/3-2019/21

Tsoulfas, G., & Mouzakitis, Y. (2021). Framing the transition towards sustainable agri-food supply chains. *IOP Conference Series. Earth and Environmental Science*, 899(1), 012003. 10.1088/1755-1315/899/1/012003

Tugrul, B., Elfatimi, E., & Eryigit, R. (2022). Convolutional neural networks in detection of plant leaf diseases. *Revista de Agricultura (Piracicaba)*, 12(8), 1192.

Turner, K., Georgiou, S., Clark, R., Brouwer, R., & Burke, J. (2004). The role of water in agricultural development. In *Economic valuation of water resources in agriculture*. Research Gate.

Tutul, M., Alam, M., & Wadud, M. (2023). Smart Food Monitoring System Based on IoT and Machine Learning. *2023 International Conference on Next-Generation Computing, IoT and Machine Learning (NCIM)*, (pp. 1-6). IEEE. 10.1109/NCIM59001.2023.10212608

Tyagi, N., Khan, R., Chauhan, N., Singhal, A., & Ojha, J. (2021). E-rickshaws management for small scale farmers using big data-Apache spark. []. IOP Publishing.]. *IOP Conference Series. Materials Science and Engineering*, 1022(1), 012023. 10.1088/1757-899X/1022/1/012023

Tzachor, A. (2021). Barriers to AI adoption in Indian agriculture: An initial inquiry. [IJIDE]. *International Journal of Innovation in the Digital Economy*, 12(3), 30–44. 10.4018/IJIDE.2021070103

Ullah, M. W., Mortuza, M. G., Kabir, M. H., Ahmed, Z. U., Supta, S. K. D., Das, P., & Hossain, S. M. D. (2018). *Internet of things based smart greenhouse: remote monitoring and automatic control. DEStech Trans.* Environ. Energy Earth Sci.

Ulvenblad, P., Barth, H., Ulvenblad, P. O., Ståhl, J., & Björklund, J. C. (2020). Overcoming barriers in agri-business development: Two education programs for entrepreneurs in the Swedish agricultural sector. *Journal of Agricultural Education and Extension*, 26(5), 443–464. 10.1080/1389224X.2020.1748669

Upendra, R. S., Umesh, I. M., Varma, R. R., & Basavaprasad, B. (2020). Technology in Indian agriculture–a review. *Indonesian Journal of Electrical Engineering and Computer Science*, 20(2), 1070–1077. 10.11591/ijeecs.v20.i2.pp1070-1077

USAID. (2018). *Land and Conflict: A Toolkit for Intervention.* United States Agency for International Development.

Valdez, A., Amba, O., Jr., R., Dimalna, H., Gomampong, A., & Manaol, N. (2023). Farmers Perceptions on Artificial Insemination (AI): A Mixed Method Design. *European Journal of Theoretical and Applied Sciences.* .10.59324/ej-tas.2023.1(3).31

Van Klompenburg, T., Kassahun, A., & Catal, C. (2020). Crop yield prediction using machine learning: A systematic literature review. *Computers and Electronics in Agriculture*, 177, 105709. 10.1016/j.compag.2020.105709

Varchenko, O. (2019). Theoretical aspects of functioning of agro-food chains and features of their development in Ukrainian. *Ekonomìka ta upravlìnnâ APK.* .10.33245/2310-9262-2019-148-1-6-20

Varol Altay, E., & Alatas, B. (2020). Performance analysis of multi-objective artificial intelligence optimization algorithms in numerical association rule mining. *Journal of Ambient Intelligence and Humanized Computing*, 11(8), 3449–3469. 10.1007/s12652-019-01540-7

Velesaca, H. O., Mira, R., Suárez, P. L., Larrea, C. X., & Sappa, A. D. (2020). Deep learning-based corn kernel classification. In *Proceedings of the IEEE/CVF Conference on Computer Vision and Pattern Recognition Workshops* (pp. 66-67). IEEE.

Vemuri, N., Thaneeru, N., & Tatikonda, V. M. (2023). Smart farming revolution: Harnessing IoT for enhanced agricultural yield and sustainability. *Journal of Knowledge Learning and Science Technology, 2*(2), 143-148..

Veni, A., & Rani, K. (2023). Improvement of Agriculture Productivity by using Artificial Intelligence and Block Chain Technology. *International Journal of Scientific Research in Science and Technology*, 445–456. 10.32628/IJSRST52310451

Vidyarthi, S. K., Singh, S. K., Xiao, H. W., & Tiwari, R. (2021). Deep learnt grading of almond kernels. *Journal of Food Process Engineering*, 44(4), e13662. 10.1111/jfpe.13662

Vikas, K., & Vishal, D. (2021). KrishiMantra: Agricultural Recommendation System. *Proceedings of the 3rd ACM Symposium on Computing for Development.* ACM.

Wale Anjali, D. (2019). Rokade Dipali. Smart Agriculture System using IoT. *International Journal of Innovative Research In Technology*, 5(10), 493–497.

Wamba, S. F., Queiroz, M. M., & Trinchera, L. (2020). Dynamics between blockchain adoption determinants and supply chain performance: An empirical investigation. *International Journal of Production Economics*, 229, 107791. 10.1016/j.ijpe.2020.107791

Wang, N., Zhang, N., & Wang, M. (2006). Wireless sensors in agriculture and food industry—Recent development and future perspective. *Computers and Electronics in Agriculture*, 50(1), 1–14. 10.1016/j.compag.2005.09.003

Wang, Y., Li, A., & Khodaei, H. (2020). Optimal designing of a CCHP source system using balanced Sunflower optimization algorithm. *Energy Sources. Part A, Recovery, Utilization, and Environmental Effects*, 1–23. 10.1080/15567036.2020.1747575

Wani, N. R., Rather, R. A., Farooq, A., Padder, S. A., Baba, T. R., Sharma, S., Mubarak, N. M., Khan, A. H., Singh, P., & Ara, S. (2023). New insights in food security and environmental sustainability through waste food management. *Environmental Science and Pollution Research International*, 31(12), 1–23. 10.1007/s11356-023-26462-y36988800

Wikipedia. (n.d.). *Agriculture in India.* Wikipedia. https://en.wikipedia.org/wiki/Agriculture_in_India

Wolfert, S., Ge, L., Verdouw, C., & Bogaardt, M. J. (2017). Big data in smart farming–a review. *Agricultural Systems*, 153, 69–80. 10.1016/j.agsy.2017.01.023

Wong, L. W., Tan, G. W. H., Lee, V. H., Ooi, K. B., & Sohal, A. (2020). Unearthing the determinants of Blockchain adoption in supply chain management. *International Journal of Production Research*, 58(7), 2100–2123. 10.1080/00207543.2020.1730463

World Bank. (2014). *The Importance of Land Tenure*. World Bank Group.

Wu, H. C., Hong, W. H., & Chiu, M. C. (2018). Comparisons with subjective and objective indexes of lifting risk among different combinations of lifting weight and frequency. *Journal of Ambient Intelligence and Humanized Computing*, 1–5.

Xia, C., Yang, S., Huang, M., Zhu, Q., Guo, Y., & Qin, J. (2019). Maize seed classification using hyperspectral image coupled with multi-linear discriminant analysis. *Infrared Physics & Technology*, 103, 103077. 10.1016/j.infrared.2019.103077

Xie, H., & Huang, Y. (2021). Influencing factors of farmers' adoption of pro-environmental agricultural technologies in China: Meta-analysis. *Land Use Policy*, 109, 105622. 10.1016/j.landusepol.2021.105622

Xie, S. (2019). *Does a dual reduction in chemical fertilizer and pesticides, improve nutrient loss and tea yield and quality? A pilot study in a green tea garden in Shaoxing*. Zhejiang Province. 10.1007/s11356-018-3732-1

Xu, P., Yang, R., Zeng, T., Zhang, J., Zhang, Y., & Tan, Q. (2021). Varietal classification of maize seeds using computer vision and machine learning techniques. *Journal of Food Process Engineering*, 44(11), 13846. 10.1111/jfpe.13846

Xu, Z., Elomri, A., El Omri, A., Kerbache, L., & Liu, H. (2021). The compounded effects of COVID-19 pandemic and desert locust outbreak on food security and food supply chain. *Sustainability (Basel)*, 13(3), 1063. 10.3390/su13031063

Yadav, A., & Agrawal, S. (2021). Mathematical model for robotic two-sided assembly line balancing problem with zoning constraints. *International Journal of System Assurance Engineering and Management,* 1-14.

Yang, T., Mei, Y., Xu, L., Yu, H., & Chen, Y. (2024). Application of question answering systems for intelligent agriculture production and sustainable management: A review. *Resources, Conservation and Recycling*, 204, 107497. 10.1016/j.resconrec.2024.107497

Yang, Y. P., Tian, H. L., & Jiao, S. J. (2019). Product design evaluation method using consensus measurement, network analysis, and AHP. *Mathematical Problems in Engineering*, 2019, 1–9. 10.1155/2019/4042024

Yawar, S. A., & Seuring, S. (2018). The role of supplier development in managing social and societal issues in supply chains. *Journal of Cleaner Production*, 182, 227–237. 10.1016/j.jclepro.2018.01.234

Yazdani, M., Hashemkhani Zolfani, S., & Zavadskas, E. K. (2016). New integration of MCDM methods and QFD in the selection of green suppliers. *Journal of Business Economics and Management*, 17(6), 1097–1113. 10.3846/16111699.2016.1165282

Yigezu, Y., Mugera, A., El-Shater, T., Aw-Hassan, A., Piggin, C., Haddad, A., Khalil, Y., & Loss, S. (2018). Enhancing adoption of agricultural technologies requiring high initial investment among smallholders. *Technological Forecasting and Social Change*, 134, 199–206. 10.1016/j.techfore.2018.06.006

Yogitha, S., & Sakthivel, P. (2014, March). A distributed computer machine vision system for automated inspection and grading of fruits. In *2014 International Conference on Green Computing Communication and Electrical Engineering (ICGCCEE)* (pp. 1-4). IEEE. 10.1109/ICGCCEE.2014.6922281

Yoon, J., Talluri, S., & Rosales, C. (2020). Procurement decisions and information sharing under multi-tier disruption risk in a supply chain. *International Journal of Production Research*, 58(5), 1362–1383. 10.1080/00207543.2019.1634296

Youssef, A. E., Kotb, Y., Fouad, H., & Mustafa, I. (2021). Overlapping gait pattern recognition using regression learning for elderly patient monitoring. *Journal of Ambient Intelligence and Humanized Computing*, 12(3), 3465–3477. 10.1007/s12652-020-02503-z

Yusuf, M., Adiputra, N., Sutjana, I. D. P., & Tirtayasa, K. (2016). The improvement of work posture using rapid upper limb assessment: Analysis to decrease subjective disorders of strawberry farmers in Bali. International Research Journal of Engineering. *IT and Scientific Research*, 2(9), 1–8.

Zagurskiy, O. N., & Titova, L. L. (2019). Problems and prospects of blockchain technology usage in supply chains. *Journal of Automation and Information Sciences*, 51(11), 63–74. 10.1615/JAutomatInfScien.v51.i11.60

Zambon, I., Cecchini, M., Egidi, G., Saporito, M. G., & Colantoni, A. (2019). Revolution 4.0: Industry vs. agriculture in a future development for SMEs. *Processes (Basel, Switzerland)*, 7(1), 36. 10.3390/pr7010036

Zareiforoush, H., Minaei, S., Alizadeh, M. R., & Banakar, A. (2015). A hybrid intelligent approach based on computer vision and fuzzy logic for quality measurement of milled rice. *Measurement*, 66, 26–34. 10.1016/j.measurement.2015.01.022

Zhai, Z., Martínez, J. F., Beltran, V., & Martínez, N. L. (2020). Decision support systems for agriculture 4.0: Survey and challenges. *Computers and Electronics in Agriculture*, 170, 105256. 10.1016/j.compag.2020.105256

Zhang, D. (2019). Detection of systemic pesticide residues in tea products at trace level based on SERS and verified by GC – MS. Springer. 10.1007/s00216-019-02103-7

Zhang, K. (2021). The design of regional medical cloud computing information platform based on deep learning. *International Journal of System Assurance Engineering and Management*, 1-8.

Zhang, M., Li, N., & Zhang, C. (2023, August). Early Warning Method of Credit Risk of Agricultural Enterprises in Guizhou Province Based on AHP. In *Proceedings of the 2nd International Academic Conference on Blockchain, Information Technology and Smart Finance (ICBIS 2023)* (pp. 6). Atlantis. 10.2991/978-94-6463-198-2_66

Zhang, D., Chen, S., Liwen, L., & Xia, Q. (2020). Forecasting Agricultural Commodity Prices Using Model Selection Framework With Time Series Features and Forecast Horizons. *IEEE Access : Practical Innovations, Open Solutions*, 8, 28197–28209. 10.1109/ACCESS.2020.2971591

Zhang, H., He, L., Di Gioia, F., Choi, D., Elia, A., & Heinemann, P. (2022). LoRaWAN based internet of things (IoT) system for precision irrigation in plasticulture fresh-market tomato. *Smart Agricultural Technology*, 2, 100053. 10.1016/j.atech.2022.100053

Zhang, K., Wu, Q., Liu, A., & Meng, X. (2018). Can deep learning identify tomato leaf disease? *Advances in Multimedia*, 2018, 2018. 10.1155/2018/6710865

Zhang, N., Wang, M., & Wang, N. (2002). Precision agriculture—A worldwide overview. *Computers and Electronics in Agriculture*, 36(2-3), 113–132. 10.1016/S0168-1699(02)00096-0

Zhang, T., Zhao, Y., Jia, W., & Chen, M.-Y. (2021). Collaborative algorithms that combine AI with IoT towards monitoring and controsystem. *Future Generation Computer Systems*, 125, 677–686. 10.1016/j.future.2021.07.008

Zhang, W. Y., Jing, T. Z., & Yan, S. C. (2017). Studies on prediction models of Den- drolimus superans occurrence area based on machine learning. *Journal of Beijing Forestry University*, 39(1), 85–93.

Zheng, H., Ma, J., Yao, Z., & Hu, F. (2022). How Does Social Embeddedness Affect Farmers' Adoption Behavior of Low-Carbon Agricultural Technology? *Frontiers in Environmental Science*, 10, 10.3389/fenvs.2022.909803

Zuidhof, M. J., Fedorak, M. V., Ouellette, C. A., & Wenger, I. I. (2017). Precision feeding: Innovative management of broiler breeder feed intake and flock uniformity. *Poultry Science*, 96(7), 2254–2263. 10.3382/ps/pex01328159999

About the Contributors

Suchismita Satapathy is an associate professor in the School of Mechanical Sciences, KIIT University, Bhubaneswar, India. She has published more than 130 articles in national and international journals and conferences. She has also published many books and e-books for academic and research purposes. Her areas of interest include production operation management, operation research, acoustics, sustainability, and supply chain management. She has filled three Indian patents, among them two published. She has more than 15 years of teaching and research experience. She has guided many Ph.D. (two completed, four ongoing), MTech (20), and BTech (49) students. Suchismita has published books such as Production operation management (Stadium Press) and MCDM methods for waste management with CRC Press, Innovation, Technology, and Knowledge Management with Springer, Soft Computing and Optimization Techniques for Sustainable Agriculture by DEGRUYTER, and many of her projects are ongoing. ORCID: 0000-0002-4805-1793 Scopus Author ID: 55085975400

Kamalakanta Muduli has about 18 years teaching/research experience (Aug 2000-till July 2010 and 1st August 2014 till date) as Lecturer, Senior lecturer, Assistant Professor and Associate Professor in Degree level Engineering colleges/Universities in, Subjects taught include Optimization Engineering, Production & Operation Management, Quality Control and Reliability, Industrial Engineering and Management, Manufacturing Technology, Maintenance Engineering, Engineering Mechanics Published 80 research papers in reputed indexed journals and presented papers in 34 national and international conferences

***

C.V. Suresh Babu is a pioneer in content development. A true entrepreneur, he founded Anniyappa Publications, a company that is highly active in publishing books related to Computer Science and Management. Dr. C.V. Suresh Babu has also ventured into SB Institute, a center for knowledge transfer. He holds a Ph.D. in Engineering Education from the National Institute of Technical Teachers Training & Research in Chennai, along with seven master's degrees in various disciplines such as Engineering, Computer Applications, Management, Commerce, Economics, Psychology, Law, and Education. Additionally, he has UGC-NET/SET qualifications in the fields of Computer Science, Management, Commerce, and Education. Currently, Dr. C.V. Suresh Babu is a Professor in the Department of Information Technology at the School of Computing Science, Hindustan Institute of Technology and Science (Hindustan University) in Padur, Chennai, Tamil Nadu, India. For more information, you can visit his personal blog at https://sites.google.com/view/cvsureshbabu/.

P. Chandrasekhar is having more than 18 years of professional experience. Dr. P.Chandrasekhar has been actively involved in research and development (R&D) activities in engineering space. He is actively involved in bringing industry orientation to the engineering education system working with several industries. Dr.Sekhar providing service to the industries by sharing his expertise and competence of his experience to solve industrial problems objectively in selected areas of engineering (PLM, CAD, CAM, and CAE),

Ayan Chaudhauri is pursuing his B.Tech in Mechatronics Engineering from KIIT University. His research interests are Supply Chain Management, Artificial Intelligence and Robotics.

Hullash Chauhan is an assistant professor at Bharti University Durg, India. He has more than one year of teaching experience in engineering college at both undergraduate and diploma levels. He has published more than 4 book chapters and 18 articles in reputed journals and international conferences. His areas of interest include ergonomics, occupational health and safety, sustainability, supply chain management, and optimization techniques. Dr Chauhan obtained his PhD in mechanical engineering from KIIT (Deemed to be University), Odisha, India. He completed his MTech with a specialization in Industrial Engineering and Management and his BE in Mechanical Engineering.

Debankur Das is pursuing B.Tech in Mechatronics Engineering from KIIT University. He is an avid reader, interested in Business Strategy, Product Development, and Supply Chain Management.

Uma N. Dulhare is working as a Professor, Department of Computer Science & Engg, MJCET, Banjara Hills, Hyderabad. She has more than 20 years teaching experience. She has published more than 30 research papers in reputed National & International Journals & as a book chapter. She is member of Editorial Board and reviewer of International Journals like IJACEA, ICDIWC and ICEOE, IIE, IJERTREW,IJDMKD, Elsevier Procedia. She was Keynote & Chair Person of International Congress on Multimedia 2014 [ICMM2014], Bangkok, Kingdom of Thailand & also 5th National Conference on National "Computer Network & Information Security" NCCNIS-2016 Vasavi College, Hyderabad . She is also the member of various professional societies like ISTE, CSTA of ACM,ASDF,IAENG & Fellow member of ISRD. She has also received a Best research paper Award in 2010, ASDF Global Award for Best Computer Science Faculty of the Year 2013 by the Lt. Governor of Pondicherry & also Best Academic Researcher of the year 2015. She honored with Outstanding Educator & Scholar Award 2016 by NEFD. Her area of interest is Networking, Database, Data Mining, Information Retrieval and Neural Networks & Big Data Analytics.

Dwijendra Nath Dwivedi is a professional with 20+ years of subject matter expertise creating right value propositions for analytics and AI. He currently heads the EMEA+AP AI and IoT team at SAS, a worldwide frontrunner in AI technology. He is a post-Graduate in Economics from Indira Gandhi Institute of Development and Research and is PHD from crackow university of economics Poland. He has presented his research in more than 20 international conference and published several Scopus indexed paper on AI adoption in many areas. As an author he has contributed to more than 8 books and has more than 25 publications in high impact journals. He conducts AI Value seminars and workshops for the executive audience and for power users.

Nitish Karn is currently pursuing Master's degree in Agronomy from Uttaranchal University, Dehradun, India. He completed his Bachelor's degree from Punjab Agricultural University, Ludhiana, India. He have published few book chapters and review articles. He is a research enthusiastic.

Kazi Kutubuddin Sayyad Liyakat has completed his B.E., M.E., and Ph.D. in E&TC Engineering and is nowadays working as a Professor & Head in Electronics and Telecommunication Engineering Department and also as Dean R&D. He is Post Doctoral Fellow working on "IoT in Healthcare applications". He has published more than 110+ articles in various Journals. Also published 11 books in the field of Engineering. He has 15 Indian Patents, 2 Indian copyright patents, 2 South African Grant Patent, and 8 UK Grant Patent. All patents are in the field of IoT in Healthcare. He worked as a Reviewer for Scopus Conferences and Journal also reviewer for IGI Global. Also work as Editorial Board Member for various Journals.

A.V. Senthil Kumar to his credit he has industrial experience for five years and teaching experience of 27 years. He has also received his Doctor of Science (D.Sc in Computer Science). He has to his credit 33 Book Chapters, 220 papers in International and National Journals 60 papers in International Conferences in International and National Conferences, and edited 12 books (IGI Global, USA). He is as Associate Editor of IEEE Access. He is an Editor-in-Chief for many journals and Key Member for India, Machine Intelligence Research Lab (MIR Labs). He is an Editorial Board Member and Reviewer for various International Journals. He is also a Committee member for various International Conferences. He is a Life member of International Association of Engineers (IAENG), Systems Society of India (SSI), member of The Indian Science Congress Association, member of Internet Society (ISOC), International Association of Computer Science and Information Technology (IACSIT), Indian Association for Research in Computing Science (IARCS), and committee member for various International Conferences.

Ashish Kumar is a highly skilled Mechanical Engineer with a diverse range of experiences in academia and industry. I hold an M.Tech. in Automobile Engineering from Shri Rawatpura Sarkar University, Raipur, and a B.E. in Mechanical Engineering from CSIT, Durg. I have excelled in various roles, including Maintenance Engineer, Trainer, Workshop Incharge, and Lecturer. I am currently serving as an Assistant Professor at Bharti Vishwavidyalaya, Durg. My achievements include developing innovative solutions like the Paddle Press Body Sanitizer Machine and publishing technical papers, showcasing my commitment to excellence. With strong technical skills and a passion for innovation, I believe I am a valuable asset to any organization.

Lopamudra Lenka is working as an Assistant Professor in Economics at KIIT Deemed to be University, Bhubaneswar, Odisha, India. She was also working as a guest faculty at National Defence Academy, Pune, India Maharashtra. Her area of research interest includes Agricultural Economics, Rural Economics, Environmental Economics, Reverse Migration & Rural Sustainability.

Puspalata Mahapatra presently working as Asst Professor (II) in the deptt. of Commerce under KIIT DU, BBSR and having more than 15 years of teaching experience in different Universities and B'Schools of Odisha. She has received her M.Com, MBA and Ph.D degree from Utkal University, Bhubaneswar, Odisha. She has published 1 Book with Kunal Publication, more than 25 Articles in different Journals and presented more than 30 papers in national and international conferences both in India and outside India. She has filled three Indian and one international patent out of which three already published. She has guided four research Scholars out of which two awarded and two are continuing. Her area of research interest is Finance, Financial service, Banking, Insurance and Agri business etc.

M. Ravi Sankar, working as an Associate Professor in Department of Computer Science and Engineering, K.L. deemed to be University, Guntur dist., Andhra Pradesh, India. He has 24 years of teaching experience. Dr. Ravi has received Excellence in Research Award and Best Senior Faculty Award. He has Published 5 Patents and he has published more than 20 articles in Scopus, SCI, WOS and International Journal., His areas of specializations are Data Mining and Artificial Intelligence.

Gopal Krushna Mohanta serves as an Assistant Professor in the Department of Mechanical Engineering at GIET University in Gunupur, Odisha. With a passion for advancing engineering knowledge, he has contributed significantly to the field through his research publications and projects. His dedication to academic excellence and commitment to nurturing the next generation of engineers make him a valuable asset to the university. Through his teaching, research, and academic initiatives, Mr. Mohanta continues to inspire students and colleagues alike in the pursuit of excellence in mechanical engineering.

Anmol Panda is an Assistant Professor in the Department of Agriculture Extension and Communication at School of Agriculture GIET University, Gunupur, Odisha. As an social science researcher he is dedicated to research in agricultural development, with publications focusing on performance of Grassroot Agricultural Extension workers and Climate Smart Agricultural practices. His expertise and passion contribute significantly to the university's agricultural initiatives, extension outreach, inspiring students and colleagues alike.

Lakshmi Prasad Panda is an Assistant Professor in the Department of Humanities at Govt. College of Engineering Kalahandi in Bhawanipatna, Odisha, India. He holds expertise in areas such as literature, culture, and social sciences. Dr. Panda has made significant contributions to academia with his research publications, covering a wide range of topics. He has also delivered talks and presentations at various national and international conferences, showcasing his deep understanding and scholarly insights. Dr. Panda's dedication to education and research serves as an inspiration to his colleagues and students alike, contributing to the academic excellence of his institution.

Chandra Prabha R received the B.E. degree in electronics and communication engineering from Visvesvaraya Technological University, India, in 2004, the M.Tech. in digital electronics communication systems from Visvesvaraya Technological University, in 2008, and currently perusing the Ph.D. from the Visvesvaraya Technological University, India. She is currently working as an Assistant Professor with the BMS Institute of Technology and Management, Bengaluru, India. Her research interests include image processing, machine learning, and deep learning.

Hakikur Rahman is an academic over 30 years has served leading education institutes in Bangladesh and abroad, and established various ICT4D projects funded by ADB, EU, UNDP and World Bank. He is also currently serving as the Director at the Ansted University Sustainability Research Institute. Before joining Presidency University, he was a Professor, CSE at the Asian University of Bangladesh. Before that he served the BRAC University as an Associate Professor (Chair, CSE Dept.) and a Post-Doctoral Researcher (Research Associate) at the University of Minho, Portugal under the Centro Algoritmi. He has written and edited over 25 books, more than 50 book chapters and contributed over 200 articles on computer education, ICTs, knowledge management, open innovation, data mining and e-government research in newspapers, journals and conference proceedings. Graduating from the Bangladesh University of Engineering and Technology in 1981, he has done his Master's of Engineering from the American University of Beirut in 1986 and completed his PhD in Computer Engineering from the Ansted University, BVI, UK in 2001; and PhD in ICT from the Empresarial University of Costa Rica in 2011.

Seema Singh got her B.Tech (Electronics and Communication Engineering) degree form Jamia Millia Islamia, a Central University, Delhi in 2002. Her M.Tech degree is in Electronics branch from Sir MVIT, Bangalore in 2007. She is Gold medalist and VTU rank holder in M.Tech degree. She is awarded the PhD degree from JNTU, Hyderabad in the year 2015 in the area of Neural Networks and Flight control systems. Currently, she is working as Professor and Head in Dept. of Electronics and Telecommunication Engineering, BMSIT&M. She has 17 years of teaching experience and 10 years of research experience. She has authored one book chapter, 30 publications in reputed International Journals and International/National Conferences. She has received award for guiding best project of the year from KSCST during 2015. Her research interest lies in the field of anomaly detection, neural networks, image processing, Control Systems, biomedical engineering. She has given several expert talks on Artificial Neural Networks and its Applications and chaired many sessions of conferences. Her research work is mainly dedicated to application of neural networks in varied fields with undergraduate, postgraduate and PhD scholars. She has filed a patent in 2018, which dealt with application of neural network in image processing field.

B. Narendra Kumar Rao is currently Working as Professor and Program Head, Department of AI&;ML in the School of Computing, Mohan Babu University, Tirupati. He is a Member of Research Advisory Committee and also Doctoral Supervisor Cluster Head in School of Computing, Mohan Babu University. His Research interests include Software Testing, Embedded Systems and Machine Learning. He has been part of several International Conferences as Convener and Conference Chair for THREE International Conferences. Has published articles in reputed Journals and Conferences. He was the convener for International Conference on Data Analytics, Intelligent Computing and Cyber Security-2021, He is editor for two Proceedings by Springer published in the year 2018 and 2022. He has been associated with IEEE, ACM, CSTA and IAENG. He has won Best Faculty Recognition, Nava Bharat Nirman Award by Information Technology Association of AP; India Servers, October, 2019 and Best Researcher Award by Integrated Research Group(IRG), Chennai for research work carried out, January, 2018. He received his Ph.D in CSE from Jawarharlal Nehru Technological University, Hyderabad.

N. V. Jagannadha Rao is working as a Professor at the School of Management Studies, GIET University in Gunupur, Odisha, India. With a wealth of experience in management studies, Dr. Rao is renowned for his insightful research contributions across various domains of management. He has authored for numerous publications in esteemed journals, presenting innovative perspectives on contemporary management issues. Additionally, Dr. Rao is a sought-after speaker, having delivered talks and lectures at prestigious national and international conferences. His expertise and dedication to advancing the field of management studies greatly enrich the academic environment at GIET University, inspiring both students and faculty members to excel in their scholarly pursuits.

Kali Charan Rath is an Associate Professor in the Department of Mechanical Engineering at GIET University, Odisha, India. With a strong focus on research, he has authored numerous publications in esteemed journals, covering diverse topics within mechanical engineering. Dr. Rath is also a distinguished speaker, having delivered talks and presentations at various national and international conferences, sharing his insights and expertise with fellow researchers and scholars. In addition to his research and speaking engagements, he has contributed to edited books in his field, further showcasing his expertise and commitment to advancing mechanical engineering knowledge. Dr. Rath's multifaceted contributions make him an invaluable asset to both his department and the broader academic community.

Anirban Roy is pursuing B.Tech in Mechatronics Engineering from KIIT University. His area of interests are Supply Chain, Technological Advancements, and Sustainability.

P.K. Sethy is an esteemed Associate Professor in the Department of Electronics and Communication Engineering at Guru Ghasidas Vishwavidyalaya (Central University, Govt. of India), Bilaspur, Chhattisgarh, India since December 2023. Prior to this, he had served at Sambalpur University (State University, Govt. of Odisha) from February 2013 to December 2023, and worked as an Engineer in Doordarshan, Ministry of Broadcasting, Govt. in India, between August 2009 and February 2013. Holding a Ph.D. from Sambalpur University, M. Tech. from IIT Dhanbad, and B.E. from BPUT Odisha. Dr. Sethy is from small village, named Kapundi, situated on the banks of the Baitarani River in the Keonjhar District of Odisha. His early education, including primary and intermediate studies, took place in Keonjhar, Odisha. Dr. Sethy is an Editor of three reputable journals and serves as an editorial board member for the International Journal of Electrical and Computer Engineering and Ingénierie des Systèmes d'Information (IIETA). Additionally, he serves as an editorial member for Automation, Control and Intelligent Systems (Science Publishing Group) and PriMera Scientific Engineering (ISSN: 2834-2550). He also holds the position of Associate Editor of Onkologia I Radiotherapy. With two patents and one copyright to his name, Dr. Sethy has been recognized for his exceptional contributions. In 2020, he received the "InSc Young Achiever Award" for his research paper on "Detection of coronavirus (COVID-19) based on Deep Features and Support Vector Machine," organized by the Institute of Scholars, Ministry of MSME, Government of India. As a Senior Member of IEEE, he actively engages as a reviewer for various journals and takes on the role of session chair in international conference

Himani Sharma currently associated with the Division of Agronomy, Sher-e-Kashmir University of Agricultural Sciences and Technology of Jammu – 180009, Jammu and Kashmir, India. She did her graduation with B.Sc. (Hons.) Agriculture from Sher-E-Kashmir University of Agricultural Sciences and Technology, Jammu, and Master's degree of M.Sc. Ag. (Agronomy) from School of Agriculture, Uttaranchal University, Dehradun (U.K.). She also published 14 articles, 4 book chapters (Scopus Indexed), 2 trainings, and attended 3 national/international seminars/conferences. She has a life member of the Indian Society of Agriculture & Horticulture Research Development (ISAHRD), Chandigarh, Punjab, India. She has also received the "Young Achiever Award" for outstanding contribution to the field of Agronomy. She has expertise in soil fertility, nutrient management, and irrigation water quality.

Sarita Kumari Singh serves as an Assistant Professor within the distinguished School of Economics and Commerce at KIIT Deemed to be University. She earned her Post Graduation degree from Utkal University, specializing in the intricate field of Accounting. Adding to her academic repertoire, she holds an M.Phil. in Commerce from the Rama Devi Women's University in Bhubaneswar and has also qualified UGC-NET in Commerce. She holds a Ph.D in Commerce from KIIT deemed to be University. Her areas of academic interest encompass a wide spectrum, spanning Direct Taxation, Economics, and Business Mathematics.

Deepak Singhal is an associate professor in the School of Mechanical Engineering of KIIT Deemed to be University, Bhubaneswar. He holds his a bachelor's degree in Mechanical engineering from Indian Institute of Information Technology (IIIT), Jabalpur. He completed his PhD in the area of Industrial engineering. His research interests include decision modelling, sustainability and supply chain management. He has published the research papers in various reputed Journals like "International Journal of Production Research", "Resource, Conservation and Recycling","Management Decision", "Sustainable Production and Consumption", Research in Transportation Economics, etc.

Sushanta Tripathy (PhD) is presently working as a Professor at the School of Mechanical Engineering in KIIT University, Bhubaneswar, Odisha, India. He completed his PhD from the Department of Industrial Engineering and Management, Indian Institute of Technology, Kharagpur. His major areas of interest include production operations management, multivariate analysis, service operations management, supply chain management and productivity management. He is a Fellow in the Institution of Engineers, India.

Index

A

B

C

D

E

F

G

I

K

L

M

O

P

Q

R

S

T

W

Milton Keynes UK
Ingram Content Group UK Ltd.
UKHW011954080824
446595UK00005B/149